An Introduction to Safety and Security Intelligence Science

"Safety and security (SS) is a critical concern in any society, especially in the digital world today. SS science has emerged as one of the fastest-growing disciplines of the 21st century. In this context, effective safety and security management (SSM) relies heavily on intelligence, which serves as the cornerstone of our digital society. One of the important tasks and functions of modern intelligence studies is to study the principles, methods and techniques of SS warning. This book combines SS science with information science in an attempt to establish a new interdisciplinary field. Based on the research results of the project, the author puts forward the idea of intelligence-led SSM for the first time. He emphasizes the necessity of applying SSI to SSM, proposes that a major challenge facing SSM is the lack of valuable SSI, and airs his concern that the importance and urgency of SSI research should be deeply understood by related professionals. The book introduces the research object and nature of SSI, constructs the content system of SSI, and expounds the basic problems and theories, major application fields and typical applications of SSI, as well as the developing trends of SSI in the era of big data. The author also makes some important and enlightening points. For example, the essence of SSM lies in the behavior of SS managers utilizing SSI to implement SSM; The concept of SSI is a new understanding of SSM from the perspective of information science; SSI is an emerging applied interdisciplinary science that studies the nature, structure, function and the laws of its generation, transmission and utilization to solve the problem of the lack of SSI, and so on. The publication of this book will further enrich the knowledge bank of SS science and information science, and provide valuable knowledge and reference for researchers, practitioners who work in related fields and readers who are interested in this field."

——Lai Maosheng, *Professor of the Department of Information Management, Peking University*

"At all times and in all countries, intelligence operation has been an important guarantee for safety and security. The general research on SSI has very important practical significance and academic value for the research on SSI in terms of the broad SS system guided by the philosophy of overall national security. The publication of this book adapts to the needs of SSI operations in the era of major SS and broad intelligence, enriches the academic system of SSI theory, and enhances the academic value of SSI research. This book has strong practical value as it aims at addressing problems; It also possesses a complete academic system featuring theoretical innovation. Its research methodology is scientific, which embodies the distinctive characteristics of interdisciplinary study. Whether you are engaged in practice, conducting scientific research, or simply interested in SSI issues, you would find this book with enduring value of reading and reference."

——Liu Yuejin, *Professor of the Department of Public Administration, School of International Relations*

"SSI is an emerging discipline formed by the intersection and integration of information science and SS science in recent years. This book opens up a new field of information science and SS science by constructing a theoretical system of SSI and proposing a new method of intelligence-led SSM. It also points out the future trend of SSI, and makes outstanding contributions to the construction and improvement of SSI. This book is well-structured, rich in content, and clear in logic. It analyzes the basic concepts of SSI science of SS, SS science, intelligence, information science and SSI, discusses the problems of SSI and related theories of SSI science, and carries out intelligence-led SSM and relevant applications. At last, it serves as an outlook for development of SSI in the era of big data. The innovation lies in the fact that from the perspectives of major SS, SSI and major SSI, this book proposes a new intelligence-led SSM method, which makes a great contribution to the innovation of SSM methods and addressing the problem of SSI absence in SSM."

——Li Gang, *Director of the Information Resource Research Center of Wuhan University of the project of "Key Research Institute of Humanities and Social Sciences in Universities" of the Ministry of Education, Changjiang Scholar conferred by the Ministry of Education*

"When the era of big data provides more diverse and larger-scale data information for SSM, it also poses a significant challenge for decision-makers to form decisive intelligence by identifying and exploring valuable insights from a vast amount of data. How to provide meaningful SSI for SS decision-making and promote smart SSM has become a major task faced by fields of SS science and information science. *An Introduction to Safety and Security Intelligence Science* constructs the theoretical system of SSI for the first time, proposes a new method of intelligence-led SSM, expounds the practical application paradigm of SSI, and discusses the development trend of SSI. The work and innovative research in this book have effectively enhanced the integration of safety and security science and information science, and have also laid a crtical foundation for the development of the new field of SSI."

——Sha Yongzhong, *Vice President of Lanzhou University, Professor of the Department of Management, Lanzhou University*

Bing Wang · Chao Wu

An Introduction to Safety and Security Intelligence Science

SCIENCE PRESS

Springer

Bing Wang
School of Resources and Safety Engineering
Central South University
Changsha, Hunan, China

Chao Wu
School of Resources and Safety Engineering
Central South University
Changsha, Hunan, China

Translated by
Diandian Yi
School of Foreign Languages
Xiangtan University
Xiangtan, Hunan, China

ISBN 978-981-96-7260-8 ISBN 978-981-96-7261-5 (eBook)
https://doi.org/10.1007/978-981-96-7261-5

Jointly published with Science Press

Translation from the Chinese Simplified language edition: "安全情报学导论" by Bing Wang and Chao Wu, © China Science Publishing and Media Press, Ltd. 2022. Published by China Science Publishing and Media Press, Ltd.. All Rights Reserved.

This Springer imprint is published by the registered company Springer Nature Singapore Pte Ltd.
The registered company address is: 152 Beach Road, #21-01/04 Gateway East, Singapore 189721, Singapore

Preface

Since human beings stepped into a risky society, various security threats, security risks, and security incidents have emerged one after another, constantly threatening human survival and the development of nations, societies, and other communities. The current world is full of changes. Man-made systems like cities are growing larger and more complex, and various uncertain security factors and risks are increasing dramatically as they are interweaving with each other. Traditional security issues and non-traditional security issues are intertwined, which brings many challenges to modern safety and security management (SSM). Against this backdrop, security has gradually become a mainstream narrative for the human society and countries around the world, and security issues have become an important real-world concern widely recognized by global communities. In recent years, the security industry has evidently flourished. For example, in terms of disciplinary development, China has established five first-level security disciplines: Security Science and Engineering (Code 0837), National Security (Code 1402), Cyberspace Security (Code 0839), Public Security (Code 0306), and Public Security Technology (Code 0838), which have formed a large group of security disciplines.

Intelligence science has been focusing on the studies of major real-world issues in national and social development. As the support and foundation for SSM, the study of safety and security intelligence (SSI) science is of great significance to promote SSM. Therefore, intelligence studies and operation in China in the new era aim to serve the security and development of the country, society, and other communities. In view of this, the study of SSI has gradually emerged and has been included in the development of intelligence studies and SS science in recent years, and has become a new growth point and point of extension in the intersection of intelligence studies and SS science.

In fact, the concept of intelligence has long been associated with SS science. To put it another way, it originated from SS science. Why? If we trace its origin, intelligence should originate from military security intelligence, national security intelligence, public security intelligence (or police security intelligence, which involves many forms of public SSI), and so on. This is also one of the key reasons why people associate intelligence with military security intelligence and national security intelligence in the first place. Additionally, SSI is a product of human needs for safety and security. Human beings began their pursuit for safety and security since their appearance on earth, which gave rise to the need for safety and security intelligence, and activities related with such intelligence naturally occurred.

In the early days, SSI, same as intelligence, had an apparently sensitive and mysterious color. It was mainly concentrated in the security field, such as national security, military security, and related to emergency management. At that time, SSI research appeared to be dispersed yet specific to certain contexts, a consensus on the understanding of SSI was yet to form. With the development of the concept of intelligence and the deepening of SSI research, its scientific nature and universal applicability are generally recognized, and its transition from dispersion to unity and from specificity to universality has been achieved. This is reflected in three aspects: from partial security to overall security, from security to security & safety integration, and from emergency management to whole-process SSM. From the perspective of SSM, SSI refers to the safety and security information (SSi) that can influence SSM (or the SSi that is useful for SSM). SSM is dependent on the support provided by SSI, which is designed to enhance SSM and has a significant impact on SS performance.

The information age (especially the era of big data) has witnessed an explosive growth of SSi in its volume; the role of SSI has been significantly reinforced. The problem of excessive useless SSi and insufficient valuable SSI becomes more and more prominent, which severely affects the effectiveness of SSM. In this context, professionals of SSI and SSM have begun to reflect on safety and security science, especially SSM. The difference between SSI and SSi has become more obvious, and it is urgent for professionals to shift their concern from the lack of SSi to the lack of SSI when the reason of SSM failure is considered. The key task of future SSM should be to solve the problem of missing or incomplete SSI in SSM, and to understand the concept of SSI and the importance of SSI research. At the same time, in the era of big data and intelligence, data information should be regarded as the basis, and intelligence (intelligence is intellectual prowess encapsulated, and intellectual prowess is intelligence uncapped) as the "network," in order to achieve intelligent management of the countries and societies. Similarly, SS big data and SSI have become the main driving forces for promoting the transformation, innovation, and progress of SSM for the aim of achieving intelligent SSM.

As mentioned above, in the era featuring high risks, big data and intelligence, the significance of SSI research and practice becomes increasingly prominent. The implementation of *The National Intelligence Law of the People's Republic of China* and *The National Security Law of the People's Republic of China*, and the proposal of "a holistic approach to national security" made by General Secretary Xi Jinping further enhanced their significance. As SSI becomes a necessary choice for innovation in intelligence studies and SS science, the rise and development of SSI research calls for a corresponding discipline as its guide. From an academic perspective, the difference between SSI and SSI science is not merely a word, but a qualitative leap in SSI research from the input of intelligence studies plus SS to the output of SS plus intelligence studies.

SSI science is a discipline born in the intersection and integration of intelligence studies and SS science. As a new branch of universal SS science and intelligence studies applicable to all SS fields or industries, as well as a science specializing SSI, it aims to solve the problem of missing or incomplete SSi in the process of SSM. At present, the status of SSI has changed from passive subsidiary to active leading. SSI helps to promote the gradual formation of intelligence studies as a first-level discipline, and boost the vertical and horizontal development of intelligence studies and SS science research in the new era. It can be said that SSI has ushered in its spring of development, as it emerges as a promising and dynamic discipline with great potential.

Nevertheless, SSI is yet to become mature. It faces numerous challenges, among which the most notable is that the fundamental theory of SSI (including the basic disciplinary theory and the theory of application) is comparatively weak. The incomplete theoretical system and the pending paradigm of practical application of SSI have led to a lack of theoretical basis, direction, and method guidance, which seriously hinders the future research, practice, and development of SSI and casts a shadow on the future development of SSI. In response, scholars of both intelligence and SS fields should accept and support the research and construction of SSI with an open and inclusive mind, while paying more attention to building an overall system of SSI, especially to composing a comprehensive academic monograph that introduces and studies SSI, which is a major task to be fulfilled in the current stage.

In order to address the problems mentioned above, to promote the development of SSI, and to cater to the needs of fellow scholars for learning, researching, and developing SSI, the author has been engaged in writing an academic monograph which introduces and studies SSI on the basis of the research done by experts and professionals. The author's original intention was to write a systematic and comprehensive monograph on SSI. Yet, during the writing process the author found that due to the complexity and vastness of the research object, it is difficult to encompass all the research content of SSI within one single book. Besides, there is currently no monograph specifically named *Safety and Security Intelligence Science*. In view of this, the author decided to write a book that is general, basic, and introductory at the level of disciplinary construction, hence the book named *An* Introduction to Safety and Security Intelligence Science.

This book is grounded in the significant research and practical needs of the disciplines of intelligence studies and SS science. The fundamental idea and goals of this book are as follows: From the perspective of major SS, SSI, and SSM, this book focuses on the whole process of SSM. Its theoretical goal is to build a theoretical framework for SSI, and its practical application goal is to enhance the support and service capabilities of SSI for SSM. In the meanwhile, it depicts and envisions the future of SSI.

From the perspective of major SS, SSI, and SSM, this book is grounded in the significant research and practical needs of the disciplines of intelligence studies and SS science. It does not only aim to develop the theoretical system and practical application paradigm of SSI with its focus on the whole process of SSM, but also encompasses the basic application and outlook of SSI. The fundamental idea and goal of this book are as follows: Rooted in major SS, SSI, and SSM, the book centers on the entire process of SSM with its objectives of building a theoretical framework for SSI (theoretical objective) and of enhancing the support and service capabilities of SSI for SSM (practical objective), so as to envision and map out the future of SSI.

As a researcher, spending a quiet time writing papers and books is to taste joy and solitude in spiritual transcendence. Seeing the fruit of such arduous labor finally take the form of an accomplished book brings a sense of accomplishment and joy to the author. This is an experience that non-writers can hardly understand. In recent years, the author has persisted in continuous cultivation in the intersection of SS science and related disciplines, and has published books of *Safety Culture Science*, *Safety and Security Information Science*, *New Branches of Safety and Security Science*, etc. As the author has accumulated a certain amount of research experience, his research and thinking in the field has gradually become more mature and systematic, leading to the completion of this book as his fourth monograph in this field.

Since SSI is an emerging field formed by the intersection of intelligence studies and SS science, this book is composed to be more of an intermediate result than a conclusion of our exploration of SSI. There are certainly shortcomings, as some theories, understandings, and viewpoints of the book are far from perfect, which welcome criticism, correction, and enrichment from researchers and practitioners. As the disciplinary system and practical application of SSI are still constantly evolving and developing, the main objective of this book is to build a preliminary framework in order to lay a theoretical foundation and propose ideas of methods and methodology for research and application of fellow scholars, contributing to the promotion of the theoretical research and practical application of SSI. It is hoped that *An Introduction to Safety and Security Intelligence Science* is an exploratory and introductory monograph. Moreover, the book is the first to include "safety and security intelligence studies" in the title. Constrained by time and the interdisciplinary nature, the author tried to collect enough materials about this discipline, yet some issues remain to be further discussed. Because of the author's limited insights and experience, there are inevitably oversights and imperfections in the book, and the author sincerely welcomes criticism and suggestions from colleagues and readers alike.

The author would like to express his gratitude to those who have offered invaluable suggestions and to the authors of the references cited in the book. As a young researcher with the background of security science, the author is a new "recruit" in the field of SSI. During his research, the author has received wholehearted support and diligent guidance from senior researchers in the field of intelligence studies and the academic journals on information science, which has brought experiences of openness, inclusiveness, and enthusiasm to young scholars in this field. The research on SSI would otherwise be fraught with difficulties if support and help of this kind is not given to scholars like the author, after all it is not easy to conduct interdisciplinary research. The author's deepest gratitude goes to anyone that has offered their assistance.

Changsha, China
February 2021

Bing Wang

Acknowledgements

I would like to express my deepest gratitude to Prof. Wang Bing and Prof. Wu Chao (Central South University, China), the authors of the original book for their insightful work, which has provided invaluable knowledge in the field. Their expertise and dedication have made this translation project both a privilege and a rewarding challenge.

My sincere thanks also go to Springer Nature and Science Press (China), and Mr. Yang Fengbo, the editor of the Chinese edition, for their trust, support, and guidance throughout the translation process. Their professionalism and attention to detail have been instrumental in bringing this work to a broader audience.

I am particularly indebted to Prof. Hu Qiang (Xiangtan University, China), a leading expert in foreign language studies, and Prof. Li Jielin (Central South University, China) and Mr. Deng Cyril Chol (Zijin Mining Group Co., Ltd., China), experts in safety and security studies, for their valuable feedback and suggestions which have greatly enhanced the accuracy and clarity of this translation. Their scholarly input has been indispensable in ensuring the technical precision of the text.

Special thanks are due to the proofreaders from the School of Foreign Languages at Xiangtan University, China, for their meticulous review and constructive comments. They are Li Jingjing, Liang Dashuang, Zhang Jing, Xiong Mei, Liu Min, Jia Peiyi, Yin Fanghui, Chen Cong, Xie Yuyuan, and Xu Zhiqiu. Their keen eyes and linguistic expertise have helped refine the translation and ensure its fluency.

This project would not have been possible without the collective effort of all those mentioned above. Any remaining imperfections are, of course, my own.

Translator
Diandian Yi

Contents

Chapter 1
Prolegomenon

1.1 Introduction

As the introductory chapter of the book, Chapter I serves to lay a reasonable foundation for readers. It defines and analyzes the five basic concepts involved in safety and security intelligence (SSI), namely, safety and security (SS), SS science, intelligence, intelligence studies, and SSI. The challenging nature of proposing and constructing SSI studies requires that scholars recognize and accept it with an open, inclusive, and innovative academic mind. Moreover, scholars are expected to conduct meticulous and in-depth research on the construction of SSI with a rigorous scientific attitude. In this regard, this chapter discusses the background and significance of carrying out SSI research and practices from the aspects of disciplinary development and practical considerations, and demonstrates the basic conditions for constructing SSI on the bases of philosophical, practical, and theoretical foundations, aiming to clarify the necessity and importance of learning and studying SSI. The chapter concludes with a systematic analysis of the research progress of the discipline of SSI.

1.2 Basic Concepts of the Discipline of SSI

There are basic concepts in every discipline. They are the fundamental questions and problems that a discipline explores, which determines the value of the discipline. In terms of SSI, it is necessary to clarify what its core basic concepts are, so as to comprehend its denotation and connotation and integrate relevant resources, theories, and methods organically. Theoretically speaking, the understanding and positioning of the basic problems of SSI (such as disciplinary attributes, research objects, research methods, etc.) should be grounded on the system of essential concepts of SSI science. If the disciplinary system of SSI is a building, the concepts of the discipline are building materials, and defining the concepts related to SSI is the premise and

B. Wang and C. Wu, *An Introduction to Safety and Security Intelligence Science*,
https://doi.org/10.1007/978-981-96-7261-5_1

foundation for understanding, learning, researching, and exploring SSI. Therefore, the basic concepts of SSI are pivotal in the construction and development of SSI.

Although there is no need for academic research and discussions to be restricted within the textual research of a certain academic concept, nor do they need to follow a certain foregone conclusion, it is necessary to clarify and unify the basic concepts of the discipline. Especially, if a disciplinary and theoretical system is to be established for smooth academic dialogues, a preliminary system of disciplinary concepts is necessary. Therefore, the construction of a disciplinary SSI system which is based on the fundamental concepts of the discipline calls for the delineation and understanding of the concepts.

It is the basis and premise to neatly define the five concepts of SS, SS science, intelligence, intelligence studies, and SSI for "SS is the meta-concept of SS science and its branches," "SSI is a comprehensive interdisciplinary outcome of SS science and intelligence studies as far as disciplinary attribute is concerned," "SSI is the research object of SSI science." Yet it is somewhat regrettable that the current definition and understanding of the above five basic concepts of SSI are still vague, resulting in biased or even erroneous positioning and understanding of the research object, research scope, and disciplinary attribute of SSI in academia.

The vagueness in defining and understanding the basic concepts of SSI is reflected in four aspects: (1) The definition of safety and security in modern SS science has not been clarified, and the full comprehension of the connotation, denotation, and attributes of security has not been achieved; (2) The concept of intelligence has not been uniformly defined (Liang Zhanping 2003; Chen Chao 2017), leading to disagreements in defining and understanding the concept of SSI; (3) Although the concept of SSI is being interpreted in numerous forms in specific security fields and industries (such as national security, military security, biosecurity, technology security, public safety, emergency management, etc.), a unified and universal definition of SSI has not yet been formed; (4) The difference (distinction) between SSI and safety and security information (SSi) has not been clearly identified, causing the understanding of safety and security management (SSM) to stagnate at the level of SSi, waiting to be transitioned to the level of SSI.

The main reasons for the existing disagreements and problems in defining and understanding the basic concepts of SSI include: (1) SS and intelligence have been two highly controversial concepts, thus it is extremely difficult to reach a consensus especially in different disciplines where the definition and connotation of the two and the understanding of their denotations vary; (2) Some researchers with their academic backgrounds in intelligence studies and management disciplines have comparatively limited knowledge of the connotation and origin of SS science, and are not quite clear about the mainstream and scientific definition of security formed in the development of modern SS science, they may not be capable enough to examine and interpret the concept of SSI from the perspective of SS science.

Theoretically speaking, as SSI is an interdisciplinary field of intelligence studies and SS science, joint participation and communication of scholars from both fields are required in order to clarify the basic concepts of SSI. With such understanding, the author with his primary academic background in SS science has fully committed to the basic theoretical research of SSI in recent years, being an earnest learner and communicator of intelligence studies in the field. The long-term effort has brought him the encouraging impression that the definitions and understandings of the basic concepts of SSI are gradually converging with each other, which is beneficial to the solidarity of the foundation for the construction, research, and development of the SSI discipline.

For this reason, this section proposes the definitions of five terminological concepts, and briefly explains their meanings and categories, based on the mainstream views and their evolving trends. Although explanations of these concepts can be sought from related literature, they still need to be summarized and analyzed for a coherent discussion in the book.

1.2.1 *Safety and Security*

The Chinese word "安全" can be seen as a synthesis of two English words, namely, "safety" and "security." Piètre-Cambacédès and Chaudet (2010) have made detailed analyses toward the difference between the two concepts (see Table 1.1).

Based on the above understanding, Piètre-Cambacédès and Chaudet (2010) constructed a framework of system safety and security based on "System & Environment-Malicious & Accidental (SEMA)," see Table 1.1 (Fig. 1.1).

Strictly speaking, any issue of system safety and security belongs to the issue of safety-security integration according to Table 1.1. From the perspective of system safety and security, safety-security integration implies that both "safety" and "security" aspects of risks are included in the risks of "system safety and security," and that attention needs to be paid to the control of overall system safety and security risks in the promotion of system safety and security (Wu Chao and Wang Bing 2018a; Wang Bing and Wu Chao 2018a, 2018b). For example, urban system safety and security (referred to as "urban SS") involves security risk factors and safety threats including natural disasters, cases of social security terrorist attacks, industrial accidents, safety issues of underground pipeline, fires, traffic accidents, stampede accidents at public places, and so on. These issues cover both "safety" and "security" risks, which

Table 1.1 Differences between "safety" and "security"

Perspective	Explanation
System and environment	"Security" relates to risks that originate from the environment and may potentially impact the system. Conversely, "safety" risks that arise from within the system itself and have the potential to affect the environment
Malicious and accidental	"Security" typically focuses on risks that are maliciously inflicted, while "safety" addresses risks that are purely accidental. For any given system, it is inevitable that both "safety" and "security" risks will be present and can sometimes interconvert

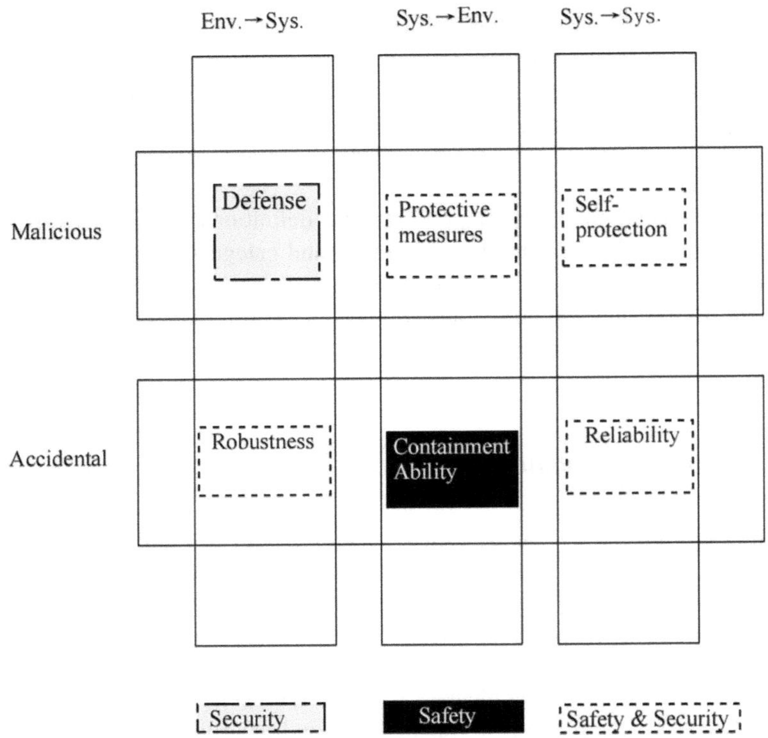

Fig. 1.1 Framework of system safety and security based on "System & Environment-Malicious & Accidental (SEMA)"

are usually intertwined and convertible. Therefore, urban SS is a typical issue of safety-security integration.

As research and development of SS science progresses to an advanced stage (or reaches the stage of system safety and security), both the academic and practical communities increasingly recognize that for a system (such as a system of enterprise), industrial system, urban system, social system, and national system), its safety and security issues inevitably involve risks concerning both "safety" and "security," which mutually influence, intertwine, and transform into each other, and are extremely difficult to separate (Wu Chao and Wang Bing 2018a; Wang Bing and Wu Chao 2018a, 2018b). It's obvious that "system safety/system security" has gradually evolved into a composite issue and concept of "system safety and security" (i.e., safety-security integration) (Wang Bing and Wu Chao 2018b).

In recent years, SS practitioners have attached more and more importance to the research on safety-security integration, and have begun to recognize and discuss the composite concept of "safety and security" (Florea and Popa 2012; Riel et al. 2018; Wang Bing and Wu Chao 2018b). For example, various SS issues within the perspective of safety-security integration (such as SS of key facilities and equipments, SS of major projects, SS of important industrial systems such as chemical industry, nuclear SS, social SS, and so on) have attracted widespread attention; Technische Universiteit Delft in the Netherlands (a world-renowned SS science research institution) has established the "Safety and Security Science Group" on the basis of the past "Safety Science Group"; in 2017, Central South University (a key SS science research institution in China) where the author works established the "Safety and Security Theory Innovation and Promotion Center of Central South University (STIPC of CSU)"; in 2018, the internationally renowned SS science academic journal *Safety Science* posted a call entitled "Editor Security Research Selection" on its journal website to invite papers on safety-security integration; in 2020, Beijing University of Technology established the Research Institute of Macro-Safety Science, here "Macro-Safety" obviously covers both "safety" and "security."

Although definitions about "safety" or "security" have been many, there has not yet been a definition integrating both. From a system perspective, the book proposes the definition of SS on the basis of the connotation of safety-security integration discussed above, that is, the state of a system being free from unacceptable adverse effects of internal and external factors (Wang Bing and Wu Chao 2017a). The logic and mathematical expression of this definition is:

$$\begin{cases} S = (0, X_0) \\ X = f(x_1, x_2) = f(x_{11} + x_{12},\ x_{21} + x_{22}) \end{cases} \tag{1.1}$$

where S is safety of the system; X is the adverse effect of internal factors on the system; X_0 is the biggest (thresh hold) of the acceptable internal and external factors on the system; x_1 is the internal factor of the system; x_2 is external factor of the system; x_{11} is the deliberate risk within the system; x_{12} is the accidental risk within the system; x_{21} is the deliberate risk outside the system; and x_{22} is the accidental risk outside the system.

Here the specific manifestation of "adverse effects" include interference or termination of the normal operation of the system, reduction or even failure of system functions, occurrence of SS incidents (including accidents) within the system, and losses within the system, etc. According to the attributes (relativity) of SS and SSM habits, "acceptable SS risks" can serve as the basic criteria for measuring and judging the "acceptance degree of adverse effects of internal and external factors." Evidently, the SS defined above belongs to the broad concept of SS: (1) In terms of its content and connotation, it includes both "safety" and "security"; (2) In terms of subjects, industries, or fields, it targets at universal safety (the most general and universal safety), which transcends all specific subjects, industries, or fields. This is because the "system" in the definition can refer to any specific subject, industry, or field including production, sector, city, country, and society, which is the so-called "general safety and security" in the SS field (which will be detailed in Sect. 3.2). Please see Fig. 1.2 for the evolving process of the SS concept.

The definition of the system is also presented here for understanding and clarifying SS: the system refers to an organic whole with specific functions formed by the

Fig. 1.2 Schematic diagram of the evolution process of the SS concept

combination of several reciprocal and interdependent components (Wang Bing and Wu Chao 2017a).

1.2.2 SS Science

With the proposal of the compound concept of safety and security and the recognition of the broad concept of SS by the academia, the academia has proposed the broad concept of modern safety and security science (modern SS science), born out of the integration of the traditional "safety science" and "security science."

SS science is a discipline that studies SS promotion (prevention and control of SS events) (Wang Bing and Wu Chao 2018c). Its research object is SS; the research purpose is to promote safety and security (prevention and control of SS events); its research path is a combination of forwarding (starting from system SS) and reversing (starting from SS events); the research content is the laws of occurrence and development of SS (SS events) and means of SS promotion (prevention and control of SS events); the research scope is the SS of all systems. Please see Fig. 1.3 for the basic problems of modern SS science in the broad sense (i.e., general SS science) (Wang Bing and Wu Chao 2017a, 2018d).

It is worth noting that in the current catalogue of discipline classification in China [*Discipline Classification and Code* (GB/T 13,745–2009)], the English for SS science is "Safety Science" with its primary research areas being production safety and certain aspects of public security, as the narrow sense of traditional security science indicates, which is different from the broad sense of modern SS science. In contrast, the SS

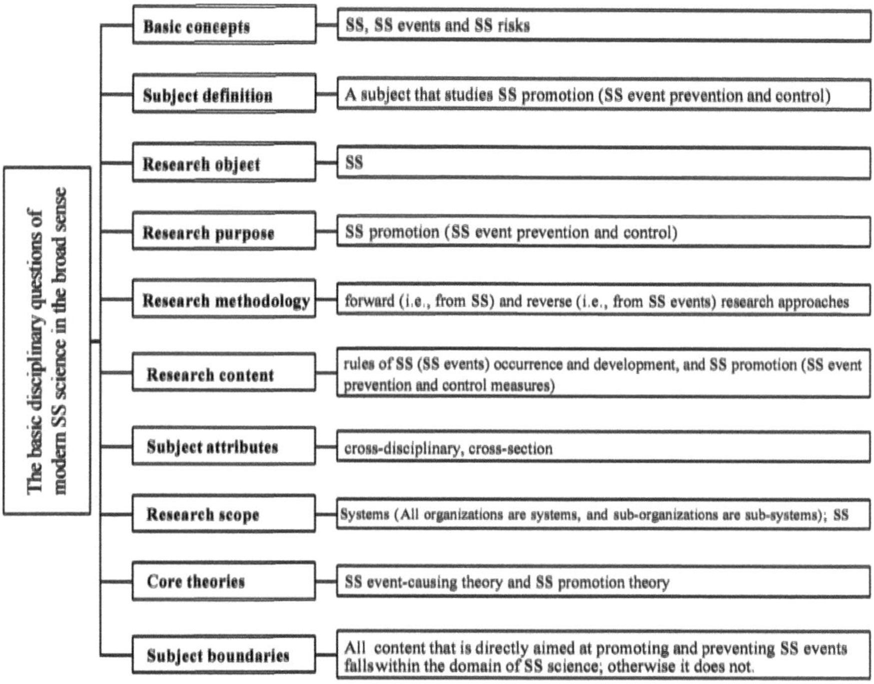

Fig. 1.3 Basic issues of modern SS science in the broad sense

science discussed in this book refers to the "broad sense of modern SS science." It belongs to general SS science, which is a vast discipline (or system), including a wide range of branches (such as studies of national security, cyberspace security, public security, production safety, engineering safety and resource safety, etc.). There ought to be more branches in this disciplinary system considering the correspondence between the system and the universality and diversity of safety and security in reality.

The discussion about the connotation of SS science shows that, the SS science is bound to develop from "safety science" or "security science" to the comprehensive concept "safety and security science" in the twenty-first century. In addition, the modern SS research practices tend to pay more heed to the idea of "prevention first," and stresses the idea that early identification and prevention of SS events is the best SSM mode to improve the level of SS in life and production, according to Wang Bing and Wu Chao (2018d). Therefore, the SS science of the twenty-first century is going to evolve and develop from the SSM mode of "focusing on post-event and mid-event control" to the full-cycle, whole-process, and all-round SSM mode which "combines the pre-event prevention and post- and mid-event control."

1.2.3 Intelligence

Seen from the inclusion relation of concepts, SSI belongs to the category of intelligence, where "intelligence" is a generic concept and "safety and security intelligence" is a subordinate concept, thus the insight into the concept of intelligence is necessary for comprehending the concept of SSI. Intelligence originated from human's social activities, a product of social development. From its broad sense, intelligence came along with humans beings, as the latter needs to frequently receive, transmit, and utilize various types of intelligence in their daily life and social production, though the early form of intelligence was rather primitive, and the content and form of intelligence exchange were simple. However, both the content and form of intelligence exchange and transmission have undergone great changes as the human society develops and progresses.

The development of intelligence has seen four major turning points. The first was the appearance of military intelligence in military actions caused by intensified contradictions between tribal alliances. The second was the emergence of intelligence exchange as a result of the birth of writing, the invention of paper-making and printing, leading to major changes in the way of intelligence transmission. The third occurred with the advancement of human civilization, science and technology, intelligence as work developed to be more and more professional as intelligence became the premise and basis of politics, economy, culture, education, science, and technology. The fourth was the socializing of intelligence which occurs with the increasing importance of the intelligence industry in national governance, national economy, and social development. In the historical development of intelligence practice, people's understanding of the intelligence concept intelligence has gone through changes as it varies from time to time, closely bound with the circumstances of intelligence operation at a particular time. Some of the representative understandings and recognition of the concept are listed and presented as follows (Liang Huiwen 2017).

The early concept of intelligence originated from and was limited to the military field (Xie Xiaozhuan and Gao Jinhu 2020). In *Etymology* (*Ci Yuan*, or *Origin of Chinese Characters*, the October 1915 edition), "情报"(qing bao, meaning intelligence) is so defined: "The various types of military reports submitted to the superior, with predictions or early signs of the enemy's military activities, and determination on the state of the enemy." In *Ci Hai* (*A Sea of Chinese Characters*, a grand dictionary of Chinese characters and words, the 1939 edition), it is explained as: "Intelligence refers to the latest reports about enemies in war time." In *Ci Hai* (the 1969 edition), it is defined: "Intelligence generally refers to all reports on the latest situation." The word intelligence has a sense of secrecy and confrontation if understood from the military perspective, which represents people's instinctive understanding and feeling of the word (Liang Huiwen 2017).

With the rapid development of science and technology in the 1950s and 60s, the number of literature records increased as most countries established institutions of scientific and technological intelligence (also known as "scientific intelligence" or "literature intelligence"), whose main tasks included compiling abstracts, catalogs, and indices, for the aim of carrying out bibliographic control management and utilization of literature intelligence. The concept of intelligence, with the nature of serialized knowledge, perfectly reflected the intelligence operation during that period. Research shows that in 1963 Tadao Umesao, a Japanese scholar, defined intelligence as all symbolized and serialized knowledge communicated between people (Zhou Bailin 1997). As the volume of scientific and technological literature expanded, computers became popularized and were applied to literature management, assisting in solving problems of large-scale collection, condensation, and serialization of knowledge. The computer-aided "serialization" evolved from traditionally treated objects like some specific books and publications to many pieces of literature, and then to numerous knowledge units (i.e., data). In the field of scientific and technological literature, the above evolution has prompted the shift of the concept "intelligence" from treating literature as a work object to treating the content of literature (i.e., knowledge) as a work object, which gives the term "intelligence" a unique sense—the transmission of intelligence as knowledge (Liang Huiwen 2017).

As intelligence develops from the military field to fields of politics, economy, science and technology, business, management, culture, education, and so on, this concept is no longer restricted to "messages or reports with confidential nature," but gradually forms closely ties with terms such as literature, data, news, information, and knowledge. For instance, the 1989 and 1999 editions of *Ci Hai* explain it respectively as: "Intelligence is the result of obtaining and analyzing the other party's situation" and "Intelligence is the result of acquiring and analyzing the other party's situation." It's worth mentioning that these two editions supplement the denotation of the concept, and divide it into political intelligence, economic intelligence, military intelligence, and scientific and technological intelligence according to the different connotations and attributes of intelligence. Sherman Kent, the father of American strategic intelligence (Kent 1951), believes that "intelligence is the specific knowledge pursued by an organization through practice." Kent stripped away the mysterious veil that has covered the work of intelligence for thousands of years, helping people know the fact that intelligence is the knowledge or information that is indispensable in life, and "in essence, the work of intelligence is the process of seeking the only best answer" (Gao Jinhuh and Wu Xiaoxiao 2014). Qian Xuesen, the renowned Chinese scientist highly praised the "knowledge theory" of the intelligence concept. At the 1983 National Conference on Defense Intelligence, he pointed out that intelligence is the knowledge activated and revitalized (Yuan Youxiong 2013). In 1988, Yang Zhengrong and Xie Hong (1988) raised the point that intelligence is the new knowledge that can affect decision-making.

With the development of intelligence studies, the "knowledge theory" of the traditional intelligence concept has evolved into the "information theory" of the concept. For instance, the American scholars Yovits and Kleyle stated that intelligence is the valuable data and information for decision-making (1993). Yang Jianlin (2020) pointed out that intelligence is the information obtained from people or groups other than stakeholders on the basis of the problems that users need to solve and the environmental constraints that users face. In other words, intelligence can be seen as information with the value of transmitting. Now both the theoretical and practical circles of intelligence agree that it is a kind of useful "information (including knowledge)." This should be seen as the essence of intelligence (Chen Chao 2017). To put in another way, the fundamental attribute of intelligence is possessing value, to be more specific, it is the useful information provided for the needs of specific objects.

The definition of intelligence has always been one of the hottest topics for intelligence theorists. Although no consensus has been made on the definition of intelligence, two universal ideas can be sought from the various viewpoints (Wang Bing and Wu Chao 2018b): (1) Though intelligence is the "processed information" "activated," its essence is still information (in terms of logical order, information comes before intelligence); (2) When it is oriented toward management practices, intelligence research aims to serve management (mainly decision-making) and solve the problem of intelligence lacking (incompleteness) in the management process. Therefore, intelligence in its general sense can be understood as referring to all information (content) that influences management (Chen Chao 2017).

1.2.4 Intelligence Science

Intelligence science as an independent research field came into being with the rise and development of scientific and technological intelligence in the 1950s. For example, in 1967, Mikhailov, director of the Soviet Institute of Scientific and Technological Intelligence, published "Intelligence Science: A New Name for the Theory of Scientific and Technological Intelligence," which stated: "The construction and characteristics of scientific and technological intelligence are presented by this new discipline: intelligence studies, though inappropriate to be called a scientific theory. In fact, intelligence studies is more than a theory, as it involves issues of research methods and history, and organizational issues, while scientific and technological intelligence refers to activities of scientific and technological intelligence" (Ma Feicheng 2013). It has been 70 years since intelligence studies started developing. In terms of modern Chinese intelligence studies, most intelligence scholars traced from the establishment of the Institute of Scientific Intelligence, Chinese Academy of Science in 1956 in reviewing the development of Chinese intelligence studies. Some scholars also believe that the establishment of the Chinese Society for Scientific and Technological Intelligence in 1978 marked the official start of the developing process of Chinese intelligence studies (Wang Yanqing 2020).

Generally speaking, intelligence studies refers to a discipline that studies the generation, transmission, and utilization of intelligence. In other words, it is a science that studies the rules, principles, means, and methods of intelligence collecting, organizing, storing, retrieving, reporting services, and analyzing (Ye Ying and Wuyishan 2012). Modern intelligence studies highly values the application of modern information technology and means for the intelligence communication process and the intelligence system to keep their best performance. They also aid intelligence practitioners to harness information techniques and means to improve the efficiency of intelligence generation, processing, storage, retrieval, communication, and utilization (Ye Ying and Wu Yishan 2012).

The understanding of intelligence and intelligence studies in the domestic and foreign intelligence community is diverse and there hasn't been a unified standard yet. Representing opinions include: intelligence studies "is a discipline that studies the theory, method, and principle of intelligence transmission," "[it] is a discipline that studies how to develop and utilize information resources, its technology and methods," "[it] is a discipline that studies how humans create, use and communicate various forms of intelligence," and so on (Ye Ying and Wuyishan 2012). Generally, it can be viewed as a discipline that studies intelligence itself, and the theory, principles, technology, and methods of intelligence operation (Yan Yimin 1994; Ye Ying and Wu Yishan 2012).

As a new discipline which started developing about half a century ago, the main research tasks of intelligence studies include: (1) revealing the nature and characteristics of intelligence and intelligence operation; (2) exploring the laws of generation and development of intelligence operation; (3) discussing the content of intelligence operation and their status and role in the work, their internal connections, and the principles and methods of each procedure; (4) studying the guidelines, policies, standardization of intelligence operation in various periods, as well as modern intelligence theory, methods, and technologies; (5) clarifying the scientific management theory and methods of intelligence operation, etc.

Being an emerging discipline, it studies the nature and characteristics of intelligence, the influencing factors of intelligence circulation, and the techniques and methods for effectively querying, acquiring, and utilizing intelligence. In other words, it is a comprehensively transversal discipline for its characteristics of an interdisciplinary or cross-disciplinary field. Its related disciplines include information science (information theory), control theory, intelligent science, data science, linguistics, semiotics, cataloging, bibliography, archival science, library science, etc. (Ye Ying and Wu Yishan 2012).

With the upcoming of an information society, the importance of intelligence studies becomes more and more apparent, especially in the era of big data and intelligence. Its role and research results are regarded as one of the powerful pillars of the information society (Ma Feicheng and Zhao Zhigeng 2019). It combines various high-tech technologies and has gradually developed a relatively complete disciplinary system and research norms. It aims to explore and reveal various laws of human intelligence practices in the future information society, as well as the important impact and value of intelligence in social, economic, and cultural development.

1.2.5 SSI

1.2.5.1 Who Will Unify the Concept of SSI

Although literature retrieval about SSI shows that results about specific research on SSI are rare, three features can be sought among the related research (Wang Bing and Wu Chao 2019a): (1) In terms of research themes, the research topics focus on "security," like national security, military security, scientific and technological security, cyberspace/information security, etc. There are also studies of "security and safety" like public security and safety, urban security and safety emergency, etc., but research results related to "safety" are extremely rare [There are only a few studies of disaster intelligence (Su Youpo and Liu Ruixing 2001). In recent years, the author has conducted some exploratory research in this aspect]; (2) Regarding the research fields, research results are scattered in specific security fields such as national SS, military SS, scientific and technological SS, and public SS, or a specific SS link (mainly emergency management). As the research orients itself toward the overall SS (or "comprehensive SS/General SS"), and the research that focuses on the whole-process SSM is relatively scarce; (3) In terms of researchers, they are mainly concentrated in fields of intelligence studies and public management with only a few SS science researchers following and studying SSI, resulting in a serious lack of SSI research made from the perspective of SS science.

At present, research on SSI are scattered, without guidance under a unified norm. A consensus is yet to be formed by researchers in different fields about the concept of SSI. This does not only lead to inadequate understanding of the concept, but also causes obstacles and difficulties for academic dialogues and exchanges about SSI between scholars of different fields. Concepts are critical as they are the summary and refinement of past understandings as well the starting point for a new understanding. They should be the logical starting point and basic unit in the process of comprehension. Therefore, in SSI research, clarifying the concept of SSI is a urgent task. The author believes that researchers in SS science need to unify the concept of SSI with joined efforts. The reasons include:

1. As a new academic concept, SSI denotes the direct intersection of intelligence studies and SS science. It is an organic combination of the basic concept of "safety and security" in SS science and the basic concept of "intelligence" in intelligence studies. Therefore, one needs to understand the concept of SSI from both perspective as both are indispensable, and the interpretation and research on SSI from the perspective of SS science calls for improvement in terms of related outcomes. Besides, a common attribute is shared by both sciences, that is, both have a broad research scope of the disciplines with complex research content (Wu Chao and Wang Bing 2018a; Liang Zhanping 2003; Bao Changhuo 2009). As SSI studies from both fields complement each other by exchanging and discussing, the SSI research will surely be deepened and enriched.
2. SSI researchers from other disciplines tend to examine SS issues of a specific field or problems of a certain SSM link from the perspective of intelligence

studies (Wang Bing and Wu Chao 2019a). The current development of SS science emphasizes the research paradigm of system SS, which calls for attention on all influencing factors of system SS from a holistic (overall) perspective (Wang Bing and Wu Chao 2019b). In the meantime, SS science researchers are generally equipped with knowledge and experience of SS science research and practice. Therefore, in fulfilling the ultimate goal of unifying the SSI concept, they tend to be more capable of understanding and interpreting the concept from an overall perspective when compared with their counterparts of other fields.

3. SSI is a new frontier of interdisciplinary research between intelligence studies and SS science. The recent years have seen the concept of intelligence studies being understood, recognized, and valued by a growing number of scholars of intelligence studies, as is shown by large amounts of SSI research on intelligence studies. Yet, SSI remains to be unfamiliar to the majority of SS science researchers, which indicates that SSI will exist as an original and attractive concept in SS science. The exploratory research results of SSI by researchers in intelligence studies will enlighten SS science researchers and assist them in blazing new trails in SS science. Therefore, SS science researchers need to approach and understand the concept of SSI, and conduct exploratory research on SSI for a deepened understanding of the issue.
4. In recent years, the work of informationalizing SSM (or "informationalizing SS") has been carried out vigorously by organizations and institutions (such as government departments and enterprises, etc.) with a large array of SSi systems and platforms being constructed. However, inadequacies remain in the practice of SSI: (1) There is ambiguity in the understanding of SSI in SSM as the comprehension about SSM stays at the level of SSi (which will be detailed in Sect. 2.3), and SSM practitioners' performance of intelligence is inadequate; (2) The coherent network of SSI in SSM has not been formed (the interconnection between SSM information systems has not been fulfilled), hence the difficulty of collaboration between subsystems for a smooth sharing of SSM; (3) An adequate recognition and respect is yet to be earned from the SSM field, so that more professional intelligence personnel and institutions would be willing to join in the intelligence network of SSM, enabling a broader expansion, penetration, and integration of intelligence operations in SSM.

In conclusion, there is large space for exploration in the "rich mine" of SSI, the intersection of SS science and intelligence studies. Researchers of intelligence studies have already taken actions, which are to be further supported and complemented by SS science workers. As the specialized research on SSI for SS science (especially SSM) is still insufficient, it is urgent that research be conducted from the perspective of SS science. The preliminary task is to interpret and define the concept of SSI from this perspective, which is hoped by all SSI researchers. As a researcher of SS science, the author recognizes in his recent efforts the urgency and necessity of SSI research for the exploration in research and practice of SSi science. In this regard, this section attempts to offer the definition of SSI and analyze its connotation based on existing research fruits.

1.2.5.2 Definition of SSI

SSI, a sub-concept of intelligence, can be understood as "intelligence related to security" in plain terms. Of course, explanation like this is far from specific, scientific, or rigorous, as the efficacy and essence of SSI are not expressed, and the connotation and characteristics of SS science are not presented. For these reasons as well as the interpretation of Sects. 1.2.1 and 1.2.2, it is necessary to delineate the concept of SSI on the basis of the theoretical level of SS science in order to achieve unity in understanding, expression, and communication for adapting to the evolutionary trend of SS science and the development of SSI concept, and averting confusion between various sub-concepts of intelligence.

SSI is a combination of the two concepts: "safety and security" and "intelligence." Considering the SS definition from its broad sense and the definition of intelligence in SSM, SSI can be defined in the broad sense from the perspective of system SS science as referring to all SSi that influences (systematically) SSM (Wang Bing and Wu Chao 2019a). In detail, it refers to all SSi that influences SSM behaviors (mainly including SS prediction, decision, and execution) of SS managers ("individuals" and the "organizational units, i.e., the SSM institutions"). Here the implications of the SSI concept are to be analyzed for a deeper understanding of its definition:

1. This definition clarifies the essence of SS—a type of SSi. What needs to be stressed here is that the concepts of "SSi" and "SSI" are not to be confused. The difference and connection between them are similar to those between "information" and "intelligence" (see Sect. 2.2 for detailed explanation). While SSi runs through the entire process of SSM work, SSI is the SSi that is directly associated with SSM problems and uncertainties in SSM and the useful (i.e., of critical value and significance) SSi for SSM.
2. This definition recognizes the value (i.e., utility) of SSI—influencing SSM. Seen from the perspective of elements of SSM behaviors, SSI mainly influences three types of SSM behaviors, namely, behaviors of SS prediction, decision-making, and execution.
3. This definition emphasizes the essence of SSI—its purpose is to serve SSM. The key value of SSI lies in overcoming problems of lacking SSI in SSM, as SSI demonstrates prominent effects of supporting, improving, and correcting SSM behaviors.
4. This definition follows the development trend of the SSI concept. There are significant advantages in defining SSI from the perspective of the system SS science, for instance, such a definition drawn from the perspective of system SS embodies the general evolution of the SSI concept, which encompasses the overall (general) SS, SS integration, and the entire process of SSM.

1.2.5.3 Connotation of SSI

According to its definition, SSI is a generalized academic term which shares common traits with "intelligence." In other words, the concept of SSI has multiple meanings.

This book holds that SSI is a kind of product, process, tool (or technique), and capability.

1. SSI Is a Product

Intelligence professionals stress that if the collected intelligence (or inoperable) proves to be unusable (or inoperable), it is not intelligence (Turban et al. 2008). Thus, as a product, SSI should be the kind of SSi (including SS knowledge) that is usable and operable, a feasible output determined by an organization according to SSM needs. It refers to all available information (including data) related to one or more aspects of SS and is generated in collecting, evaluating, organizing, analyzing, integrating, and interpreting. These kinds of information are directly or potentially significant to the SSM of an organization (like the formulation and implementation of SS plans and policies or management of SS risks). In short, SSI is the product generated from the processed (i.e., analyzed and interpreted) SSi.

2. SSI Is a Process

Firstly, SSI is a process of acquiring operable and useful SSi, through planning, collecting, processing, and analyzing SSi from internal and external environments of an organization for the aim of assisting SSM and improving an organization's SS level. There are 5 stages in the process:

(1) SSI planning, like focusing on the SSI needs of SS managers of an organization and the urgent SS problems that the organization needs to solve.
(2) SSi collection, like gathering SSi from various sources inside or outside an organization.
(3) SSi processing, like classifying SSi according to certain rules from different perspectives, and obtaining key SSi from physical and electronic SSM records with various means and techniques.
(4) SSi analysis, like converting SSi into operable SS decision-making information.
(5) SSI reporting and dissemination, like reporting and conveying SSI to personnel of an organization who are authorized and responsible for taking actions on the result of SSi analysis.

According to Wang and Wu (2019) and Brummer et al. (2006), the SSI process is cyclical with a series of related steps (Fig. 1.4). SSi analysis is the critical step to transform information of raw SS data into the SSi that supports SSM.

As a process, SSI refers to the process that transforms fragmented SS data information into operable information or knowledge associated with organizational SS goals and SSM suggestions, improving the organization's SSM abilities with such information or knowledge.

Secondly, SSI is a process driven by SSi (including SS data), which combines SSi storage and collection with management of SS information and knowledge, so as to provide input for the SSM (especially SS decision-making) process.

Thirdly, SSI is a process driven by software and technology as SSi processing relies on software and information technology. This process aims at making more

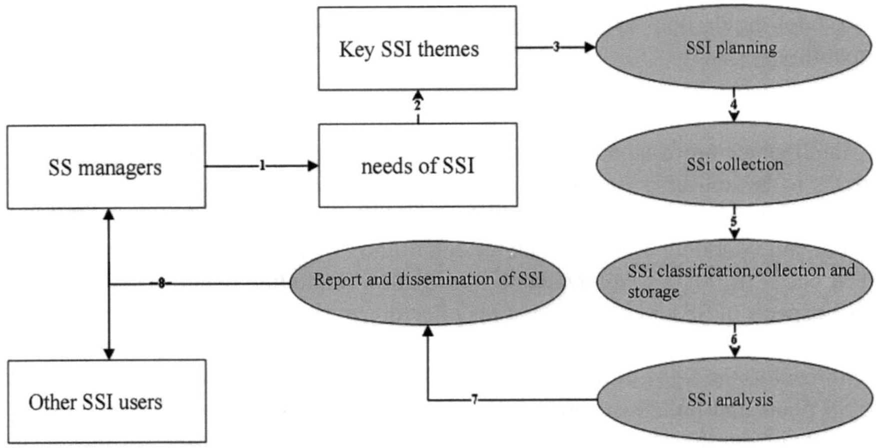

Fig. 1.4 Cycle of SSI process

effective SSM decisions by analyzing raw SSi from multiple sources and extracting insights from it.

Fourthly, seen from the perspective of SSM, SSI is a process providing support for SSM and a reflection of the SSI-based SSM process (or "intelligence-led SSM"). In other words, SSI is an important component of the SSM process.

3. SSI is a tool (or technique)

Firstly, as an effective tool (or technique), SSI is able to transform SS data into SSi, which is further transformed into an SS decision, thereby promoting SSM actions. SSI includes data warehouses, SS analysis (including SS risk analysis) tools, and SS data/information/knowledge management. Specifically, as a tool (or technique), SSI consists of a series of architecture and technologies (such as databases, data warehouses, data mining, etc.), which convert raw SS data into useful SSi to support SSM. In the meantime, various information technologies are essential supporting technologies for SSI, for example, the SSI system adopts the technology of machine learning to identify relevant information (the relationship between the matters that change in organizational SSM without an apparent cause), thereby intelligently predicting SSi, which is beneficial for managers of organizational SS to better understand the development and alteration of organizational SS-related factors and their interrelationship. In recent years, the rapid development of information technologies such as database management, data warehouse, big data mining, etc., and their application in SSM have promoted the application of SSI in SSM (Fig. 1.5). It can be concluded that as a tool (or technique) for processing and analyzing collected SS data information, SSI requires a multitude of tools developed from disciplines of SS science, data science, intelligence studies, computer science, and artificial intelligence studies.

Secondly, SSI is a powerful tool and technique for analyzing the cause-and-effect relationship of SS incidents and diagnosing SS risks as it provides a method driven by SS data information, which can effectively associate the strategic SS goals and

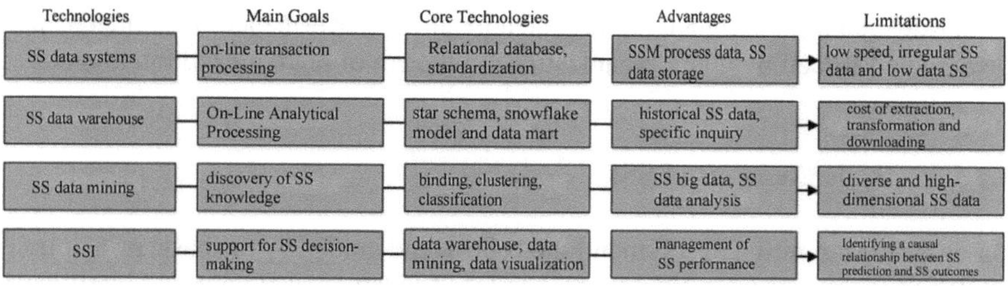

Fig. 1.5 Comparison of various information technologies applied in SSM

policies of an organization with tactical SS procedures and front-line operational SSM activities.

Thirdly, SSI technologies need to be considered for improving the quality and efficiency of SSM if SS managers of any organization wishes to achieve success in organizational SSM. SSI technologies can assist SS managers to make swifter and more effective SS decisions with highly accurate and valuable SSi.

In summary, SSI is a set of tools and techniques that enables an organization to convert SS data information into timely and accurate SSi which is needed in the SSM process, and transmit it to the right personnel in its most appropriate form.

4. SSI Is a Capability

Firstly, SSI is an organization's capability to collect and process SS data information. Secondly, SSI is an organization's ability to solve SS problems as SSI focuses on SS problem-solving, which is also the direct goal and task of SSM. Thirdly, SSI is an organization's capacity to recognize and predict SS risks and changes and take timely measures. This capacity covers foresight and insights into SSM. For instance, SSI can identify imminent SS changes in an organization, which may be positive for they represent opportunities for SS improvement, or negative for they represent SS threats or challenges. SSI also reflects many other capacities, such as:

(1) SS learning capacity (such as acquiring new SSi and understanding the latest progress in SS research);
(2) The capacity to adapt and transform the organizational SSM environment;
(3) The capacity to understand SSM factors (such as hazards, incidents, insecure behaviors, SS culture, and SS resources) and take appropriate actions with this understanding.

In summary, from the perspective of SSM, the new term SSI of SS science and intelligence studies represents the modern thought and means of SSM, which integrates methods, processes, tools, techniques, and capabilities to transform raw SS data into SSI products (i.e., meaningful and operable SSi) in order to support SSM activities. By identifying SS risks, new opportunities of SS promotion, hazardous factors and potential SS threats, SSI could reveal fresh SSM insights and optimize SSM processes. Therefore, it can play a significant role in promoting organizational

SS advancement and enhancing organizational SS performance. It is increasingly favored and valued by many organizations in the era of big data and intelligence.

1.2.5.4 Basic Elements of SSI

Based on the definition of SSI mentioned above and guided by the basic principles of intelligence services (Yan Yimin 1994) and the fundamental elements of intelligence (Yao Leyuan and Fan Wei 2014), four basic elements of SSI are formed with the characteristics and practicalities of (system) SSM operation (Wang Bing and Wu Chao 2019e). They are accurate SSi, appropriate SS scenarios, qualified SS managers, and reasonable SS costs, as explained in Fig. 1.6.

As can be seen from Fig. 1.6, the process of converting information into intelligence in SSM would be unbalanced if a basic element of SSI is considered in isolation. The four elements of SSI ought to be taken into consideration for collecting and utilizing SSI in SSM, that is to say, the SSI products can only be produced and fully utilized by SS managers when the accurate SSi is used as the basis and elements of appropriate SS scenarios, qualified SS managers, and reasonable SS costs are adequately conjoined. In this way, the supporting role and values of SSI can be exerted in SSM.

1.2.6 Conclusions

Basic concepts of a discipline are the foundation and starting point for understanding and acquiring disciplinary knowledge and conducting related research. This section defines and analyzes the five basic concepts involved in SSI, namely SS, SS science, intelligence, intelligence studies, and SSI, targeting at clarifying the fundamental concepts of SSI. The above concepts are also the primary ones involved in this book,

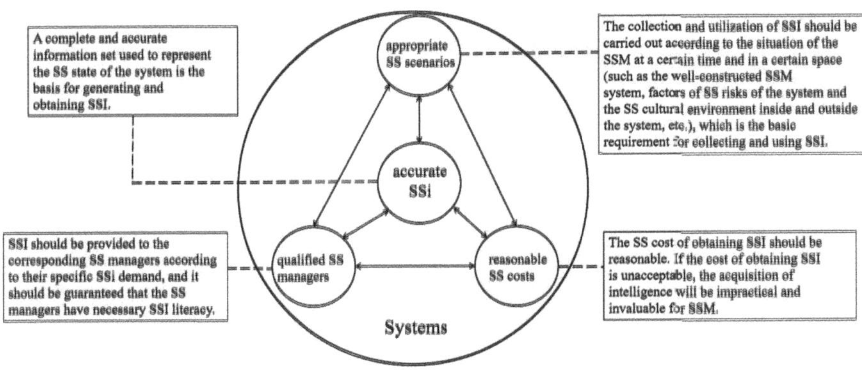

Fig. 1.6 Four basic elements of SSI

thus comprehending these concepts serves as the basis for learning the following chapters of this book.

1.3 Background and Significance of SSI Science

Human beings exist in a society of risks (or even high risks) where various SS threats, risks, and incidents emerge one after another, posing a menace to the human society (Jin Qian 2017; Wu Chao and Wang Bing 2018a). Thus the issue of SS has become one of the thorny problems which are highly concerned by the society.

Intelligence science has always highlighted the study of significant social issues, and advocated and called for interdisciplinary research with other related disciplines (hence the emergence of branches of intelligence studies, such as public SS intelligence, science and technology intelligence, medical intelligence, competitive intelligence, and military intelligence) (Bao Changhuo 2009). The research orientation of SSI (such as national SSI, cyber-SSI, military SSI, science and technology SSI, and emergency intelligence) was born in response to this trend, and has gradually developed into a core emerging disciplinary branch following the traditional branches of intelligence studies like national intelligence, competitive intelligence, public SS intelligence, military intelligence, and science and technology intelligence. Moreover, from the perspective of intelligence studies, the direct task of SSM is to collect, analyze, process, and transmit SSi (Wang Bing and Wu Chao 2018d), thus SSI operation is the key to the success of SSM, and research on SSI is of high theoretical importance and practical value for improving SSM capabilities (Wang Bing and Wu Chao 2019c).

For this reason, in recent years research on SSI has expanded into intelligence studies and SS science and become a new growth point in the intersection of the two disciplines (Wang Bing and Wu Chao 2018a). It is clear that the important value of SSI itself and the existing foundation and practice of research on SSI are driving the intersection and integration of the two disciplines, which will inevitably give birth to a new branch—SSI Science. The construction and development of the discipline of SSI science is unstoppable.

Theoretically, the birth, rise, and development of SSI science should be rooted on reasonable background with definite importance. Therefore, in terms of the construction of SSI science, the primary task is to analyze and demonstrate the background and significance of the construction of SSI science profoundly and meticulously, with the expectation of clarifying the necessity, importance, and scientific feasibility of the construction of the discipline. At present, SSI science is still in the stages of "brewing" or "accumulation," or in its "infancy." In order to help the academia accept and embrace SSI science, colleagues in SS science and intelligence studies should be more concerned with the question of "why should we carry out the construction of SSI science?" as they accept SSI science in an open and inclusive attitude and support its disciplinary development. With this in mind, this section has centered around the analysis of the construction background and significance of SSI science, which are divided into the content of academic background and importance, and practical background and significance.

1.3.1 *Disciplinary Background and Importance*

1.3.1.1 Objective Requirements of the Disciplinary Development of Intelligence and SS

The name of SSI science can be explained neatly as: SSI science is a specific field of applying the theory, methods, and technology of intelligence studies to SS science (especially SSM), aiming to abstract and summarize relevant principles and laws in the practical work of SSI. It is clear that SSI science is a new interdisciplinary subject formed by the direct intersection and integration of intelligence studies and SS science, or SSI science is a branch discipline extending from intelligence studies and being closely related to SS science. Therefore, the original disciplines of SSI science are intelligence studies and SS science. Members of this science "family" are closely linked, and mutually dependent, promoted, and enriched in the general and theoretical development of both (Zhao Rongying 2017). Therefore, research of related branch disciplines of intelligence studies and SS science should be highlighted and strengthened, as they create a breeding ground for SSI science. According to the needs and trends of disciplinary development, the construction of SSI science has inevitably become an important opportunity for the expansion and deepening of intelligence studies and SS science.

Firstly, the construction of SSI science is a requirement of the development of intelligence studies. As both sciences study the characteristics and laws of procedures in the process of intelligence production, transmission, and utilization, they possess the same nature. So far, intelligence studies has more than 15 sub-disciplines (branch disciplines), whose development will further improve the disciplinary development of their original disciplines (Zhao Rongying 2017), which holds true for intelligence studies. In the meantime, as the construction of SSI science benefits the broadening the research scope, content, and application and practice of intelligence studies, it will contribute to the deepening of the research and application value of intelligence studies.

Secondly, the construction of SSI science is a requirement of the development of SS science. Both sciences share the same research purpose—to provide support and basis for SS promotion (Wang Bing and Wu Chao 2018a). As modern SS science is a disciplinary category with a large group of disciplines, major groups, and field groups (Fig. 1.7), there are broad application fields and infinite application prospects for SSI science. In recent years, SS-related disciplines and majors have been officially acknowledged. For example, five first-level disciplines of the SS category have been established in China, they are Safety Science and Engineering (Code 0837), National Security Studies (Code 1402), Cyberspace Security Studies (Code 0839), Public Security Studies (Code 0306), and Public Security Technology (Code 0838). What

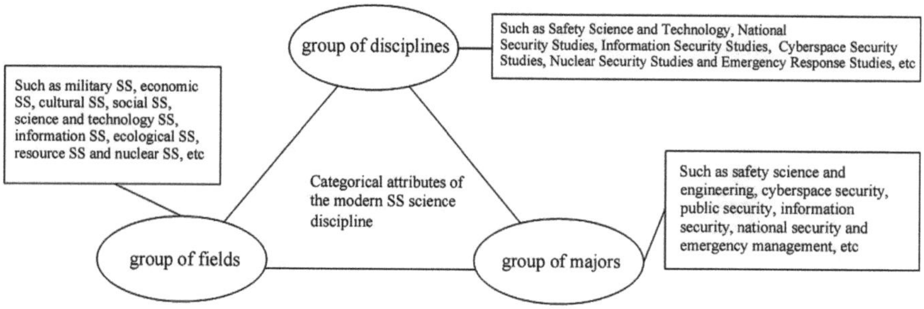

Fig. 1.7 Disciplinary attributes of modern SS science

also have been established are a system of branches of general SS sciences (such as SS science, SS philosophy, SS informatics, SS systematics, SSM science, SS pedagogy, SS cultural studies, and SS law) which are universally applicable and a system of branches of SS science application targeted at specific departments/fields/entities/industries, which have contributed to establishing the discipline for SSI science as a legitimate one. The construction of this discipline is conducive to further broadening the connotation and research scope of SS science, adding fuel to the future development of SS science.

SSI science is a natural outcome of the developing SS science at the stage of "Security 4.0." It originates from the industrial society (specifically, issues of industrial SS), and evolves and changes with the development of human industrial production (Wang and Wu 2020). The advancement of human society shows that starting from the eighteenth century, the history of human society can be seen as a history of industrialization, thus SS science existed concomitantly with the industry, progressing with the advancement of industrial development. According to different stages of industrial development (Wang Feiyue 2015), the modern SS science has undergone four major transformations (Wang Bing and Wu Chao 2019d): (1) "Industry 1.0" corresponds to "SS 1.0—science of SS experience"; (2) "Industry 2.0" corresponds to "SS 2.0—technical SS science"; (3) "Industry 3.0" corresponds to "SS 3.0—system SS science"; (4) "Industry 4.0" corresponds to "SS 4.0—computational SS science." Computational SS science is the product of Era "SS 4.0," as it ushers in a new era when new paradigms and directions of SS science are designed for a new course of development. It becomes a new field as well as a hot topic for the current SS science R&D (Fig. 1.8). In this era, the focus of SSM is intelligence-led smart prevention and control, which makes SSI an increasingly important aspect in SSM, thereby giving rise to the emergence of SSI science.

["4E + C" refers to the five SSM measures, namely, SS enforcement, SS education, SS engineering, SS economics, and SS culture; Mode of "6P SS science" refers to the mode of SS science which integrates six features, requirements, and targets: predictive, preventive, personalized, participatory, precise, and public.]

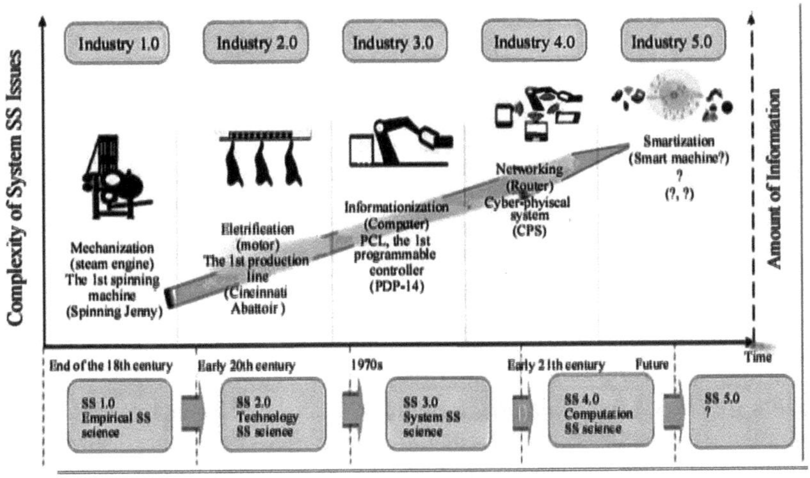

Discipline features	empirical	technological	systematization and informationization	coupling and intelligent	?
Disciplinary focus	accident disaster/natural disaster	emergencies	Safety or Security	Safety &Security	?
Focus of SSM	random control based on experience	technology-based key prevention and control	comprehensive prevention and control based on the "4E+C" means	intelligent prevention and control based on intelligence	?
Disciplinary code	afterwards	beforehand+interim	full-cycle, full-process, and all-round	mode of 6P SS science	?
Methods of SSM	experience-oriented	problem-oriented	"Risk+information"-oriented	intelligence dominated	?
Disciplinary methodology	"+SS"		"SS+"&"+SS"		?
Driving force of the discipline	forced		active		?

Fig. 1.8 Development process of modern SS science

1.3.1.2 Requirements for the Development of SSI as a Discipline

The rise and development of SSI activities and research call for the support and guidance from SSI studies which takes intelligence as the object of study. As SSI research and practice are being conducted worldwide in recent years, a good many experts, scholars, and practitioners at home and abroad have made extensive research and practical exploration of SSI, which greatly enriched the theoretical connotations of SSI studies. However, the SSI research and practical fruits are far from being systematic, as the perspectives that researchers from different disciplines take are manifold, resulting in diversified starting points and destinations in their studies and evidently different understandings about the nature of SSI. In the meantime, the current research on SSI mainly focuses on the surface (such as the direct introduction to the definition, research content and achievements of SSI) and its application; the in-depth research and exploration into the disciplinary construction are still inadequate. Therefore, it is necessary to comprehend the importance of SSI in order to understand the nature and inherent rules of SSI from the theoretical level and from the height of disciplinary construction, and recognize the significance of guiding and promoting SSI research and practice from a macroscopic perspective.

1.3.2 Background and Significance of Application

1.3.2.1 Responding to and Serving the Needs of the Major Strategy of a Holistic Approach to National Security

This is an era where strategies dominate competition (Zhao Rongying 2017). As human beings stride into the twenty-first century, it becomes clear that the key to an organization's ultimate success in a secured course of development is whether it is capable of formulating its own SS strategy. A valid and accurate SS strategy can be used to enhance an organization's (such as an enterprise, government, society and country) SS capacities, thereby promoting the growth of the organization. Thus SS strategy management as a new concept in SSM has been positioned at the central position in modern SSM. For this reason, aiming at creating a new horizon for national security, General Secretary Xi Jinping first proposed the theory of major national security and strategy of "a holistic approach to national security" at the first conference of the National Security Commission of the Communist Party of China held on April 15, 2014. The Fifth Plenary Session of the 19th CPC Central Committee which was held in 2020 proposed "for building a peaceful country at a higher safety level through coordinated efforts in development and maintaining peace." The strategy of security—the national security strategy—has taken shape. In view of this, how to respond to and serve the needs of the major strategies concerning national security, the Peaceful China Initiative and peaceful development with the support of SS resources has become the top priority.

As a key SS resource in SSM, the focus of SS strategy research is to collect, analyze, process, and utilize SSI in order to generate applicable ideas and goals about SS strategy. In other words, the formulation and implementation of SS strategies based on SSI is a major task of SSI research. Therefore, conducting SSI research is not only the basis for implementing the major strategies of national security, the Peaceful China Initiative, and peaceful development, it is also the primary task.

In summary, the development of the SSI discipline is both necessary and critical. It can provide knowledge services for the three major strategies (namely, national safety and security, the Peaceful China Initiative, and peaceful development) on the one hand, and fulfill the needs of reform, innovation, and high-quality peaceful development in the new era of socialism with Chinese characteristics on the other hand. In addition, it is conducive to consolidating the theoretical foundation of the strategies of the three major strategies. It also provides theoretical support for the feasibility and scientific rationality of formulating and implementing these strategies.

1.3.2.2 Requirements of Addressing SSM Challenges in the Information Age

In general, the main features of the era in which SSI was born are: overlapping and intertwined SSM challenges such as the complexity of SS threats/risks in age of digital information, the severity of the SS situation, the explosion of SSi and the lack of valuable SSI, and the transformation of SSM models. The formation, rise, and research development of SSI constitute the practical demands for proactively responding to the complicated and volatile SS threats/risks in the information age. They are also the inevitable outcomes of constantly adapting to the development and changes of the security of the information society, the necessary paths to solving the prominent problem of "lack of valuable SSI" in SSM in the information age, and the definite choices for fulfilling the value of SSI in the information age.

1.3.2.3 The SS Situation is Becoming Increasingly Severe in the Information Age

As the information age develops, the flow of information steadily increases, the trend that the society and a majority of organizations are undergoing the process of informationizing and networking is becoming prominent, and more and more SS issues take on an "informational" look (Wang Bing and Wu Chao 2018d). In this age, the SS risks caused by information in some sense are severer than the nuclear security risks of the industrial age, and sudden cyber-attacks have become one of the most serious types of threats to national security (Wu Shizhong 2013; Wu Handong 2017). Generally speaking, the SS challenges that confront our age can be reflected in three aspects:

1. SS risks are more interdependent, uncertain, unpredictable, and complicated (Wang Bing and Wu Chao 2018c). As a medium, link, or connector, information enables SS issues of virtual and realistic societies intertwined with each other. It blurs the boundaries of these issues in political, economic, military, social, informational, and manufacturing domains (Wang Bing and Wu Chao 2018a), exacerbates the overlapping and transformation of various SS risks and significantly increases the probability of a "butterfly effect" which triggers a massive chain reaction. It also makes the interweaving of traditional and non-traditional SS factors increasingly confusing (Wang Bing and Wu Chao 2018a). What's more, it leads to the growth of the number of predictable and unpredictable SS threats and risks, and the linkage effects of various SS threats and risks becomes more pronounced.

2. In the era of big data and intelligence, SS challenges for humans are on the rise (Wu Shizhong 2013). To be specific, current SS approaches are no longer sufficient to meet the SS (especially information SS) needs of the big data era (Wu Shizhong 2013); SS risks of the artificial intelligence era are emerging, but the potential social SS risks and negative impacts of artificial intelligence and robot technology cannot be determined (Wang Bing and Wu Chao 2018a); High technology often comes with high SS risks. The fear of unemployment due to artificial intelligence, incidents of robots attacking humans, the existence and spread of weapons of mass destruction—such as nuclear, chemical, biological, and radiological weapons—all pose more serious existential threats to us.
3. Currently, the Chinese society is undergoing a transitional phase, characterized notably by heightened and proliferating contradictions, SS crises and risks (Wang Bing and Wu Chao 2018a). For instance, the class nature of existential SS risks has become evident. The allocation and possession of SS resources are marked by irrationality and disparity, particularly the issue that "SS resources are concentrated at the top, while SS risks gather at the bottom" has become progressively more obvious. In detail, it is observed that in contrast to the vulnerability of the impoverished to SS risks, the affluent tend to experience fewer adverse effects and may even reap benefits from such risks. This phenomenon indicates that SS risks not only fail to dissolve class distinctions but also widens the gap of wealth and intensify class stratification. As China develops to be more industrialized and urbanized, various man-made systems such as urban centers are growing in scale and complexity, with an expansion in the scope of production and operation. A multitude of uncertain SS factors and risks are surging, interweaving with one another or even emerging in a comprehensive manner. Furthermore, driven by the escalating demands and expectations for safety and security from the government and the public, as well as the promotion and widespread adoption of the concept of "cost-effectiveness and high efficiency," the pressure of expectation and cost in SSM is further intensified, which in turn fosters the transition from traditional SSM to a modern SSM model—an intelligence-led SSM model. The paradigm of this model makes SSI the nucleus of SSM, thereby promoting the SSM performance and the allocation of SS resources.

In summary, the myriad of SS issues in the current information age present numerous challenges to SSM. With the growing diversity and complexity of SS factors, the severity of the SS situation, and the relentless pressure on SSM, there is a heightened focus on how to effectively address the intricate and grave SS landscape by efficiently collecting, analyzing, processing, and utilizing SSI. Clearly, under such SS conditions, the emergence of SSI studies is not only timely but also imperative. The development and establishment of SSI are essential practical responses to the emergence of a list of SS challenges, risks, and threats that characterize the information age. Based on Bao Changhuo's (2009) perspective, intelligence studies should focus on the assessment, analysis, early warnings, prevention, and management of significant events, threats, and risks. In light of the aforementioned SS situation, it is imperative to address the assessment, analysis, early warnings, prevention and management of significant SS events, risks, and threats from the vantage point of intelligence studies. Intelligence science ought to be the central disciplinary pillar of SS science, and SSM is in need of the intervention and support of intelligence studies. It is fitting to establish a common branch that integrates intelligence studies and SS science, that is, "safety and security information (SSi)" studies.

1.3.2.4 Problem of the "Deficiency of Valuable SSI" is Becoming More Prominent in the Information Age

In the context of SSM in the information age, the deficiency or incompleteness of SSI is a pervasive problem, often cited as the root cause of SSM failures. This is a new understanding emerging from the SS academic and practical fields (Wang et al. 2017). Wang Bing and Wu Chao (2018c) have emphasized this issue, stating that "SSI is the indispensable pathway to SS, and a threshold that must be crossed." It is widely recognized that the most effective SSM strategy is the one that is grounded in the most accurate and comprehensive SSI. The success or failure of SSM is directly influenced by the collection, management, and utilization of SSI (Wang et al. 2017). Thus, in recent years, the SS community is diligently advocating a new concept and methodology in SSM—Evidence-Based Safety and Security (EBS) management (Wang et al. 2017), aiming at addressing the dilemma of SSI deficiency in SSM.

In fact, intelligence in management is commonly understood as follows: intelligence refers to all information (content) that influences management decisions, primarily those related to management actions (Chen Chao 2017). This understanding shows that, in precise terms, "SSi (the SSi that influences SSM)" is in fact "SSI." Therefore, it can be concluded that what is truly lacking in SSM efforts is not SSi, but SSI. For instance, the delay in disaster intelligence was the primary reason for the unsatisfactory rescue efforts following the Great Hanshin-Awaji Earthquake in 1995 in Japan (Xu Jingbo 2008).

In the information age, it is imperative to highlight and promote research on SSI science in order to enhance capabilities of SSI operation in SSM. Currently, SS issues are becoming increasingly complex and interconnected, and new information technologies such as the Internet of Things, mobile internet, big data, cloud computing, and even blockchain have been extensively integrated into SSM. The volume of SSI is growing exponentially, and SSM has truly entered the era of big data, where the role and value of SSI are becoming more prominent. However, the issue of "the overflowing of non-essential SSI and the deficiency of valuable SSI" is becoming evident, which seriously affects the improvement of SSM capabilities. One of the major reasons is that SS managers have not paid sufficient attention to SSI in SSM, which leads to the underperformance of SSI, such as the production of low-quality SSI products and the untimely transmission of SSI. What's more, the human society is facing heightened risks. Jin Qian[1] posits that a high-risk society is not merely "one with escalating dangers," but rather one that is increasingly vigilant about "future SS

[1] "Characteristics, Dangers, and Responses of Risk Society", *Legal Daily*, February 22, 2017.

concerns." To enhance our predictive and preemptive control over future SS risks, it is imperative to explore and employ the capabilities of SS prediction and early warnings and its potential.

1.3.3 *Conclusions*

This section demonstrates and discusses the background and significance of developing SSI science from both disciplinary and practical perspectives. It is evident that the development of SSI is underpinned by profound contextual conditions and carries significant implications. In recent years, the field of SSI has become a new growth point and point of extension in the fields of intelligence science and SS Science. The establishment of SSI as a scientific discipline is both crucial and urgent.

1.4 Foundation of SSI Science Development

A solid foundation is essential for the establishment, research, and development of any academic discipline. This section probes into the fundamental conditions necessary for the construction of SSI science from three perspectives: philosophical, practical, and theoretical. The goal is to elucidate the foundational principles and prerequisites that underpin the development of SSI science.

1.4.1 *Philosophical Foundation*

A critical foundation for the establishment of SSI science is its utility in SSM. The fundamental assumption for SSI's role in SSM is that the efforts of SSI can be seamlessly integrated into SSM processes. This raises a core philosophical question regarding the establishment of SSI science: Is the integration of SSI and SSM possible? Philosophically, the basis for integrating two or more entities is their shared commonalities. Thus, identifying and elucidating the commonalities between SSI and SSM is a prerequisite and foundation for their fusion. There are four main aspects in "the converging points" or the commonalities between SSI and SSM (Yao Leyuan and Fan Wei 2014; Tian Shuicheng and Jing Guoxun 2016; Wang Bing and Wu Chao 2018a).

1.4.1.1 Centering on SS

Both SSI and SSM highlight the idea of "centering on SS." SSM covers a spectrum of management practices aimed at the prevention of SS risks (or the enhancement of

SS), centering around safety and security. In the same vein, SSI adheres to "centering on SS" (in other terms, the idea of "centering on SS" essentially distinguishes SSI from other forms of intelligence). SSI operations revolve around the identification of SS risks, the prevention of such risks, the surveillance of hazard sources, the approaches to promoting SS, and the execution of SS enhancement initiatives.

1.4.1.2 Concern for the SS Environment

Both SSM and SSI advocate for the idea of "the SS environment is an open-up system." They are both concerned with the influence of internal and external factors on the system's (organization's) SS profile (this includes the organization's adaptive response to shifts in the external environment, as well as the SS threats, challenges, and opportunities for SSM that arise from changes within the environment involved). The focus of this approach is to enhance understanding and facilitate effective interventions in the factors that contribute to and shape the progression of SS incidents.

1.4.1.3 Consideration for the SS of the Entire System (Organization)

SSI, aligned with the system's (organization's) SSM objectives, endeavors to thoroughly gather and provide SSI that can enhance the SSM across the entire system (organization). Modern SSM also emphasizes the management of SS risks that could adversely affect the system (organization). Both approaches are rooted in a holistic view of the system (organization), prioritizing the SS operation and development of the system (organization) and its associated subsystems (sub-organizations).

1.4.1.4 Similar Operational Characteristics

SSM and SSI share numerous commonalities. For instance, (1) Planning and implementation are two critical steps in the processes of both SSM and SSI; (2) Both SSM and SSI are ongoing and iteratively evolving endeavors; (3) They both concentrate on "future SS," highlighting a "prevention-oriented" philosophy within the SSM; (4) Both SSM and SSI aim for the long-term sustainability, SS operation, and development of the system (organization). These shared characteristics are critical for the integration of SSM and SSI efforts.

Furthermore, SSI operation is more closely aligned with the practical methods and tasks of SSM operation, whereas the latter emphasizes comprehensive planning and design at the macro-level from a strategic and systemic (organizational) SS perspective. Thus, SSI operation should be considered as a core element of SSM. Conducting research related with SSI helps SSM operations become more scientific and improve SSI practices in SSM, so as to enhance SSM capabilities.

1.4.2 Practical Foundation

SS science (especially SSM) has become more informationized with the ongoing evolution of information technology and its prevailing application in SS science. There has been a surge in the "informationization" of SSM, as it is recognized that efforts of informationization can significantly enhance SSM, refine SS performance, and promote SSM capabilities. For instance, research outcomes in the domain of SSM informationization are emerging rapidly; furthermore, the promotion of informationized SSM has been designated by the Chinese government as one of the pivotal and prioritized areas for the development of SS initiatives [In the field of workplace SS, both *The Opinions of the CPC Central Committee and the State Council on Promoting the Reform and Development of Workplace Safety* (issued in 2016) and *The 13th Five-Year Plan for Workplace Safety* (issued in 2017) have emphasized the need to promote the informationization of SSM]. This underscores that the informationization of SSM has become a symbol and objective of contemporary SSM, as it provides a solid foundation and creates favorable conditions for SSI operations. SSI operation based on SSM informationization has garnered increasing attention.

In effect, the notion of intelligence and its operations have long been part of the domain of SS science (arguably, they originated from SS science). This is primarily because the intelligence concept and its operational framework stem from subconcepts such as "military intelligence (including military SSI)," "national SSI," and "public SSI/police intelligence," along with their respective intelligence activities. This is also the reason why intelligence is often linked to military and national security. From another vantage point, SSI has always been a product of human safety needs in the extensive history of SSI operations. People have been attentive to safety and security since the dawn of humanity, hence the emergence of SSI requirements and relevant SSI operations. Since the ultimate goal of SS is "prevent problems before they occur," SSI, with its predictive and early-warning capabilities, would naturally give rise to SSI operations. From this perspective, SSI has been intricately involved in SSM operation, and its importance and status within the informationization of SSM have become increasingly prominent. It is believed that high-quality SSI operation is a necessary requirement and the ultimate goal of modern SSM, particularly within the context of SSM informationization. This is specifically manifested in the following six aspects:

1. SSI operation is an essential requirement for actively adapting to the evolving methods of SSM and the advancement of SSM informationization.
2. SSI operation is an essential requirement for addressing the issue of incomplete or missing SSi within SSM.
3. SSI operation is an important requirement for taking the "first move" in SSM, or a "prevention-oriented" SSM approach. It enhances the proactive nature of SSM and effectively leverages the predictive, early-warning, and preventive value of SSI to avert "unnoticed" SS incidents before they occur.
4. SSI operation is an indispensable requirement for fostering the ongoing development of SSM informationization and to achieve the 6P paradigm and objective

of modern SSM, namely "Predictive, Preventive, Personalized, Participatory, Precise, and Public" (Wang Bing and Wu Chao 2018d) in the information age, particularly in the era of big data and artificial intelligence.

5. SSI operation is an essential requirement for implementing mainstream SSM methods characteristic of the information age, such as "EBS methods" and "methods of SS knowledge management."
6. SSI operation is an inevitable requirement for breaking through the "SS information barriers" and eliminating the "lonely islands of SS information" between SS-related departments or institutions, so as to achieve the integration of SS resources, sharing and use of SSI, and collaborative governance of SS.

In conclusion, the advancement of SSI operation, grounded in the informationization of SSM and driven by its practical demands, has established a solid practical foundation for the emergence and evolution of SSI studies. This represents one of the key external driving forces in the inception, growth, and development of SSI studies. There are the practical dimensions of SSI operation:

1. The development and implementation of SSM informationization serve as a crucial foundation for SSI endeavors. As operations of informationizing continue to progress and expand across various sectors, SSM informationization has emerged as a timely necessity. In recent years, China has taken vigorous efforts to advance SS informationization. For instance, in the domain of workplace safety, during the period of the "11th Five-Year Plan" of China, the program of the National Workplace Safety Information System (the initial phase of Project "Jin'an") was implemented. In alignment with the "2006–2020 National Strategy on Development of Informationization," *The 12th Five-Year Plan for the Informationization of Workplace Safety* was specifically developed and implemented in 2011. To date, a national system of safety supervision information and a platform of safety emergency information have been established and operated in China. However, the development of SSM informationization is an ongoing process, necessitating the continuous enhancement of the informationization of foundational SSM operations and the integration of the latest information technologies to elevate the caliber of informational SSM. This, in turn, lays a more robust foundation for the advancement of SSI endeavors.
2. SSI operation is integral to the entire SSM process. The implementation of SSI within SSM demonstrates that SSI operation constitutes an intelligence system fundamentally supported by SSM informationization, centered on SSI research and analysis, and aimed at serving and supporting SSM operations as its core objective. This system elucidates the essence and critical elements of SSI operations: the supply and application of SSI are the foundation and safeguard; SSI analysis is the central challenge; and leveraging SSI to serve SSM is the ultimate goal and aspiration. Consequently, SSI operation is not only a fundamental aspect of SSM but also a basic tool for enhancing SSM capabilities. Guided by the objective laws governing SSM and intelligence operation, and in collaboration with SSM personnel and intelligence professionals, it employs the theories and methods of SSM and intelligence studies to integrate and share SSI (data)

resources, and plan, collect, and process SSI. Thereby, it produces valuable SSI products for SSM (based on the platform of SSM informationization) to provide a basis and intellectual support for SSM.

1.4.3 Theoretical Basis

Theoretical considerations suggest that the emergence and development of any method of management are benefited by a certain philosophy of management, and the proposal of the SSM approach is no exception. For instance, building upon SSM principles such as "safety/security first," "preventability and controllability of SS incidents," "SS standardization," "risk management," "systematic SS," "SS culture," and "evidence-based SS," a variety of specific SSM methods have been developed to practice these concepts. In recent years, with the advent of the "intelligence-led SSM" concept, relevant intelligence-led SSM methods have been proposed and applied.

Broadly speaking, the existing classical SSM paradigms can be categorized into four types when examining the initiating point of SSM: problem-oriented SSM (Fig. 1.9), risk-oriented SSM (Fig. 1.10), statistics-oriented SSM (Fig. 1.11), and intelligence-led SSM (Wang Bing and Wu Chao 2018a). The first three models are dependent on the support of SSI operation and are intricately linked with intelligence-led SSM methods, suggesting that they can all be assimilated into the intelligence-led SSM paradigm. For detailed analysis, please refer to Figs. 1.9 to 1.11. In essence, the intelligence-led SSM paradigm represents a holistic approach to SSM.

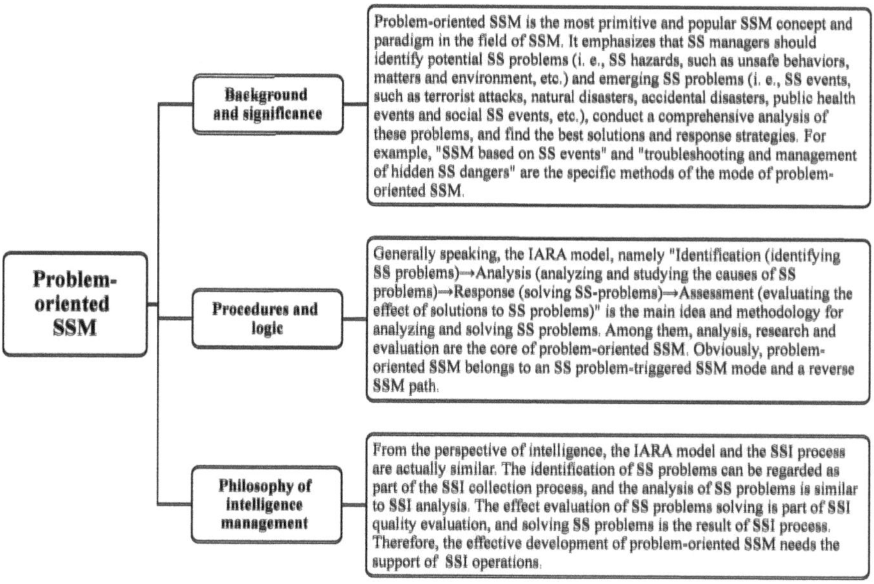

Fig. 1.9 Problem-oriented SSM paradigm

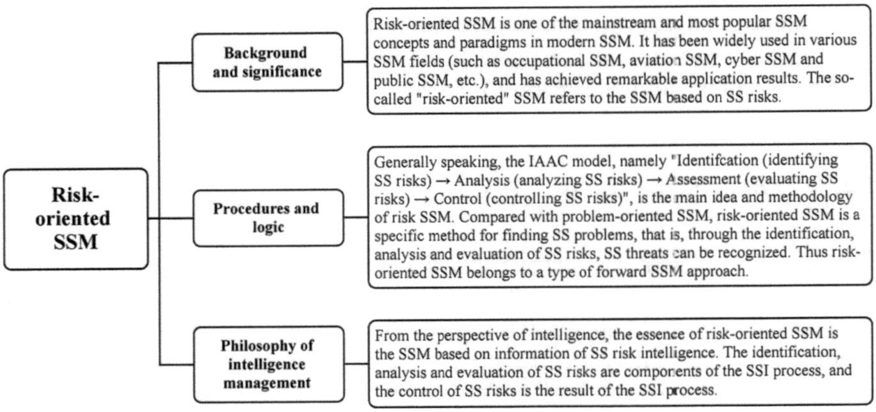

Fig. 1.10 Risk-oriented SSM paradigm

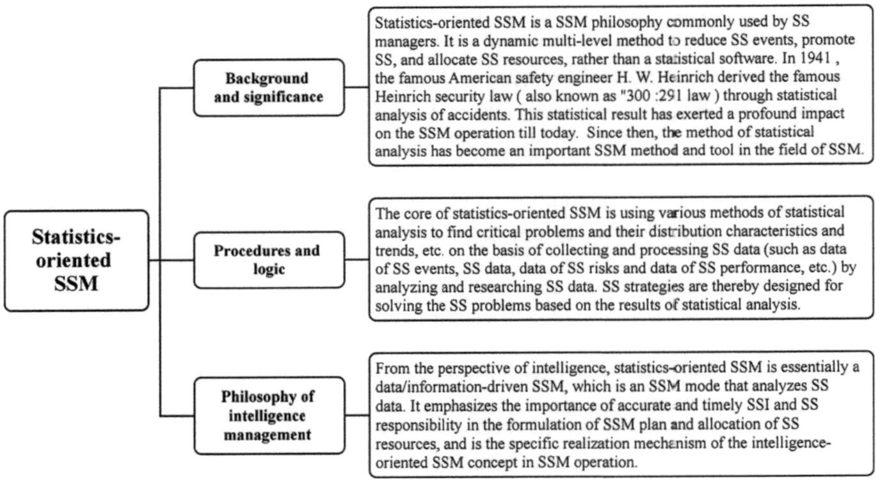

Fig. 1.11 Statistics-oriented SSM paradigm

The concept of intelligence-led SSM has been introduced and applied in practice (Liska 2015). For instance, the intelligence-oriented SSM paradigm has been piloted in the fields of information security, public security, and emergency management, yielding significant results. In fact, across various domains of SSM, specific implementation methods of the intelligence-led SSM paradigm have been proposed and employed, with the most notable being the method of "Evidence-Based Safety and Security Management" and approaches to SS knowledge management. The realization of SSI's value hinges on actively promoting and embracing the concept of intelligence-oriented SSM (Liska 2015). The crux of the intelligence-oriented SSM concept lies in the execution of effective SSM, which involves accurate SS prediction, decisions and actions, while SSM relies on precise, timely, and comprehensive SSI, the production of which in turn is contingent upon the support of SSI operations (includes planning, SSi collection, SSi processing, SSI analysis, and communication, and is supplemented by the general management activities of leadership, organization, coordination, control, and evaluation) (Liska 2015).

SSi (data) analysis and SSI are pivotal to SSM. Intelligence-led SSM represents an original paradigm and philosophy within the field of SSM. It is important to recognize that intelligence-led SSM is an analysis-driven approach, emphasizing in-depth, systematic SSI analysis and the optimization of SS resource allocation and strategic policy management based on SSI. This approach markedly differs from the foundational principles of problem-oriented, risk-oriented, and statistics-oriented SSM. Consequently, SS managers must learn to collaborate with intelligence professionals. They should participate in SSM activities as well as intelligence analysis of SS issues and encourage intelligence professionals to devise relevant strategies for addressing SS challenges. Intelligence-led SSM can only be fully realized within an organization by integrating intelligence as an essential component of the organizational SSM philosophy.

1.4.4 Conclusions

From a theoretical standpoint, this section analyzes the underpinnings for developing SSI research (its philosophical, practical, and theoretical foundations), on the basis of the practical outcomes of existing SSI research. The objective is to elucidate the fundamental prerequisites and the viability of developing SSI research, reinforcing the basis for conducting the research.

1.5 Progress of Studies in SSI Science

As SSI emerges as a foundational and potent tool for managing or governing SS risks in the era of big data, intelligence, and SS 4.0 (referring to the computational SS science) in the high-risk society, it becomes a prominent research topic in intelligence studies and SS science in recent years (Wang and Wu 2020; Wang 2020). For instance, Yan Hui (2020) reviewed the prominent research topics in the field of library information studies and archive management in China for 2019 and identified SSI as one of them. The ascendance and evolution of SSI call for guidance from an academic discipline, which leads to the emergence of SSI science. It represents the culmination of the profound integration of intelligence studies and SS science, marking an inevitable stage in their development. SSI science has now evolved into a significant academic field situated at the intersection of intelligence studies and SS science.

In recent years, the field of SSI science has seen extraordinary growth, and there has been advancement in the studies of SSI science. Theoretically, a comprehensive understanding the progress within a discipline is a prerequisite for ensuring its healthy and scientific development. Yet there has not been a systematic review and in-depth analysis of the recent progress in SSI research to date. From a disciplinary standpoint, this book addresses general issues of SSI research, drawing on representative domestic and international literature of SSI research. It provides a systematic review and thorough analysis of the progress in SSI research, with the aim of fostering future research and promoting sound and rapid development of SSI.

Following a process of classification and analysis, it has been determined that existing studies in SSI science predominantly concentrates on three key areas: the development of the SSI science discipline, the execution of SSI operations, and the management of SSI activities.

1.5.1 Research on the Construction of the SSI Science Discipline

In China, the trajectory of a discipline's development and research is inherently linked to its inclusion within the national directory of discipline classification. Government attention and research fund are typically allocated in accordance with this directory. In recent years, China's intelligence and SS communities have been actively working to expand and reinforce their respective disciplines. The establishment of the SSI discipline as a significant branch within the fields of intelligence studies and SS science garners considerable attention. Therefore, in terms of dedicated and systematic research and practice concerning the development of the SSI discipline, China stands at the forefront, signifying an attitude not commonly found among researchers and practitioners in other countries.

In recent years, the SSI community in China has given significant attention to the development and research of the SSI science discipline, achieving notable outcomes. Generally speaking, the construction of this discipline primarily concentrates on three areas: the development research of SSI science guided by the holistic approach to national security, the construction research of SSI science which is oriented toward national security, and the construction research of universally applicable SSI science which is oriented toward major SS.

1.5.1.1 Development Research of SSI Science Guided by the Holistic Approach to National Security

On April 15, 2014, the concept and approach to national security with Chinese characteristics—namely, the "Holistic Approach to National Security"—was formally proposed by General Secretary Xi Jinping at the first meeting of the National Security

Commission of the Communist Party of China. Since then, the evolution of intelligence under the auspices of this principle has garnered attention within China's intelligence community. Seen from a disciplinary standpoint, SSI studies represents the confluence of this principle with intelligence studies, and the principle serves as the strategic guidance and theoretical foundation for the construction and innovative advancement of SSI with distinctive Chinese characteristics. Thus, the research on intelligence development within the framework of the principle is intrinsically linked to the establishment of the discipline of SSI science.

Zhang Qiubo and Tang Chao (2015) have pointed out that the development of intelligence studies should contribute to the expansion of research domains, the enrichment of theoretical and methodological frameworks, and the cultivation of a more productive disciplinary landscape, all underpinned by the holistic approach to national security, to better serve the general national SS and development objectives. Guided by this principle, SSI science should concentrate on overarching national SS and non-traditional SS issues. Xie Xiaozhuan (2019) has identified that intelligence studies oriented toward national SS primarily encompasses three major subdisciplines: national SSI studies, military intelligence studies, and public SSI studies. Xie posits that, under this principle, the cognitive constraints that have historically influenced the development of intelligence studies should be addressed, with a focus on resolving the issues of insensitivity, isolation, narrowness, fragmentation, and excessive secrecy that have characterized past intelligence studies. The development of intelligence studies should aim to establish and develop a unified, large-scale theoretical system through a broad, interdisciplinary, and integrative approach. With these research efforts, the development of SSI science should transcend the confined perspectives of specific fields and industries, fostering interdisciplinary integration across various sub-fields or industries of SS. It is essential to construct a unified system of major SSI science that aligns with the requirements of modern governance of risk society and the requirements of the holistic approach to national security.

1.5.1.2 Research on the Construction of SSI Science Oriented Toward National Security

Enhancing the capacity for national SS risk governance necessitates the backing of intelligence, rendering national SSI a critical domain of interest within both spheres of intelligence and national governance. Research indicates that systematic and specialized national SSI research and practice originated during World War II, with early national SSI studies primarily concentrating on military (overseas-oriented) SSI and regime (domestic-oriented) SSI (Gao Jinhu 2019a). Foreign national SSI research predated domestic efforts, and the disciplinary framework of national SSI studies in Western nations, particularly the UK and the US, have reached a moderately mature stage. In contrast, formal national SSI research in China commenced after the policy of "Reform and Opening up" was implemented (Gao Jinhu 2019a; Xie Xiaozhuan and Gao Jinhu 2020). This research predominantly centers on military SSI, of which the level and perspectives are relatively macroscopic, emphasizing strategic-level military SSI operations. This emphasis results in an apparent disparity between research and military SSI operation at the tactical and battlefield (operational) levels, though it tends to align more closely with macroscopic research themes of national SSI studies (Gao Jinhu 2019a).

In 2018, China's Ministry of Education promulgated the document of *Recommendations on Enhancing National Security Education in Primary and Secondary Schools*, which mandated the establishment of national security studies as a first-level discipline. Therefore, since 2018 the development of this first-level discipline has become a contentious topic within the academia and the field of higher education. In this context, national SSI science which is dedicated to providing intelligence services and support for national SS governance as a branch of national security studies merits significant attention. Despite of a long history of research on national SSI within the fields of intelligence studies and national security studies, there is a dearth of research on the disciplinary construction of national SSI studies (Shang Pu 2018). To address this research gap and to foster future research and development in national SSI studies, numerous scholars have begun to focus on the research on its disciplinary construction which is oriented toward national safety and security, guided by the holistic approach to national security.

Shang Pu (2018) elaborates on the research object of national SSI, delineating its characteristics—which are predominantly political, practical, scientific, humanistic, and ethnic—as well as its disciplinary system, comprising ontology, epistemology, methodology, and practice theory as four distinct sub-disciplines. He also outlines the research methods, such as the Marxist philosophy, the experimental method, the method of analytical synthesis, the method of comparative reference, the method of survey, and the method of specific case analysis. He also identifies the discipline's iconic communication platform, its practice system, and its educational system as three crucial external elements in the construction of national SSI studies. He notes that national SSI studies is an interdisciplinary field that intersects with national SS studies and intelligence studies (national intelligence studies). Gao Jinhu (2019a) highlights the primary areas and significant topics within national SSI studies, including foundational theories of national SSI, such as its definition, characteristics, functions, classification, and constituent elements, as well as warning errors of national SSI, strategic deception, warning intelligence of national SS, national SSI analysis, the national SSI management system, and national SSI supervision and legal frameworks. He also points out five major shortcomings in national SSI studies: the blurring of disciplinary boundaries, weak theoretical research, being confined to the English-speaking countries, insufficient integration and interdisciplinary research, and inadequate participation of local researchers. The future direction and objective for national SS studies are to address these five key deficiencies in national SSI studies. Furthermore, he posits that the work of national SSI constitutes the research object of national SSI studies, and the ultimate goal is to elucidate the principles governing national SSI operations and to establish a theoretical and methodological framework for guiding the practice of national SSI activities (Gao Jinhu 2019b).

1.5.1.3 Research on the Construction of Universally Applicable SSI Science Oriented Toward Major SS

In recent years, the field of international SS science has officially entered the era of "major SS," also known as "general SS" (Wang Bing and Wu Chao 2019b). It is important to note that the understanding and recognition of SS science should not be confined or influenced by the disciplinary and specialty classifications within the country (Currently, SS science and engineering disciplines are categorized as scientific and engineering fields; however, internationally, SS science is recognized as a quintessential interdisciplinary and comprehensive field encompassing natural sciences, social sciences, and engineering technology). Additionally, the understanding and recognition of SS science should not be constrained by its origins within the domain of workplace safety in the country, as internationally, it extends to the broader domain of "security" and "safety." The general SS paradigm poses two significant requirements for the research and development of SS science from a systemic perspective (regions, nations, cities, enterprises, or ecological, military, social, resource, and political systems). First, system SS research should shift from the focus on partial SS—encompassing the specific SS of various fields, industries, and subjects within the system—to an overall system SS that includes the SS of all fields, industries, and subjects involved, aligning with the essence of "a holistic approach to national security" in the context of national SS. Second, system SS research should move from a single-layer focus on "security" or "safety" to emphasis on "Safety and Security integration," reflecting the complete meaning of the term "安全" in the Chinese context (Wang Bing and Wu Chao 2019b). Following the mechanism and thought process behind the formation of interdisciplinary fields, the cross-integration of SS science and intelligence studies can lead to the development of a universal, or "generally applicable" SSI studies oriented toward major SS.

In recent years, scholars in the field of SS science, notably Wang Bing and Wu Chao (2018a, 2019b, 2019c, 2020a), have conducted extensive research on the development of universal SSI studies oriented toward major SS. Firstly, they differentiate between the establishment of a new academic major, which is primarily based on societal needs, and the establishment of a new discipline, which necessitates a thorough examination of its background and rationale. After in-depth analysis and research into the background and foundation of SSI science, they identified the disciplinary background, including the objective needs for the advancement of intelligence and SS disciplines and the imperative for the standardization and scientific progression of SSI, as well as the practical background, which encompasses the operational demands of SSI operation geared toward SSM. They assert that the foundational prerequisites for the construction of SSI science include philosophical, practical, and theoretical underpinnings, such as the alignment between SSI and SSM, SSI operation supported and promoted by the informationization of SSM, and the SSI management philosophy inherent in and permeating the classical SSM paradigm (Wang Bing and Wu Chao 2018a).

Secondly, from the vantage point of major SS, they have provided a comprehensive elucidation of the significance, disciplinary characteristics, research objects, content, foundational principles, and branches of SSI science. They have clarified the fundamental issues within the discipline of SSI science and have established a macro-level architecture of disciplinary system for SSI science (Wang Bing and Wu Chao 2019b). Furthermore, to dispel confusion and potential misdirection in the evolution of SSI studies, they have delineated the discipline's origin (a product of "SS + "), the connotations of its meta-concepts (including SS, intelligence, and SSI), the discipline's identity (distinguishing SSI science from SS information studies with its own disciplinary legitimacy), and the perspective of disciplinary construction (based on views of "major SS" and "major SSI," and oriented toward SSM) (Wang Bing and Wu Chao 2019c). Additionally, in response to contemporary demands, they have analyzed the transformation and progress of SSI science within the context of the big data environment (Wang Bing and Wu Chao 2020a).

1.5.1.4 An Overall Review and Prospect of Research

As the review above indicates, considerable progress has been achieved in the disciplinary development of SSI science in China in recent years, which can effectively foster further disciplinary research and advancement in SSI science. Based on the analysis presented, it is evident that its disciplinary development in China, whether it is associated with national SSI studies oriented toward national safety and security, or major SSI science oriented toward general safety and security, should be guided and promoted by the holistic approach to national security; a consensus has been reached on this issue. What's more, it is imperative to delineate the relationship between national SSI studies and major SSI studies in order to facilitate their mutual reinforcement, coordinated evolution, and integrated growth. The relationship between national SSI studies and major SSI studies, categorized according to their disciplinary attributes, types, and scientific affiliations, is illustrated in Fig. 1.12.

As Fig. 1.12 illustrates: (1) SS encompasses various subjects, including national SS, social SS, urban SS, and enterprise SS. Consequently, national SS studies, which focuses on national SS, should be considered as a significant branch of SS science, which takes SS as its research object (Wang Bing and Wu Chao 2019d). (2) SSI is the outcome of the convergence and integration of intelligence studies and SS

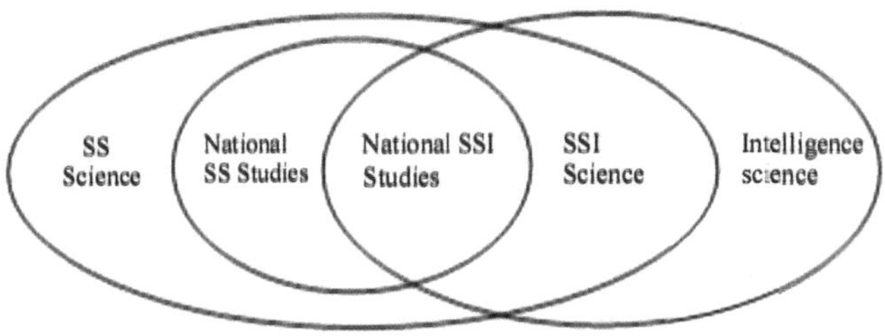

Fig. 1.12 Relationship between intelligence science, SS science, SSI science, national SS studies, and national SSI studies

science, representing a shared branch of both disciplines. (3) National SSI, an important branch of SSI, is the product of the intersection and integration of national SS studies—a branch of SS science—and intelligence studies (more specifically, SSI). It is evident that the theories and methods of SSI are applicable to national SSI and can offer a theoretical foundation and methodological guidance for the research, practice, and development of national SSI. National SSI is a specific application branch of SSI, akin to the leaves of the main stem, representing a horizontal extension and development of SSI. Furthermore, the research and practice of national SSI have preceded those of SSI, and the theories and methods of national SSI can be elevated and sublimated into those of SSI, thereby enriching the theoretical framework of SSI.

In fact, both SSI and national SSI have garnered attention and recognition within domestic academia. For instance, the call for papers for the academic conference "New Era, New Mission, New Action — Intelligence and Intelligence Studies Work Development Forum (2019)," co-hosted by the China Society for Scientific and Technical Information, the China Information Society of Social Sciences, the China Society for Defense Science and Technology Information, and the School of Information Management at Central China Normal University, included a special topic on "SSI," which encompassed national SSI, information SSI, science and technology SSI, enterprise SSI, urban SSI, social SSI, cultural SSI, military SSI, anti-terrorism SSI, public SSI, and so on. Clearly, this topic falls under the research content category of SSI oriented toward major SS, with national SSI being just one of the many areas of interest. The "Huashan Intelligence Forum," hosted by *The Journal of Intelligence*, has consistently focused on the research and development of SSI. Notably, the theme of the sixth Huashan Intelligence Forum held in 2019 was "Research Topics of National SSI."

It is evident that the relationship between "national SSI" and "SSI" is not one of conflict, but rather one of mutual enhancement and complementarity. In China, the academic background of researchers significantly influences their focus within SSI research. Scholars with a background in SS science primarily concentrate on developing the SSI discipline oriented toward major SS, and their efforts have led to the formation of an SS-oriented academic school of SSI science. In contrast, researchers with a military, public SS, or defense background focus on developing national SSI disciplines, leading to the formation of an academic school of national SSI studies which is oriented toward national SS. Looking ahead, these two academic schools should aim to strengthen dialogue, cooperation, and exchange, continuously refining their consensuses to promote coordinated and mutual development. This collaborative approach will help to unify and expand the SSI research community, pooling efforts toward the construction and advancement of the SSI discipline and striving collectively to elevate the field to greater heights.

Furthermore, there are three primary issues currently confronting the development of the SSI discipline. Firstly, a consensus on the fundamental concepts and systems within the discipline has yet to be established. The author concurs that the three basic concepts of SS, intelligence, and SSI, as defined by Wang Bing and Wu Chao (2019c), should serve as the foundation for constructing and developing the basic conceptual framework of the SSI discipline. Secondly, the subsystem and methodology of SSI have not been thoroughly examined, and these areas require focused subsequent research. Thirdly, the future developmental model for SSI remains unclear. The author posits that the future development of SSI should adhere to a "horizontal-vertical" model, which involves enriching and perfecting the disciplinary subsystem through cross-integration while simultaneously strengthening the overarching and general theoretical and methodological research of SSI, as well as in-depth studies within each sub-discipline. This approach is essential for deepening the scope and depth of SSI research and for expanding the potential for SSI's development and inquiry.

In addition, the evolution of SSI must be in sync with the disciplinary and practical contexts of its original disciplines, including SS science and intelligence studies. This includes adapting to new characteristics of SS science, such as the emergence of SS risks or issues and the overlaying and complexity of SS risks, as well as new features in intelligence studies, like the integration of intelligence and computation. SSI must respond to the development of the era—the era of big data and intelligence—in its advancement and innovation.

1.5.2 Research on SSI Operations

SSI operations, or "SSI businesses," cover a spectrum of activities aimed at acquiring, analyzing, storing, utilizing, and disseminating SSI to maximize its potential and effectiveness. SSI operations constitute the main content in foundational application and practical scenarios. Current scholarly endeavors in SSI operations primarily encompass three areas: analysis of SSI, research on SSM-oriented SSI operations, and investigation into national SSI operations.

1.5.2.1 Research on SSI Analysis

SSI analysis stands at the center of SSI operations. For instance, Fingar (2011) pointed out the critical role of intelligence analysis in mitigating uncertainties in his discussion of SSI analysis and national SS. As research on SSI progresses, several noteworthy findings have emerged, particularly in three key areas:

1. Research on comprehensive models or methodologies for SSI analysis: Wang Bing and Wu Chao (2019), from an SSM perspective, proposed two SSI acquisition and analysis approaches—"SS risk" and "SSM"—and introduced the "Risk-Management (R-M)" model and methodology for SSI acquisition and analysis. Gao Jinhu (2020) noted the distinction between intelligence analysis and information analysis, emphasizing the complexity of national SSI analysis, which conventional methods cannot adequately address. He advocated for the development of a new intelligence analysis methodology grounded in falsificationism and cognitive psychology, and suggests that SSI analysis should be guided by

the principles of complexity science and systems science due to its inherent complexity.

2. Research on specific SSI analysis methods: Guided by SSI analysis methodologies, practical SSI analysis involves the application of specific methods. For example, Atkinson et al. (2010) proposed analysis of border defense SSI based on news taping; Xue Mei (2011), Cao Wen (2017), and Dai Lina (2016) explored the use of spatial modeling, multi-layer heterogeneous complex networks and social networks in SSI analysis respectively; Kim et al. (2014) discussed SSI analysis methods for cyber security. SSI analysis methods are often borrowed from other disciplines as there are few proprietary methods. Xiao Lianjie et al. (2019) conducted a targeted study of specific methods of SSI analysis supported by deep learning techniques, as shown in Fig. 1.13. Their research identified commonly used methods of SSI analysis, including the method of social investigation (field investigation, expert consultation, sampling survey, and document investigation), classification (theme or text), comprehensive evaluation, induction, simulation, and cluster analysis.

3. Research on SSI analysis of new information and intelligent technologies such as big data and artificial intelligence: The advent of new information and intelligent technologies like big data and AI has presented fresh opportunities, methods, and tools for SSI analysis. Pramanik et al. (2017) discussed the potential application of big data analysis in SSI within a crime analysis framework and proposed a big data analysis framework for SSI. Ding Xiaowei and Su Xinning (2019), using financial SSI analysis as an example, pointed out that "blockchain, big data,

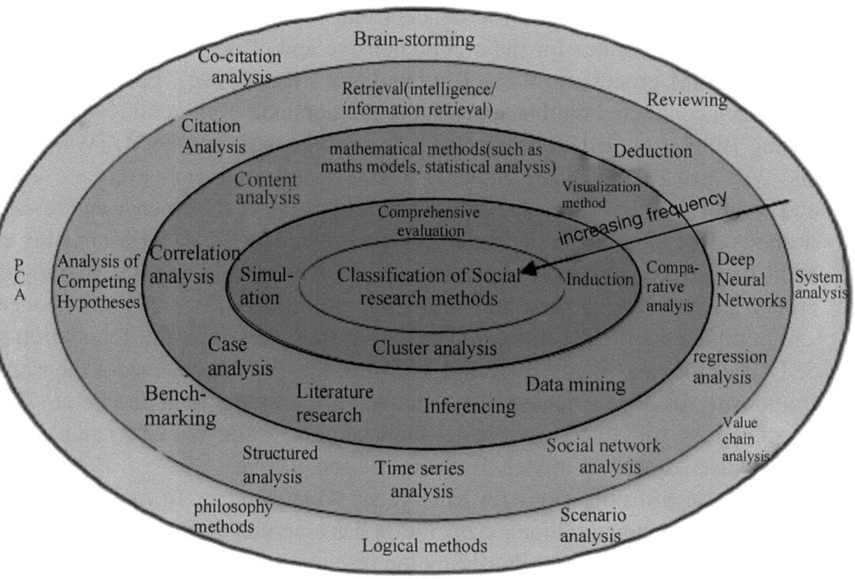

Fig. 1.13 Specific methods of SSI analysis

and artificial intelligence combined with intelligence analysis" offer a new and effective pathway for financial risk prevention and control. They proposed an SSI analysis concept based on blockchain-trusted big data artificial intelligence to enhance capabilities of financial SS risk governance. Benzidane et al. (2016) proposed an SSI analysis processing method on the basis of loud computing, which includes big data processing. Li Min et al. (2019) constructed a system of industrial information SSI analysis based on big data, while Liu Jingjuan and Huang Dan (2013) discussed information SSI analysis on the basis of big data.

1.5.2.2 Research on SSM-Oriented SSI Operations

The ultimate goal of SSI studies is to provide effective SSI services and support for SSM endeavors (Barnes et al. 2007; Wang 2020; Wang Bing and Wu Chao 2019). Consequently, research on SSI operations should be SSM-oriented (Asthana and Nirmal 2008). By adopting a systemic perspective and integrating the objects and content of SSM, a comprehensive SSM cycle model can be constructed, as depicted in Fig. 1.12. The content, process, or means of SSM encompass both normal SSM and emergency management (i.e., anomalous SSM) (Yan Ye 2019). Based on this analysis, existing research on SSI operations for SSM can be categorized into the following two types:

1. Comprehensive Research on SSI operations within the SSM Framework: Since 2019, significant strides have been made in this domain. Initially, Wang Bing and Wu Chao (2019e) elucidated the efficacy and value of SSI within SSM, devising an "SSI—SSM" interaction paradigm. Drawing inspiration from Liska (2015) and others, they introduced the concept and methodology of intelligence-led SSM, which advocates for the comprehensive and effective application of SSI to guide SSM processes (Wang Bing and Wu Chao 2019g). Empirical studies have been conducted on this approach in the domains of mineral resource SSM (Wang Bing et al. 2019), urban SSM (including smart city SSM) (Wang Bing and Wu Chao 2020b; Guang Xialei et al. 2020), and emergency event prevention and control (Wang Bing et al. 2020). They also explored the essence and research framework of intelligence-led smart SSM propelled by big data (Wang Bing and Wu Chao 2020c). Zhang Ya Lei et al. (2019) emphasized that SSI literacy is a crucial attribute for SS personnel, initiating preliminary research on the concept and components of SSI literacy. The SSI system serves as the foundation for SSI operation, and Wang Bing and Wu Chao (2019h) established a theoretical framework for SSM-oriented SSI systems. Furthermore, Trim and Peter (2003) examined the role of SSI in disaster management, while Wu Chao and Wu Lin (2020) discussed SSM models from an SSI perspective.
2. Research on SSI operations for Emergency Management: In recent years, the study of emergency intelligence operation has emerged as a prominent theme in research on SSI operation. Preceding the research on SSI operation in the SSM process, this area boasts a good number of findings which are systematic and rich in content. Schlegelmilch and Albanese (2014) explored how intelligence

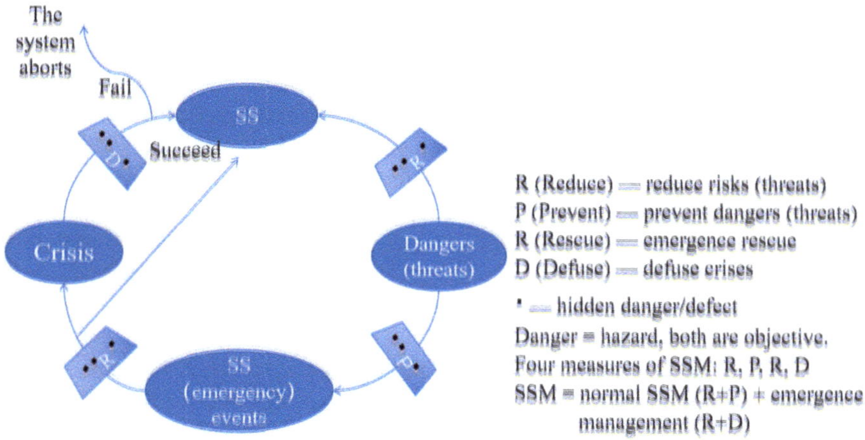

Fig. 1.14 Cycle model of system SSM

can foster business innovation and management. Zhao Fazhen (2020) conducted a comprehensive literature review on emergency intelligence operation in China, revealing that existing research mainly involves the process of emergency intelligence operation, systems, decision-making promoted by emergency intelligence, emergency intelligence operations in the big data environment, and specific research on various emergency types. As indicated in Fig. 1.14, emergency management includes both responses to SS events and crisis management. Barnes et al. (2007) have investigated the intricacies of embedding intelligence flows and other critical information within the risk framework to facilitate early warnings in crisis management for newly emerged threatening scenarios. Deng Kaifan (2018) explored the SSI and the mechanism of early warning in crises in Russia. Wang Keping (2012) analyzed the nexus between competitive intelligence and corporate crisis warning. Zou Ping et al. (2018), through a comprehensive review of studies, found that research in this domain predominantly focuses on early warning in crises and management of risks in crises.

1.5.2.3 Research on SSI Operations for National Safety and Security

The operation of national SSI is a crucial component of SSI endeavor which is oriented toward SSM, specifically tailored to address national SS governance. It represents a vital area of inquiry within the field of national SSI studies, which is a specialized branch of SSI research. Gao Jinhu (2019b) posits that national SSI operations essentially constitute the primary subject of national SSI studies. The domain boasts an extensive history of scholarly endeavors, with contributions from both domestic and international researchers dating back several decades (Rankin 1986; Johnson 2012; Weller 2010). Here an overview of recent advancements in China's research in this area is provided, which covers five key dimensions: First, foundational research on national SS is being conducted, encompassing the essence, characteristics, status, systems, and institutional frameworks of national SS endeavors. For instance, Gao Jinhu (2019b) offers an analysis of the significance and traits of national SSI operation. Dulles (1965) and Asthana and Nirmal (2008) discuss the role and functions of national SSI agencies in national SS operations. Jiang Huanhui (2015) examines the evolution of the relationship between national SS and national SSI operations, while He Hui and Liu Shengxiang (2017) dissect the system of SSI operations in France. Second, research on national SSI operations are being conducted under the holistic approach to national security, including the exploration of the development and deepening of national SSI operations against this backdrop (Zhang Jianian and Ma Feicheng 2018; Yang Guoli and Li Pin 2018), and research on national SSI operations in specific SS domains, such as anti-terrorism SSI operations (Li Heng and Liu Zuoxinhui 2018) and cyberspace SSI operations (Jiang Xi 2019). Third, there is a focus on the transformation and development of national SSI endeavors in the era of big data. Wu Chengyi and Tang Xiaohong (2020) analyze the changes and challenges that national SSI endeavors face in the era of big data. Fourth, research is being conducted on national SSI operations for national SS governance. Zhang Jianian (2017) investigates the integration and countermeasures of national SSI and national SS strategy resilience from the perspective of national SS assurance. Fifth, recognizing the critical nature of the SS of national intelligence itself, Sun Ruiying and Ma Haiqun (2019) analyze the system of national SSI operations with Chinese characteristics, and Chen Ming and Wang Qiaobao (2018) provide additional research contributions about this critical area.

1.5.2.4 Review and Prospect of the Overall Research

In summary, while research on SSI operation has made progress, there are shortcomings and areas for further concern.

1. Research on SSI analysis: Firstly, the field is in dire need of tailored models of SSI analysis, as the current methodology appears to be rather one-dimensional. It is imperative that a comprehensive SSI analysis model is constructed, which would be informed by extracting and refining the common threads from the various specific methods of SSI analysis. Secondly, the methods of SSI analysis which are employed are largely borrowed from other disciplines or adapted from analyses of different types of intelligence. It is needed to develop a proprietary method of SSI analysis that is attuned to the unique characteristics and requirements of SSI analysis, guided by an overarching model or methodology of SSI analysis. Thirdly, there is insufficient empirical validity of certain SSI analysis methods, and the effectiveness, scientific validity, and applicability of certain SSI analysis methods need to be enhanced. It is imperative to strengthen the empirical validation of these methods. And efforts should be made to integrate existing SSI analysis methods by reconstruction (aiming to complement

the strengths of various methods) and propose new, effective, and applicable SSI analysis methods. Lastly, the integration of SSI analysis with emerging information and intelligent technologies, such as big data and artificial intelligence, is still in its infancy. The potential value of these technologies for SSI analysis is underutilized, and their impact on SSI analysis has not been sufficiently analyzed. Research in these aspects needs to be done (for instance, on the integration of SSI analysis and computational SS science).

2. Research on SSI operations oriented toward SSM: Firstly, while the spotlight has primarily been on emergency intelligence operations (the theoretical and methodological frameworks have been set up, and the applied empirical research has started), there's still a noticeable gap when it comes to SSI operations that are oriented toward the entire SSM process. The essence, attributes, and foundational patterns of SSI operation have not been clarified, and the SSI services and products have not been adequately concerned. Secondly, the research on SSI operation across the SSM spectrum is fragmented, and a cohesive theoretical and methodological system is yet to be constructed. The SSM process and the SSI operation need to be studied in their entirety, and the research content of SSI operation needs to properly planned for establishing a mode of SSI operation, and a system that effectively support the SSM process from start to finish. Thirdly, the current research is predominantly theoretical and qualitative, with a significant absence of quantitative analysis and empirical validation. Attention should be paid to quantitative research and applied empirical research on related research fruits. Fourthly, there's a dearth of research on SSI operations that employ big data and intelligent technologies, thus the needs of SSI operation in the era of big data and intelligence cannot be properly met. Fifthly, and perhaps most critically, there is insufficient research on SSI operations concerning the prevention and resolution of major and emerging SS risks (including unconventionally major SS accidents). Attention should be paid to this aspect and relevant research needs to be carried out.
3. Seen from the research on SSI operations of various types, it is evident that the field remains relatively narrow, and there is an urgent need to enrich, expand, and deepen the scope of research. The primary issues include: (1) In terms of the varying interpretations of SS, current research predominantly concentrates on the aspect of "security" (such as military security, cyber security, and financial security), while research dedicated to the aspect of "safety" (including urban safety, workplace safety, aviation safety, and medical safety) and the aspect of "Safety and Security" is limited. (Patriarca et al. 2019; Bruenisholz et al. 2019; Wang Bing and Wu Chao 2019a; Gao Min 2019) (2) With regard to research domains, existing studies are primarily centered on traditional areas of interest within intelligence studies, such as military security, counter-terrorism security, scientific and technological security, and cyber security. There is a significant lack of attention and research in other areas, such as biosafety and biosecurity, public SS, cultural SS, resource SS, and nuclear SS (Barnes et al. 2007; Asthana and Nirmal 2008; Cates 2015; Wang Bing and Wu Chao 2019a). (3) Regarding the subjects of research, the focus is largely on national SSI operations, with a

noticeable lack of research on the SSI operations of societal, urban, and corporate SS.

In essence, while strides have been made in research on SSI operations, there are still problems to address, including the clarification of fundamental issues, the need for directive research from the methodological level, the universal applicability of research fruits, research scopes, and empirical validation. Future research needs to be targeted at resolving these issues on the basis of existing research fruits.

1.5.3 Research on SSI Management

SSI management represents a critical management endeavor designed to address the SSI requirements of SSM. This activity is primarily characterized by a structured approach encompassing planning, control, and coordination. Essential for the efficient development and comprehensive utilization of SSI resources, SSI management stands as a central pillar within the broader field of SSI studies. In the context of China, research on SSI management is predominantly concentrated in three key areas: research on the mechanism of SSI systems, research on SSI laws and regulations, and research on SSI management systems.

1.5.3.1 Research on the Mechanism of SSI Systems

The construction of the mechanism of SSI systems is endeavor of SSI management on the strategic level. It determines the overall direction and tasks of SSI management, and is the key to the success of SSI management. At the same time, an efficient and perfect mechanism of SSI systems is extremely important for enhancing the capability of SSI services and supporting SSM operations. In recent years, scholars in China have begun to pay attention to and conduct research on the mechanism. The representative research results mainly include: First, the research on the model of the organizational structure of national SSI. This model clarifies the correspondence between national SSI categories and various government agencies on the basis of the content of the holistic approach to national security and the actual settings of government agencies in China (Zhang Jianian and Ma Feicheng 2015a); Second, the research on the operation process of national SSI (Zhang Jianian and Ma Feicheng 2015a); Third, the research on the legislation and supervision mechanisms of SSI; Fourth, the research on the system and mechanism of the national SSI education and training (Zhang Jianian and Ma Feicheng 2015a); Fifth, the study of the structure and operational mechanism of the US national SSI system from multiple perspectives, aiming to provide reference for the construction and optimization of the national SSI system in China (Zhang Jianian and Ma Feicheng 2015b); Sixth, the study of approaches to the reform of China's national SSI system. The approaches include fostering a strong intelligence-led strategic awareness, complying with and applying the holistic approach to national security, setting up specialized national SSI agencies, strengthening the top-level design of the mechanism of national SSI systems, strengthening the legislation and institutional construction of the national SSI system, emphasizing national SSI education and training, and advocating and fostering a high-quality national SSI culture (Gao Jinhu 2019c); Seventh, the study of the mechanism of national SSI transmission of the US, which found that the US national SSI system has formed a "dual-wheel drive" system of national SSI agencies and integration of tasks, and full-coverage horizontal and vertical communication networks have been built for its national SSI transmission (vertically, three levels of leadership, coordination, and execution have been covered; horizontally, effective coordination among different agencies has been achieved), and effective coordination of national SSI has been basically fulfilled (Fan Bing 2019); Eighth, the study of the system of national science and technology SSI and its construction, such as clarifying the system framework of national scientific and technological SSI, establishing the model of "dual-function cycle of SS/Surveillance," and proposing corresponding countermeasures for the construction of the system of China's national scientific and technological SSI (Zhang Jianian and Ma Feicheng 2016).

1.5.3.2 Research on SSI Laws and Regulations

The development of SSI legal framework serves as the foundational legal basis and safeguard for regulating SSI management practices and reinforcing SSI initiatives. There are mainly four aspects in the current research on SSI laws and regulations in China: (1) Research on the construction of the mechanism of the national intelligence law from the perspective of a holistic approach to national security. This has led to the preliminary formation of a structural framework featuring "three tiers" (encompassing international intelligence legal relations, inter-agency intelligence legal relations, and state-citizen intelligence legal relations) and "three statuses" (including intelligence cooperation, confrontation, and defense), which offer valuable insights for the development of the national SSI legal and regulatory mechanism (Jin Haiting 2018). (2) Research on the development of national SSI legal system. Zhou Zhibo (2020) conducted the research on the development of the national SSI legal system guided by the constitutional clauses of "national SS." He emphasized that the constitution of a nation should serve as the fundamental guidance in the development and refinement of the national SSI law, thus the national SSI legal system should be constructed in accordance with the spirit inherent in the constitutional clauses of "national SS." (3) Research on the development of public SS legal and regulatory systems. The focal points and pathways for constructing the public SSI legal system have been clarified (includes designing and establishing a value orientation that balances national SS with the protection of human rights within the overarching framework of *The State Security Law of the People's Republic of China*, and adopting an international perspective), advocating for forging a new paradigm of public SSI legal system (Tang Chao 2017). (4) Ma Haiqun and Zhang Bin (2019) have scrutinized the trends of thematic diffusion of national SSI-related laws and regulations, expecting to inform the future development of national SSI laws and regulations at both central and local government levels.

1.5.3.3 Research on SSI Management System

In the era of information where the computer technology advances swiftly, the governance of SSI and its operations are contingent upon a robust SSI management system. Theoretically, the SSI management system—alternatively termed the "SSI management platform"—envisions a comprehensive assembly of all elements of SSI, bridging from the origin to the end-users, typically SS managers, through its channels of circulation and transmission. In essence, an SSI management system is an artificial construct designed for the aggregation, processing, manipulation, conveyance, and application of SSI, that is, for the efficacious governance of SSI itself and its associated operations. As such, the SSI management system constitutes a significant domain of inquiry within the broader spectrum of SSI studies. Chinese scholars have embarked on preliminary yet exploratory research into SSI management systems, traversing both theoretical and practical application domains. In the theoretical aspect, scholars such as Wang Bing and Wu Chao (2019h) have articulated a general theoretical model for an SSI management system in the context of SSM. They posit that the SSI system represents an organic synthesis of the SSM system and the intelligence system, with its fundamental architecture encapsulated by the concept of "one organization, one intelligence source, two institutions, two major systems, and three networks." From a practical standpoint, with the ambition of achieving unified collection, automated application, and closed-loop control of all SSI, Xiao Peng et al. (2017) have crafted and implemented a centralized control platform for the intelligence of power grid information and SSI. Similarly, Xiang Qian (2019) has engineered a campus network SSI management system, tailored to the nuances of SSM within educational institutions.

1.5.3.4 Research Review and Prospect

In conclusion, the field of SSI management research presents several areas that require further refinement and deepening. Firstly, there is a pressing need for a consensus on the overarching programmatic and guiding models for SSI management (including the comprehensive systems of SSI management and of the SSI laws and regulations). It is imperative to establish a sound framework for SSI management research and practice that can effectively guide these endeavors through comparative analysis and scientific verification, leveraging complementary strengths. Secondly, the scope of research in SSI management has not encompassed all critical elements, including key determinants of SSI culture and SSI management, addressing these research deficiencies is a matter of urgency. Thirdly, while SSI management is inherently a highly practical discipline, much of the existing research remains theoretical and exploratory in nature. There is a need to enhance the practical guidance and operational applicability of research to real-world SSI management practices. Building on existing discussions, it is essential to investigate specific key areas, deepening the research and ensuring that it takes root, flourishes, and yields tangible results. Fourthly, the majority of research has been theoretical, with a dearth of empirical validation of research findings. This creates a discernible gap between research and practical application. Future research should, therefore, place a stronger emphasis on applied empirical studies to continuously refine and enhance theoretical outcomes. Moreover, research should be directed toward SSM, recognizing the inherently practical nature of SSM and the ultimate goal of SSI studies—to provide effective SSI services in support of SSM practices. Fifthly, the research has predominantly focused on national SSI management, with a relative neglect of the universality of the research and a lack of attention to SSI management in other contexts, such as urban and corporate settings. This oversight results in a paucity of theoretical guidance for the implementation of grassroots SSI management efforts. It is essential, therefore, to balance the universality and specificity of research, fostering a comprehensive advancement of SSI management research and practice across various domains.

1.5.4 Conclusions

From a disciplinary standpoint, this section probes into the research advancements in SSI within China. It conducts a thorough review, highlighting the shortcomings and charting a course for future scholarly pursuits. This assessment is grounded in the typical research domains that are reflected in the SSI literature authored by Chinese scholars, encompassing the development of the SSI discipline, operational research, and management studies. It is imperative to acknowledge that SSI is a field of study in a state of rapid evolution, one that will continue to transform with the vigorous development and innovation within the broader realms of SS science and intelligence studies. The future holds boundless opportunities and significant potential for SSI research. Besides, the trajectory of SSI's growth is replete with both opportunities and challenges, which necessitates a concerted effort among SSI professionals to enhance dialogue and communication. Adhering to the principles of general SS and intelligence, and with a focus on SSM, the SSI professionals need to align their efforts with the evolving demands of intelligence and SS sciences. It is through such collaborative and synergistic endeavors that they can hope to foster the continued progression and maturation of the SSI field.

1.6 Chapter Summary

SSI science stands as a symbolic branch within the realms of intelligence and SS disciplines. This chapter provides a comprehensive outline of the foundational concepts of the discipline—encompassing SS, SS science, intelligence, intelligence studies, and SSI—as well as the context, significance, underpinnings, and prerequisites for the development of the discipline. It also reviews the progress made in research to date. Despite the absence of a universal consensus on the fundamental concepts of SSI science within scholarly circles, this chapter endeavors to establish a set of widely accepted concepts based on prevailing perspectives and the evolutionary trajectory of related ideas. It becomes evident through this exposition that the establishment of SSI science is backed by compelling academic and practical imperatives, along with a robust foundation. SSI emerges as an innovative branch within the intelligence and SS disciplines that warrants construction. It is poised to be the most crucial new branch in the twenty-first century and for the foreseeable future. The development of SSI science is an ongoing process. In recent years, the field has witnessed rapid growth and yielded substantial research outcomes globally, with particularly notable advancements in China. Nevertheless, the research in SSI has its shortcomings, including a lack of overarching perspective, a comparatively narrow scope, fragmentation, and a deficiency in both theoretical depth and foundational theory. The future direction of SSI science research should be focused on constructing a comprehensive theoretical framework, guided by the general principles of SS and intelligence, and oriented toward SSM. This endeavor is essential for the maturation and advancement of SSI science as a discipline.

Chapter 2
Fundamental Issues in Safety and Security Intelligence

2.1 Introduction

Based on the concept of SSI defined in Chap. 1, this chapter seeks to analyze and answer several critical and fundamental questions about SSI from the theoretical perspective, in order to provide readers with a general picture and in-depth knowledge of the concept of SSI. The origin and evolution of this concept are to be explored from the theoretical perspective of SS science, then the characteristics and functions of SSI are to be analyzed, and finally the reasons for classification and specific classification of SSI are to be discussed.

2.2 Origin of the Concept of SSI

As far as the origin of the concept of "SSI" is concerned, it should be a product born with the introduction of the concept of "intelligence" into the field of SS science, and an outcome of the application of related theories and methods of information science in this field. It is also an academic concept emerging from the intersection of information science and SS science. Theoretically, the concept of SSI originates from a new understanding of SS science (especially SSM) from the perspective of information science, or a new understanding of SSM "from SSi to SSI."

2.2.1 Re-Examining SSI Deficiency in SSM

The phenomena of SSi deficiency (or incomplete or asymmetric SSi) have widely existed in SSM, accounting for the fundamental cause of failures in SSM. This is a fresh recognition as well as a new consensus reached in the academic and practical

B. Wang and C. Wu, *An Introduction to Safety and Security Intelligence Science*,
https://doi.org/10.1007/978-981-96-7261-5_2

circles of SS science in recent years. Wang Bing and Wu Chao (2018b) believe that "the road to safety and security must pass through SSi." The most meaningful and best approach to SSM is to carry it out with the best SSi. How SSi is collected, managed and harnessed determines the outcome (Wang et al. 2017).

In recent years, Evidence-Based Safety and Security (EBSS) Management (Wang et al. 2017), a budding SSM concept as well as method, is being actively advocated and practiced in the SS domain, aiming at addressing the dilemma of SSi deficiency in SSM effectively. The essence of "evidence" in EBSS is SSi, and more strictly and accurately, SSI (Wang Bing and Wu Chao 2018d). In terms of management, "intelligence refers to all the information (or content) that affects management (mainly management decisions)" (Chen Chao 2017). According to this understanding as well as Wang Bing and Wu Chao's cognition of the nature of "evidence" in EBSS Management (Chen Chao 2017), the above-mentioned SSi (namely, the SSi that affects SSM) refers to SSI from the perspective of intelligence perspective, ("SSi" and "SSI" cannot be confused, and the difference and connection between the two are similar with the difference between "information" and "intelligence," which will be analyzed in detail in Sect. 2.2.1), because according to the theory of intelligence transformation (Hua Berlin and Zheng Yanning 2012), SSi is originally "objective," "infinite," and "useless," while SSI is oriented toward SSM and directly affects SSM. Therefore, what is truly missing is "SSI" rather than "SSi" in terms of SSM. For example, delays in communicating disaster information have caused poor results of disaster relief in the Great Hanshin Earthquake in Japan in 1995 (Xu Jingbo 2008); in China, the SARS event in 2003, the "7.23" particularly serious railway traffic accident on Ningbo-Fenshuiguan Railway in 2011, the "11.22" Qingdao Oil Pipeline Explosion accident in 2013, and the "8.12" explosion accident in Binhai New Area of Tianjin in 2015 with inadequate post-emergency responses, were all closely related to the failure of SSI research and judgment, and the untimely transmission or failure of SSI (Lin Xi and Yao Leye 2014).

2.2.2 *Reflections on Intelligence in SSM*

From an intelligence perspective, SSM should be a process of upgrading the "SSi chain" progressively. It is generally believed that the information chain is composed of five key elements: "Facts → Data → Information → Intelligence → Wisdom" (Li Houqing et al. 2015). With regard to the elements of the SSi chain, there are also SS scholars [such as Huang et al. (2018a)] who have provided descriptions similar with the elements of information chain: the upstream of information chain is oriented toward physical attributes, while the downstream toward cognitive attributes (Liang Zhanping 2003). From the principle of the information (SSi) chain, it can be seen that SSi is mostly directed "downward"—close to the lower link of the SSi chain, while SSI is at the higher link of the SSi chain, and should be directed "upward" toward SSM (Li Gang and Li Yang 2015). Therefore, as far as SSM is concerned, SSi is originally "useless"—only when SSi (including knowledge of safety and security) is "activated" and converted into SSI will it affect and assist SSM (that is, play a certain role). In short, SSM without SSI is like water without an identifiable source. With this understanding, an important addition can be made about the previous assertion "The road to safety and security must pass through SSi, and SSI is a necessary asset for realizing safety and security."

It can be known from the theory of information chain that "being intelligence-oriented" is the basis and premise of achieving the level of "intelligence (wisdom)." If the understanding remains at the level of "SSi" of the SSi chain, then "smart (or intelligent) SS," the goal pursued by the SSM informatization ("SS informatization" for short), will become "out of the question" and "out of reach." At present, although numerous SS databases or SSi systems and platforms have been established at various levels, including the state, government, society, cities, and enterprises, they have not yet played a near perfect role in SSM. The main reason is that SS researchers and practitioners have not truly comprehended the difference between "SSi" and "SSI." As a result, their focus of knowing and working remains at the level of "SSi" instead of reaching the level of "SSI," and the SSi has not yet been upgraded to SSI to provide services for SSM.

It is important that SSI serve as the support, key point, and prerequisite of SSM. Obviously, the above-mentioned drawbacks in understanding and research practice will seriously hinder the process of safety and security informatization and the effective exertion of its utility, and the practice of safety and security informatization is liable to become a "safety and security (information) database," an "empty shell" or a dispensable "decorative item" like a "vase" that exists with a sheer name. More specifically, the development of SSi resources has registered remarkable achievements, but the problem of "emphasizing storage over use" has existed for a long time and has become increasingly prominent. What's more, the large-scale SSi resource database built cannot quickly and effectively provide immediate and effective solutions to specific problems of safety and security. In other words, the resource-based understanding of SSI prevails, which tends to contain its functions and fail to provide sufficient long-term service for SSM.

It is also of importance to examine the role of SSI from the perspective of safety and security managers (referred to as SS managers). According to Brookes (1980)'s knowledge equation, "$K(S) + \Delta I = K[S + \Delta S]$, where: $K(S)$ is the original knowledge structure, ΔI is the amount of intelligence absorbed, and $K[S + \Delta S]$ the terminal knowledge structure," which indicates that SSI would fundamentally improve the "quality" of SSM. Based on SSM and the knowledge equation of Brookes (1980), and starting from the SS manager (subject), $K(S)$ is the original safety and security knowledge structure of the SS managers, and ΔI refers to the safety and security knowledge that is multiplied by absorbing SSI (the main body that becomes SSI is SSi). $K[S + \Delta S]$ refers to the safety and security knowledge structure of SS managers supported by SSI. Theoretically, in this case the SSM behavior of SS managers are to become more reliable, scientific, and accurate.

With the increasing complexity and interweaving of SS issues, especially the extensive application of new technologies such as the Internet of Things, cloud computing, and the Internet in the safety and security domain, the amount of SSi has registered a growth spurt, ushering in an era of big data of the domain. In this era, either role or significance of SSI has not been reduced. On the contrary, the problem of "the profusion of useless SSi and the lack of valuable SSI" has become more and more evident, which seriously affects the predictability, decision-making, and execution of SSM. The following three reasons may account for the problem:

1. Most of the safety and security informatization only remains at the level of SSi, which leads to its dissociation from the SSI-related work.
2. The lack of strong intelligence awareness among safety and security researchers and practitioners, as well as a vague understanding and positioning of SSI, has led to SSI failures.
3. Some safety and security professionals' limited capabilities of intelligence analysis, together with the fact that intelligence professionals (including professional service agencies that provide intelligence) have not yet been well integrated into the safety and security domain or cooperated well with safety and security professionals to provide SSI services, cause unsatisfactory quality and poor timeliness in yielding SSI products.

In general, there is need for safety and security researchers and practitioners to upgrade SSi to the level of SSI from all dimensions of SSM, extract SSI in the process of SSM, and grasp the "principal contradiction" in SSM—to sublimate SSi (data) into SSI resources with SSM utility. Only in this way can the power of prediction, explanation, and supporting of SSi (including SSI) for SSM be activated to generate more value, and can the SSM system be established more readily with SSI as its core. Obviously, the SSM research and practice paradigm guided and centered on "SSI-SSM" is essentially a process of SSi cognition, SSI analysis and application, advanced and accurate SS prediction, scientific and effective SS decision-making, and timely SS implementation. To some extent, this is a kind of innovation and transformation of the SSM research and practice paradigm. But for the time being, there are still many deficiencies and confusions in the SSI-related work in SSM. Therefore, it becomes urgent to upgrade the focus of SSM from "universal safety and security variables (SSi)" to "targeted safety and security variables (SSI)," which proves to be the critical but difficult point of SSM from the intelligence perspective.

Recent years have witnessed SSI, especially security intelligence, receive extensive attention in the domain of SS science, especially in the security (such as military security, national security, scientific and technological security, information security and public security, and so on) domain, and a series of research and practical exploration with great theoretical and practical significance have been conducted, which has proved to be a favorable beginning for successful transition from SSi to SSI and for further deepening of the understanding, research, and practice of SSI. Currently, SSI research is still in the preliminary exploration stage. In SSM (especially safety and security informatization), SSI still exists in a dissociated state lacking valid integration, calling for integration of SSM with ideas of SSI so as to elevate SSi as the goals of SSI for the value of the latter to be truly unlocked and appreciated.

2.2.3 Conclusions

Seen from the perspective of SS science theory, SSI is an important concept in the field of SS science and information science. The concept of SSI stems from an original understanding of SS science (especially SSM) from the angle of information science, that is, the new understanding of SSM "from SSi to SSI." In short, SSI is the sublimation of SSi research. SSI provides guidelines needed for innovating the theory of SS science (especially safety informatics) and expanding the scope of information science. It is a pity that although SS science has developed rapidly in recent years, SS science researchers and practitioners haven't yet paid enough attention to SSI, the high-end link of the SSi chain, which tends to be cast into oblivion. It is urgent to understand the concept of SSI and the importance of SSI research so as to attach more importance to SSI research.

2.3 Distinctions and Relations Between SSI and Related Concepts

The SSI as a concept is easily misunderstood in the safety and security domain. The most common mistake is to consider "SSI" as synonymous with "SSi." Moreover, "SSI" is sometimes confused with "safety and security analysis" in the SS field. In order to understand it correctly, this section will explore the relationship, especially differences, between SSI and the two existing concepts (namely, "SSi" and "SS analysis") in the safety and security field according to the meaning of SSI (see Sect. 1.2.5) and the origin of the concept of SSI (see Sect. 2.2, this chapter).

2.3.1 SSI and SSi

According to the definition and basic elements of SSI, SSi proves to be the core of SSI as it provides the foundation and lifeblood for the production of SSI. However, SSI differs from SSi in that it requires certain form of processing (such as analysis), of which the goal is to derive specific meaning from piles of SSi. In other words, SSI covers wider objectives rather than mere gathering of SSi. Moreover, the success of SSI is not measured by the SSi collected, but by the timeliness, efficiency, and accuracy in supporting SSM activities (such as SS decision-making).

As a matter of fact, the way of defining SSI has helped clarify the differences between SSi and SSI early in Sect. 1.2.5. While SSi is factual (Kahaner 1996) and numbers, statistics, scattered SS data about an organization or a system (Wang et al. 2019), SSI is not reams of SS database print-outs. It is the SSi that has been filtered, distilled, and analyzed. In other words, there would be no SSI without the processing of SSi, as SSI refers to the information that is useful for SSM after the gathered SSi

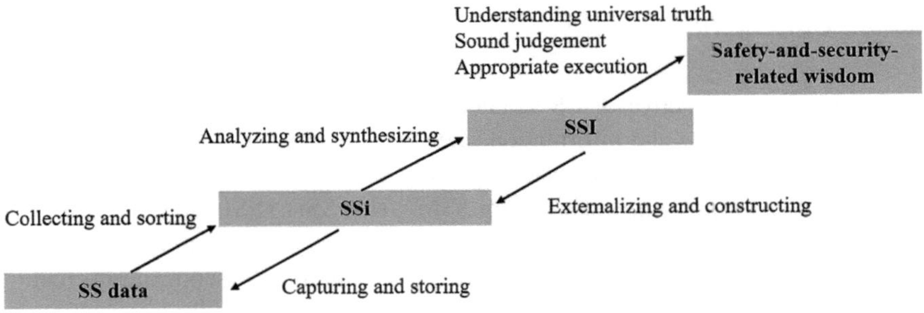

Fig. 2.1 SSi cycle

(or data) has been processed. In even simpler terms, SSI is processed SSi. It must be handled by specially trained intelligence professionals for SSi to become useful. Therefore, SSI requires precise recognition of the distinction between SSi and SSI. SSI, not SSi, is what SS managers need in order to guide and drive SSM activities of an organization.

In addition, SSI managers have a role in enriching SS data throughout the SSi cycle—to transform safety-and-security-related into exploitable SSI, which can be used by SS managers like safety decision-makers. The SSi cycle (Huang et al. 2018b) as shown in Fig. 2.1, is similar with the general information cycle (Liew 2013). According to Liew (2013) and Huang et al. (2018a; 2018b), it can be explained as: (i) SS data is a set of perceptible symbols representing raw SS facts (including observations); (ii) SSi is organized (namely, collected and selected) SS data (answering the following basic questions: What? Who? When? Where?); and (iii) SSI is processed SSi (aims to answer three basic questions: Why? How? For what purpose?).

According to Fig. 2.1, three conclusions can be made: (i) SSI is not about truth, but a further explanation of it could reveal its proximity to safety-and-security-related wisdom; (ii) strictly speaking, the direct foundation and support of SSM is SSI, not SSi; and (iii) safety-and-security-related wisdom is immediately extracted from SSI, and is the direct driver of smart SSM. According to the third conclusion, SSI serves as the prerequisite and foundation for smart SSM. In this sense, the SS manager should pay more attention to SSI rather than SSi in the SSM practice. Of course, SSi is also very critical for SSM. In a word, it is the primary purpose of SSI to bring value to SSi (or data) and to transmit SSi.

2.3.2 *SSI and SS Analysis*

As SS analysis is a scientific approach to SS problem-solving, it relies on deductive and inductive reasoning for defining SSM requirements and forecasting safety threats (Rozenfeld et al. 2010). Therefore, its concept largely focuses on analyzing potential safety hazards or threats, SSM capabilities, strengths and weaknesses of SSM, current

SS situation, trends, and so on in an organization (Rozenfeld et al. 2010). Or in other terms, SS analysis aims to address questions like "What safety phenomena or problems have appeared and what are happening in the organization?" "Why do SS problems occur in the organization?" "What are the key SS risks or threats of the organization?" and so on. For example, as a kind of typical SS analysis, the accident analysis involves a wide range of analytical work in terms of accident pattern, cause, type, position, time-series, frequency-distribution, and statistics. Both of SSI and SS analysis play key roles in SSM. However, according to the definition and aspects of SSI in Sect. 1.2.5, SSI is different from SS analysis.

With SSI being its product, SS analysis is an essential part of the SSI process, Both of SS analysis and production of SSI need to be based on SSi (Rozenfeld et al. 2010; Wang and Wu 2018b), which becomes the currency of both SSI and SS analysis. Lowenthal (2009) believed that "all intelligence is information, but not all information is intelligence," which is similar with the relationship between SSI and SSi analyzed in Sect. 2.3.1. The same could be said about SS analysis in that all the SS analysis results are SSi, but not all SSi is the result of SS analysis. With this premise, the distinction between SSI and SS analysis can be reasonably explained. Briefly, SS analysis and its added value bring about critical differences between SSi and products (result) of SS analysis. In other words, without SSi analysis, there is no SSI.

Of course, while SSI and SS analysis are not synonymous, much in common exists between them. The common ground between SSI and SS analysis is based on the same objective, which is to provide support to SSM (Rozenfeld et al. 2010; Wang and Wu 2019g). For example, both SS analysis and SSI are designed to support SS managers to make an informed safety decision. SSI allows SS managers to arrive at an informed safety decision. In a similar manner, the result of SS analysis provides SS managers with SSi to formulate SSM policies and strategies. Moreover, SSI can offer support for SS analysis.

2.3.3 Conclusions

Although SSI is closely related to SSi and SS analysis, comparative analysis reveals that SSI is different from SSi and SS analysis. It refers to useful SSi for SSM and is the product of SSi being processed by (safety) analysis. The clarification of the differences and connections between SSI, SSi, and SS analysis is of great significance for recognition and in-depth understanding of the concept of SSI and the research and development of SSI science.

2.4 Trends of the Evolution of the Concept of SSI

A good number of concepts (especially those in the fields of social sciences, such as "culture," "information," and "intelligence") are undergoing a dynamically evolving process as people's understanding continues to deepen with the ever-changing development of social needs and environment. This is the same case for the concept of SSI. With the ongoing evolution of people's SS demands, major contradictions of SS, SS situation, SS tasks, SS problems and risks, the concept of SSI is being constantly transformed. Here, a brief summary and analysis of the evolution trends of the concept of SSI is provided to introduce a concept of SSI that is better adapted to and cater to the current needs of research and practice in SS science.

2.4.1 *Fields: From Partial Safety or Security to Overall (Comprehensive) SS*

On April 15, 2014, Xi Jinping, General Secretary of the CPC central committee, put forward the great national security theory and concept of the "Holistic Approach to National Security" at the first session of the National Security Commission of the CPC. The conference aims to establish a system of national security that integrates political security, homeland security, military security, economic security, cultural security, social security, technology security, information security, ecological security, resource security, and nuclear security. In fact, from the perspective of system security, the conditions, states, and attributes of security, if combined with any system (such as the national system, cultural system, and ecosystem), can produce a series of secondary security concepts featuring "xx security." Based on security science and guided by overall national security, the safety system can be divided into three tiers based on the size of different systems (according to the Chinese safety and security scholars Wu Chao et al. (2018): it can be divided into macro-system, medium system, and micro-system). Chinese safety and security scholar Luo Yun (2018) has roughly divided the field of SS science into three first-level scopes (categories) as follows:

1. Macro-System Security (referred to as "Big Security"), that is, the field of national security. It mainly includes political security, homeland security, military security, economic security, cultural security, technological security, information security, ecological security, and other secondary security areas.
2. Medium System Safety and Security (referred to as "Medium Safety and Security"), that is, the field of public safety and security. It mainly includes food safety, disaster prevention and mitigation, fire safety, traffic safety, nuclear safety, social security, anti-terrorism security, campus security, and other secondary security areas.
3. Micro-System Safety (referred to as "Minor Safety"), that is, the field of production safety (in China, it is often called "Workplace Safety"). It mainly includes secondary safety areas such as occupational safety, occupational health, and

industrial safety (mine safety, construction safety, chemical safety, metallurgy safety and transportation safety, etc.)

In view of different SSI needs of safety and security management in the above three primary security fields, SSI can be classified into national security intelligence, public security intelligence, and production safety intelligence. The specific explanations are as follows:

1. National SI (security intelligence): According to the research, national SSI (including military SSI) is one of the earliest fields that attract attention and generate studies from the information science community. Because of this historical background, for most people in the field of non-intelligence studies or without an intelligence studies background, the understanding of intelligence has still been restricted to national security or military (including military security) fields. Since General Secretary Xi Jinping proposed the "Overall National Security Intelligence" in 2014, the study of "National Security Intelligence" in China has received widespread attention from the academic community (Zhang Jianian and Ma Feicheng 2015a). The academic community has raised some secondary level "xx SI" concepts which belong to the category of national security intelligence. Database searching of Web of Science and CNKI show that studies are mainly concerned with military SSI, information SI (specifically involves network SI or "cyberspace security threat intelligence" and cyberspace SI), science and technology SI, and so on.
2. Public SSI: The concept of public SSI was proposed at a rather early time. Tang Chao (2017) believes that in China, public SSI has long been regarded as a synonym for public security intelligence since the public security departments have long undertaken most of the tasks of maintaining public security. Judging from the above definition of public safety and security, this understanding is not entirely accurate and it is somehow limited. Meanwhile, the academic community has raised the secondary level "xx security intelligence" concept, which partially belongs to the category of public security intelligence. By searching the databases of Web of Science and CNKI, it is found that the secondary concepts mainly include nuclear SSI, food SS risk intelligence, disaster intelligence (including earthquake disaster intelligence and natural disaster intelligence), fire prevention intelligence, anti-terrorism intelligence, and so on.
3. Production SI (safety intelligence): Documents show that the concept of "production safety intelligence" as a specific referent for the domain has not yet been formally proposed in the field of production safety. The main reason could be that researchers in the traditional domain of production safety are generally educated with a background in science and engineering and they are short of knowledge in humanities and social sciences (especially information science) to some degree. Their disciplinary background determines that their understanding of safety and security informatization is mostly focused on the application of information technology in safety management or stays at the "safety information" level. Clearly, this perception needs to be changed and upgraded to the level of "safety intelligence," as detailed above.

It should be an important trend of evolution for the SSI to move from partial security to overall (comprehensive) security. Since General Secretary Xi Jinping proposed the "Holistic Approach to National Security" in 2014, Chinese safety and security scholars have come to realize that since security issues in specific fields are mutually intertwined and transformed (such as public security and production safety), and the concept of "overall (comprehensive) security" needs to be specially noted if safe development is to be sought. The so-called holistic (overall) security, according to the author, means building a safe and secure country and society in an all-round way. To that end, the key lies in achieving "overall" development, which requires that development of safety and security in all areas be equally fulfilled without any avoidable weakness. Importance needs to be attached to improving the weakness of overall safety and security to ensure that all the fields and items are covered in time. In fact, this is also an essential trend of worldwide development of SS science. Obviously, based on those requirements of holistic (overall) security, the current SSI research is merely centered on serving local security management services, which calls for transition toward holistic (overall) security development.

2.4.2 Connotations: From "Security" to "Integration of Safety and Security"

Coined with two related connotations, the Chinese phrase "安全" can be rendered into a combination of two English words "safety" and "security." The differences between "safety" and "security" have been detailed and explained in Sect. 1.2.1, from which we can conclude that for a specific system, its security will inevitably involve risks concerning both "safety" and "security," and sometimes mutual conversion may undergo between the two. Given this thought, security scholars have begun to pay wide attention to research on safety and security integration in recent years.

The connotation of the SSI concept is yet to transition from security to safety and security integration. It can be known from Sect. 1.5 that SSI research has drawn wide attention from the field of security. However, in the domain of safety there is little attention paid to SSI, which still tends to focus on collecting security information, a preliminary stage of the domain development (the specific reasons have been explained in Sect. 2.2). From the perspective of system security science, SSI research aims to promote the overall security of service system. Manifestly, the connotation of the current concept of SSI is still evidently reflected in the "security" aspect, and that of the "safety" aspect is critically insufficient, hence the failure of reflecting the connotation of safety and security integration.

2.4.3 *Links: Overall Process from Emergency Management to Security Management*

Intelligence is of particular importance to emergency management due to the contingency, emergency, and urgency of the latter. In view of this, emergency intelligence is considered as the next focal point of information science research (Yao Leye and Fan Wei 2014). Indeed, there have been a great number of research achievements of emergency intelligence. China's emergency intelligence research has been highly valued and strongly supported by the state and the government. For example, *The Law of the People's Republic of China on Response to Emergencies*, which came into effect in November 2007, clearly stated the emphasis on "intelligence cooperation" in emergency management. Besides, there are many research objects related to emergence intelligence among the National Social Science Fund Projects and the Youth Projects of National Social Science in recent years (for instance, "Research on the Construction of Intelligence Engineering Service Mechanism for Emergency Management" on the project list of 2018).

In fact, from the perspective of SS science, divided by process, SSM encompasses "normal security management (with prevention as its focus)" and "non-normal security management (or emergency management, focusing on emergency management after an accident)." SSM should focus more on "normal security management." At the same time, both kinds of management ask for the support of intelligence. However, it is regrettable that there is currently limited SSI research for the overall (or link) SSM, as the current research mainly is centered on SSI research for emergency management. However, in some specific fields, research of this kind has been carried out, such as the mode of "intelligence-led policing" in public security and the method of "intelligence-led information" in information security. Therefore, the concept of SSI should be transformed from emergency management to the overall SSM in the future.

2.4.4 *Conclusions*

With considerations of the evolution and development of security science from a diachronic perspective, SSI has evolved "from decentralization to unification" and "from concreteness to universality." Specifically, the evolution is embodied in three aspects, namely "from partial security to overall security," "from "Security" to "SSI" and "from emergency management to SSM."

2.5 Characteristics of SSI

Characteristics of SSI are many. For example, SSI possesses some basic properties of intelligence such as providing knowledge and being transmittable because it is a subset of intelligence. This section, centered on security management, will summarize significant features of SSI based on the literature (Zhao Rongying 2017; Liang Huiwen 2017; Wang Zhe and Chen Qinghua 2004; Shi Yanqin et al. 2017; Jin Sheng 1996; Qin Tiehui 1991). It will also interpret SSI's special and significant characteristics on security management with a view to security management.

2.5.1 *Providing Knowledge*

Knowledge serves as the summary and reflection of people's subjective world toward the objective world. With the development of human society, new knowledge is being produced at every minute. People can acquire useful knowledge through reading books and newspapers, listening to the radio, watching TV, attending meetings, making visits, and other activities. In the SS field, the safety knowledge that is useful for prevention and control of SS risks and for SSM is the SSI needed by people. In other words, SSI is the specific SS knowledge that people need to have for SSM and security actions. It follows that SS knowledge is the essential content of SSI. Without certain knowledge concerning security, it cannot be transformed into SSI and is thus utterly valueless. Therefore, knowledge is one of the main attributes of SSI. The content of SS knowledge contained in the SSI is exactly the value of the SSI, and the amount of SS knowledge contained in the SSI determines its value.

1. Scope of SS knowledge of SSI: It refers to the breadth of the scope of security knowledge in the content of SSI. The wider the security knowledge of SSI, the wider the scope and field that can be utilized, the greater the opportunity and possibility to be harnessed, the more adequate the conditions for realizing the value of SSI, and the greater the expected security benefits, so the higher the potential security management value will be.
2. Amount of SS knowledge of SSI: It mainly refers to the quality of SS knowledge content of SSI, which is mostly reflected in three aspects: the advanced nature and novelty of SS knowledge in SSI; the reliability and applicability of SS knowledge in SSI; and the elaboration of SS knowledge in SSI. Generally speaking, the value of SSI is directly proportional to its advanced nature, novelty, reliability, applicability, and elaboration of the content of security knowledge.

2.5.2 *Being Transmittable*

Being transmittable is one of the basic attributes of SSI. In order to demonstrate its value, it must be transmitted, as it can never come into being if not transmitted and communicated for people to use. The essence of SSI transmission is to activate the fixed security information (including SS knowledge) for realizing the potential value of SSI. As far as SSM is concerned, only by transmitting security information to security managers can it provide possibilities to implement intelligence-led SSM.

In theory, the transmission of SSI refers to the transference of SSI from one point to another in time or space. Transmission in time is named as storage, while transmission in space is named as propagation (communication). It follows that SSI can be transmitted beyond the limitation of time and space. At the same time, the energy consumption of simultaneous transmission of SSI is relatively low, which can be made even more advantageous by the technological development of modern information communication and of information storage. Compared with the energy consumed by the transportation of material products, the energy that is required by SSI transmission is much lower. Generally speaking, the timing, speed, and completeness of SSI transmission play a decisive role in the value of SSI (in other words, timeliness is one of the attributes of SSI). It can be shown in the following three aspects:

1. Timeliness of SSI transmission: SSI can be fully leveraged only when it is transmitted to its users when it is most needed. If it is sent too early, there are inadequate conditions for SSI users to absorb and utilize so that the intelligence can't be made full use of, and the utilization effect of SSI will be largely undermined. And if is too late that things have changed, the intelligence will turn meaningless for its users. With the delay of SSI, its value can also decline. Thus the more timely the transmission of SSI is, the more fully the potential value of SSI can be achieved.
2. Promptness of SSI transmission: The shorter the time used in the transmission from the SSI source to the users, the more advanced and original it will become, the more efficient it will be, and the more fully its value is to be realized.
3. Completeness (systematicity) of SSI transmission: Only when SSI is systematic and complete can it produce maximum effectiveness. The scattered, disorderly, and incomplete security information will inevitably result in difficulties in use if transmitted to the users, and its value will also be distorted and deformed. Therefore, the more complete and systematic the security information, the more fully its worth is to be actualized.

Besides, SSI's nature of being transmittable indicates that it can be shared. In theory, it can be shared by multiple cognitive subjects (SSI users). It is true that SSI can be received, perceived, and used by multiple users in certain time and space. However, it should be noted that some SSI (especially "security" intelligence) is usually confidential and adversarial. Therefore, some pieces of security intelligence are restrictively shared, though they can be shared theoretically.

2.5.3 *Effectiveness (Value) and Practicability*

People produce security information, and communicate and transmit SSI in order to fully use it, which contributes to the constant improvement of its effectiveness (value) and practicability. SSI provides services to its users. As the users call for SSI, effectiveness (value) and practicality become essential indicators for measuring the quality of SSI service work. In effect, effectiveness (value) and practicality are the most critical attributes of SSI and the essential characteristics that distinguish it from SSi. Seen from the meaning of SSi (refer to Sect. 1.2.5), SSI is a kind of SSi in its essence, but it is different from SSi. While the latter is valuable and significant for SSM which it affects, SSI aims to serve SSM in that it can effectively support, optimize and correct SSM behaviors of security managers. The key value of SSI lies in solving the absence of SSI in SSM. From an economic perspective, SSI is also a valuable resource, as SSI is extremely crucial to SSM for it is the basic resource of SSM, which is also the significance and value of researching SSI. The SSM process is thus the process of realizing the utility of SSI. It can be inferred that SSI is effective (valuable) and practicable. In terms of SSM, its effectiveness (value) and practicability lie in the improvement of the SSM performance.

Only the users absorb, utilize and put into SSM practice can the value of SSI be finally realized. Thus the matching degree between the application of SSI and its users' needs becomes the key to whether the value of SSI can be realized and to what extent. It is mainly reflected in two aspects:

1. Pertinence of SSI: In the transmission of SSI, the SSI must be matched with the absorption capacity of its users, which generally refers to their ability to identify, understand, digest, and apply SSI. As far as security managers are concerned, how well they can take in SSI mainly depends on their SSM ability, knowledge level, and intelligence literacy.
2. Effectiveness of SSI: In the transmission of SSI, the intelligence transmitted must make its users more informed. If the SSI transmitted cannot alter the knowledge structure and trigger the creative thinking of SSI users, the potential value of SSI will never be fulfilled. Therefore, the better the transmitted SSI satisfy its users' needs of solving critical SS problems and adapt to their requirements, the higher the effectiveness of SSI, and the greater the economic and social benefits produced by SSI, hence the higher the value of SSI.

2.5.4 *Strategic in Nature*

The value of SSI is represented not only by the collection and processing of security information but also by the in-depth analysis and research. SSI strives to reveal the fact hidden behind the security data and information in order to draw a comprehensive picture of the security environment where the organization or system is located. In this process, the analysis and research of SSI have become part and parcel of the process.

SSI analysts cannot copy ready-made answers, nor can they always analyze security issues in traditional, conventional, and repetitive methods. That is because the objects of SSI analysis—security issues and environment—are complex and uncertain. Therefore, analysts are required to incorporate more intellectual activities, including analytical reasoning, SS risk analysis, innovative thinking, and SS forecasting ahead of time. Because of this, SSI can be referred to as "SSi plus intelligence" engineering, which integrates security with intelligence. By exploring new ways of thinking, SSI can be produced and applied in assisting SSM with new ideas being discovered, new safety knowledge created and new ideas for SSM generated.

As can be seen from the above, in SSI activities scattered information and materials about SS risks or environment need to be transformed into interconnected security information that is accurate and practical. Such analytical SSI can enable SS managers to clearly recognize and understand the strengths, weaknesses, opportunities, and challenges of SS risks and management so as to lessen uncertainties significantly, improve the effectiveness, and bring great security benefits to SSM.

The introduction of SSI has injected new vitality into traditional SSM and SSM informatization. SSI expands the latter from rigid safety data management to the lively collection and analysis of intelligence which is centered around the organization's safety goals, and from a generalized daily management system of safety data information to a smart SSI system that highlights "functions of SSI collection and analysis." The introduction of SSI will increase intellectual participation and provide services and support for SSM with more valuable SSI products.

2.5.5 Predictability

SSI can be used to symbolize, reflect, and thus be used to predict the SS state of an organization or a system and how it is going to change. Therefore, SSI is predictive in nature. SS prediction is the logical starting point of SSM, the core method to achieve "prevention beforehand," and the precondition to realizing accurate SS prediction. In other words, SSi should be collected in a timely and precise manner for SSM services to guarantee the safety needs of an organization or a system. SSI must be predictive as SSM activities to ensure the safety and security of a system or an organization take place in the future. A piece of lagging SSI does not have any practical significance.

SSI has an evident character of predicting and foreseeing. For the SSM of an organization or a system, SSI should contribute to solving present and existing SS problems and long existing problems about SS strategy. The purpose of utilizing SSI is to serve SSM and for seizing its future. Therefore, extensive collection of information about SS situation and SS risks are required. In addition, based on in-depth research, SSI is required to predict the future SS situation of organizations or systems and formulate SS strategies to improve their security capability.

2.5.6 Purposefulness

SSI activities are purposeful, targeted, and pertinent. The basic goal of SSI activities carried out by SS subjects is to collect, analyze, and study all the aspects of their SS situation and information, provide SS subjects with highly targeted SSI services, and assist them in formulating effective SSM programs and strategies to protect their safety and security.

2.5.7 Smartness

Seen from the perspective of intelligence studies, smart SS stems from SSI, because according to the information chain, intelligence is the smartness encapsulated, while smartness is the intelligence opened up. Wang Feiyue pointed out that there is a "natural" inherent connection between intelligence and smartness. By the same token, SSI and smart SS are closely related. Safety intelligence is the ability to obtain and apply SSi (including SS knowledge) and skills, while SSI is the SSi that serves specific SS goals. Therefore, SSI and smart SS are inseparable. At the same time, the English term "safety and security intelligence" corresponds to "安全情报" in Chinese, while the term "intelligence" has both the meaning of "情报 (intelligence)"—information that is gathered by the government or the army, and the meaning of "智能 (smartness)"—the ability to think, reason, and understand. As a result, the word "intelligence" is sometimes translated into "智能 (smartness)" in Chinese, where the well-known "人工智能 (artificial Intelligence)" is an obvious example.

In addition, the smartness of SSI is fully reflected in SSI-related work, namely the production, delivery, and evaluation of SSI products. The manufacture of SSI products is the process of applying safety intelligence and generating smartness. The delivery of SSI products should take SSI into account and seek to reflect it, that is, to consider the requirements of SSI users for the number and form of SSI products. The intelligence of its research work is exhibited in the condensation and organization of SSI materials, in the strategic requirements of SSI conclusions, and in the design and management of its delivery methods and channels as well.

2.5.8 Conclusions

In this section, based on SSM, some salient features of SSI are summarized, which mainly include features of providing knowledge, transmissibility, effectiveness (value) and practicality, strategic in nature, predictability, purposefulness, and smartness. The practical significance of the above features of SSI in SSM has also been analyzed.

2.6 Importance of SSI

2.6.1 Fundamental Functions of SSI

Being the premise and foundation, and the vanguard and adviser of SSM, SSI can be likened to "the ears and nose, or the pioneer and adviser" of SSM. Specifically, the root cause for SSM failures is the lack of SSI, and the value of SSI is to solve problems of security information deficiency in systematic SSM. Thus, according to the classification of SSM activities (i.e., safety prediction activities, safety decision-making activities, and safety implementation activities), the major functions of the following three most basic activities of SSI can be outlined (Wang, Bing, and Wu, Chao 2019a).

1. SSI Can Serve as a Support System for Safety Prediction (Early Warnings)

"To be able to detect in a precise and timely manner" is the prerequisite for ensuring effective SSM. This successful security forecasting (early warning) requires comprehensive collecting, understanding, and attaining various security information and analyzing the collected information to obtain effective intelligence. Then based on SSI, advanced, correct, and scientific security forecasts can be made ahead of time. In other words, SSI helps to identify organizational safety threats and improvement opportunities. The time left for security managers' response can be extended by ensuring earlier security warnings. As a result, SS risk control and prevention will be advantageous for precautions.

2. SSI Can Serve as a Support System for Safety Decision

"Good, fast and economic decision" is the basic requirement for developing an SSM program. It refers to successful safety decisions, that is, to promptly make scientific, reliable, effective, and economical safety decisions which are based on the best SSI after comprehensive research and evaluation of various safety forecast intelligence.

3. SSI Can Serve as a Support System for Safety Execution

"Timely, comprehensive and effective prevention and control" is seen as the ultimate objective of SSM response. To successfully implement such a principle, decision programs (mainly referring to various security measures, including emergency management measures) should be carried out effectively, according to relevant safety decision intelligence. The aim is to nip all dangerous events in the mud or minimize their adverse consequences through effective emergency management measures.

2.6.2 Extended Functions of SSI

Based on its basic functions, some extended functions of SSI can be proposed as follows:

1. SSI Can Be Used as an Important Safety Learning System

SSI can provide covert safety knowledge, which is inseparable from learning. Thus it is an important safety learning resource (Wang, Bing and Wu, Chao 2019a). It can provide specific clues, ideas, grounds, and references for SSM and help SS managers access new safety ideas and advanced SSM methods. It also enables them to draw upon safety-and-security-related experience, allowing them to identify opportunities to reform and innovate SSM from SSI. Thus, organizations can find the best improvement solutions for SSM. It also enables them to continuously introduce new SSM strategies and technologies to improve the organization's performance.

2. SSI Is the Basis of Consulting Service for SSM

In recent years, consulting service for SSM has gradually become one of the pillar industries in the global field of security. It is based on the collection, analysis, and utilization of SSI. Therefore, the essence of consulting service is the collection, analysis, and utilization of SSI, regardless of the form. Consulting service institutions provide users with SSM services, which is ultimately the collection and analysis-based integration of SSI, and systematic SSM solutions and knowledge of SSM specifically tailored for users. Based on SSI collection, analysis, and utilization, the consulting service for SSM is to use, transmit, and disseminate SSM knowledge, promoting the proliferation of SSM knowledge to help users improve their levels of SSM.

3. SSI Promotes the Advancement of SSM Informatization

Informatization has become the mainstream of SSM development in this era and is also an essential reflection of SSM capabilities. The so-called "SSM informatization" refers to utilizing the general theory and methods of information management, with modern information technology supporting as the core technical tools, and with overall consideration in the collection of an organization's internal and external security information. The aim is to effectively organize and configure security information resources and conduct SSM activities of informatization to enable SS managers to promptly grasp the general safety status of the organization and obtain security information (i.e., SSI) that is useful for SSM. In this way, SSM strategies can be carried out effectively. The above shows that the development of SSM informatization cannot be fulfilled without practical support of SSI from the very beginning. SSI has a decisive role in promoting SSM informatization. The appropriate understanding and use of SSI are conducive to helping organizations seek the right development goals of SSM informatization, and then promote the virtuous circular development of both SSI and SSM informatization.

4. SSI Can Create Incalculable Socio-Economic Benefits

SSI is incredibly vital and significant to national and social security development. It can serve as an effective contribution to the safeguarding of national and social security and to the prevention and reduction of the occurrences of security incidents, thus lowering the losses generated, which can be seen as a reflection of the potential and immeasurable socio-economic benefits of security information.

2.6.3 Roles SSI-Related Work Can Play

After analyzing the functions of SSI, this part is dedicated to the roles SSI operation can play in SSM. Competent SSI is the primary manifestation and basic guarantee of SSM, and strengthening SSI-related work serves as a sharp tool in promoting the SSM ability. In summary, engaging in SSI-related work can improve SSM abilities from the following five aspects:

1. Work of SSI Promotes Timely and Accurate Prediction and Early Warnings of SS Risks

Accurate and timely risk prediction and early warnings are the key points in SSM, just as the saying reminds us that to prepare for a rainy day is to prevent against potential problems. SSI research aims to effectively predict SS risks (Zhang, Jia-Nian and Ma, Fei-Cheng 2016). The prediction mechanism of SS risks refers to the analysis of research objects to clarify the current SS situation and effectively prevent SS risks. At the same time, SSI helps to monitor, discover, and predict the development and changes of various factors of SS risk in real time, and increase the response time of SS risk prevention by ensuring earlier risk warnings to give full play to SS risk prevention. Actions can be taken through early detection, warning, and prevention before it is too late.

As can be seen from the above, SSI operation serves as the basis for achieving "smart and sensible" SSM. Seizing SSI efficiently and accurately is the key to ensuring proactive SSM. Therefore, correct and timely prediction and early warnings of SS risks rely on right, prompt, and efficient SSI monitoring, collection, and analysis. With the continuous improvement of SSI-related work, the prediction and early warnings of SS risks are going to be conducted in a more timely and accurate manner.

2. Work of SSI Makes SSM More Scientific

SSI is indispensable for scientific SSM. While scientific prevention and control prove to be the underlying requirements of SSM, work of SSI is conducive to making scientific and correct SSM decisions. Intelligence serves as the premise and basis of decision-making as it is placed at the forefront of the decision-making process (Simon 1959). Similarly, scientific SSM decisions cannot be made without the support of high-quality SSI, which is the key to SSM decisions. In fact, the decision-making of SSM proves to be a process of collecting, analyzing, processing, and applying SSI, avoiding bias and even errors in the understanding and judgment of SS risks, seizing the best and appropriate timing for SSM, identifying the SSM situation and prominent problems, making correct, scientific, and reliable judgments of SSM, and developing an implementation plan of strategies and tactics of SSM, so as to avoid incompleteness and mistakes made about SSI which prove to be the main reasons for unsuccessful SSM.

3. Work of SSI Enhances the Precision of SSM

In security management, the key to accurate governance is "to figure out the realities," or the work of SSI. Precise governance is a critical method of SSM improvement. The types, quantity, and severity of SS risks vary from place to place and from link to link, so does the focus of SSM. At the same time, SS risks will continue to exhibit new characteristics, development trends, and even SS risks over time, so SSM should be flexibly adjusted according to the real-time changes in SS risks. This shows that SSM cannot take a one-size-fits-all approach. Safety risks need to be accurately analyzed and judged, in order to identify the critical matters, major SS risks, and prominent issues. This aims to realize precise governance by taking a targeted approach according to various locations, links, and situations. By accurately controlling various types of SS risks and problems, work of SSM will be well-targeted and handled with precision. "To figure out the realities" is the first stage of accurately identifying problems and precise governance. In SSM, only after figuring out the objective reality can we accurately identify major SS risks, seize the direction and focus of SSM, and implement specific policies and governance oriented toward problems and risks.

4. Work of SSI Encourages the Synergy of SSM

SSM is a complex systematic project which involves multiple elements, factors, subjects, departments, levels, and links. Hence, the actual implementation and execution of SSM requires multi-sectoral, cross-regional, or multi-level communication, collaboration, joint prevention, and control. Thus the synergy of multiple parties (including joint prevention and control) is crucial in SSM. Intelligence information serves as the bridge and link to connect the synergy of various subjects. Therefore, the synergy of SSI becomes the basis and support of SSM. Only with the interoperability and smooth transmission of SSI can efficient synergy of SSM be guaranteed.

5. Work of SSI Makes SSM More Intelligent

To achieve intelligent SSM is the most direct requirement of the era of big data, digitalization, informatization, and intelligence, of Industry 4.0 (or era of intelligence where information technology is harnessed for promoting industrial revolution, added by the translator) and of Safety 4.0 (or era of safety where safety and security are guaranteed with technology of intelligence, added by the translator) (Wang and Wu 2020). Intelligent SSM should consciously comply with the general trend of social informatization, digitalization, and intelligence. According to the information chain (i.e., "Facts → Data → Information → Intelligence → Wisdom") (Liew 2013), intelligence is encapsulated wisdom, and wisdom is intelligence opened up and also proves to be the essential characteristic of intelligence (Wang Feiyue 2015). Because of this, intelligence should be the cornerstone and focus of realizing intelligent management. Similarly, SSI is the basis for achieving intelligent SSM. The collection, analysis, processing and application of SSI and intelligent decision-making system incorporated into the SSM system is conducive to building an intelligent SSM system.

2.6.4 *Conclusions*

From the perspective of SSM, we have introduced SSI and important functions of its performance. Its fundamental functions can be summarized as follows: to serve as a support system for SS prediction (early warning), SS decision, and SS implementation. Its extended functions include: to be used as an important safety learning system and the basis of consulting service for SSM, promotes the advancement of SSM informatization and create incalculable socio-economic benefits. At the same time, the primary roles of SSI are to achieve more timely and accurate prediction and early warnings of SS risks and to make SSM more scientific, precise, synergetic as well as intelligent.

2.7 Types of SSI

There are different types of SSI according to various criteria of classification (Wang Bing and Wu Chao 2019b). As far as the form of SSI is concerned, there are static SSI (such as SSI related to population, geographical location, and time) and dynamic SSI (such as the evolution trajectory of SS events, the development and change of SS risks, the interrelationship of SS factors and the changing trend of SS situations). It should be noted that the classification of SSi is not unique, and with the research and practice of SSi science and SSM science, other methods of classification may be generated to meet different needs of SSM.

2.7.1 *Classification of SSI in Terms of SSM*

SSI is generally utilized to serve SSM. With regard to the characteristics and reality of SSM, SSI can be classified according to the levels and links in SSM (Wang Bing and Wu Chao 2019b).

1. Classification of SSI According to Links in SSM

SSI includes normal SSI and abnormal SSI (or "emergency intelligence") in respect of the links in specific SSM served by SSI [including "normal SSM (with the focus on prevention)" and "anomalous SSM (i.e., emergency management, of which the focus is on post-emergency)"].

2. Classification of SSI According to Levels of SSM

SSM requires SS managers to carry out a series of interrelated, continuous SSM activities over time. In the long run, they need to conduct strategic activities (such as deciding on potential SSM inputs and SSM strategies, policies, systems, and capacity building) with methods specially chosen or adapted in order to meet SSM goals. In the mid-term, SS managers need to carry out tactical activities (such as formulating SS plans, procedures, resources, and prevention and control measures to guide and optimize SS risk prevention and control). However, as it is insufficient to trigger the operation and activities of the daily SSM system with mere description of tactical SSM activities, SS managers need to adopt specific methods or measures to implement and execute tactical SSM requirements or decisions. That is to say, in the short term, SS managers need to conduct SSM affairs which involve specific operation, so as to execute daily SSM operation instructions. In fact, Hale (2003; 2005) and Winge et al. (2019) also expressed similar views. They believe a solution to SS problems (covering 3 levels, namely operation, tactics, and strategy) is a reliable SSM system structure with a cycle involved. In addition, Hale (2003) pointed out that the time dynamics for the three levels of SS problem-solving are different, ranging from the first level of seconds to days, to the second level of weeks and months, and to the third level of three to five years.

Therefore, SSM can be divided into three levels: strategic, tactical, and operational. Strategic SSM refers to the long-term security decision-making process where SS managers make SS investments and long-term decisions about the SSM system. Tactical SSM refers to the mid-term SS decision-making process where SS managers update the SSM system. Operational SSM refers to the short-term SS decision-making process of SS managers, mainly for daily on-site SS risk control. Since SSI is oriented toward SSM, and each level of SSM requires corresponding biological SSI support, SSI can be categorized into three different types, namely strategic, tactical, and operational SSI. In summary, a pyramid model of SSI for different levels of SSM can be constructed, as shown in Fig. 2.2.

In Fig. 2.1, different types of SSI are layered, with strategic SSI at the top of the pyramid. Strategic SSI aims to help senior SSM personnel understand the overall SS situation of the organization and consider how to effectively improve the SS level of the organization by appropriate means. Strategic SSI should focus on long-term trends in organizational SS risks, technology, culture, or potential SS threats. It

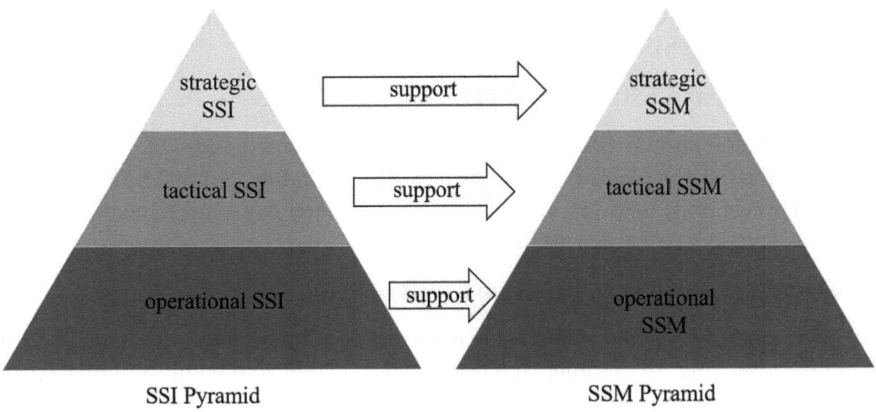

Fig. 2.2 Pyramid model of SSI for different levels of SSM

is a forward-looking SSM concept, highly dependent on SS forecasting and assessment—predicting future organizational safety and security based on past SSM actions or expected SS capabilities. Effective strategic SSI requires that SS managers (or more specifically, SS analysts) have profound professional knowledge, considerable insight and rich experience in SSM, as well as willingness to understand and adapt to highly dynamic changes in the environment of SSM.

If strategic SSI is designed for the leadership of the organizational security, tactical SSI is provided for the middle management of SSM in the organization. With its concern about SS risks and threats, tactical SSI attempts to answer questions of what and when by providing SSI about safety and security plans, procedures and technology. It is the assessment of an organization's direct SSM capabilities, mainly focusing on SS risk assessment (including hazard identification), weaknesses, strengths, opportunities, and challenges in SSM. When an organization's tactical SS risk assessment and SSM analysis are effectively conducted, those who directly control SS risks are allowed to allocate critical resources for crucial SS tasks in the most effective way, and are enabled to manage on-site SS risks at the proper time according to SS plans and procedures which have been appropriately established.

Operational SSI addresses daily SSM activities of all forms within the organization. It is oriented toward on-site SS risk managers. It is real-time or near real-time SSI, usually derived from technical means and delivered to members of the organization for direct on-site activities of SS risk management. Yet operational SSI is immediate and has a short lifespan. The immediacy of operational SSI requires SS managers (or more specifically, SS analysts) to be able to access security data and information collection systems in real time, and to convert security data and information into SSI within a short time and under high pressure.

The above analysis shows that strategic SSI focuses on the organization's SSM strategy and long-term SS investments, goals, and policies, tactical SSI on the organization's SS plans, procedures, and resources, and operational SSI on all of the organization's daily and on-site activities of SS risk management. In terms of SSM and SS enhancement, operational SSI differs largely from strategic and tactical SSI in aspects which include the level of specification and the timeliness of SS data and information. Operational SSI requires a constantly updated real-time SSI system. Strategic and tactical SSI may be real-time, but they can also be based on weekly, monthly, or yearly SSi. At the same time, these three types of SSI interact and influence each other: The need for operational SSI is driven by tactical SSI of which the need is driven by strategic SSI. Therefore, SSI provides a data-driven method to associate the organization's strategic SS goals with tactical SS policies and specific operational SS actions.

2.7.2 Classification of SSI Based on Categories of SS Concepts

Seen from the categories of safety and security concepts, SSI can be classified according to different extensions, fields, and research objects that it involves, as shown in Table 2.1.

2.7.3 Classification of SSI Based on Sources of SSI

Safety and security intelligence can be divided into three categories based on the basic sources of SSI (Fig. 2.3) (Wang and Wu 2019):

1. SS risk intelligence: It refers to intelligence specifically targeted at SS risks, which are obtained through SS risk identification, analysis, and assessment. SS risk intelligence helps organizations identify and understand potential SS threats and provides strong support for organizations to reduce SS risks.
2. SS strategy intelligence: It refers to intelligence about SSM strategies, including measures employed in management (such as SS engineering, enforcement, education, economy, and culture), strengths, weaknesses, opportunities and challenges in management, and conditions including human resources, materials and financial support for management. SS strategy intelligence can be helpful to organizations in identifying and understanding strengths and weaknesses of their SSM, predicting future trends and changes of SS, and discovering potential factors which might be useful in improving SSM and enhancing SS capabilities.
3. SS Science and technology intelligence (i.e., SSI related with science and technology): It refers to the scientific and technological evidence of SSM, including results of both basic and applied researches, academic literature and publications, criteria, norms, procedures, and so on in SS science. SS science and technology intelligence contributes to collecting and utilizing scientific research evidence for the aim of assessing costs/benefits of current and the latest SSM methods and measures in order to provide guidance for SSM practices. Similar to the SS research evidence in evidence-based SSM, it can be used by organizations to achieve better SSM performance. In addition, it is helpful in narrowing the gap between SSM research and SSM practices.

2.7.4 Conclusions

It is of utmost importance to discuss the classification of SS as classification is a prerequisite for science. This section focuses on three ways of classifying SSI: one is oriented toward SSM, the other is based on the category of SS concepts, and the

Table 2.1 Classification of SSI based on categories of safety and security concepts (Wang Bing and Wu Chao 2019b)

Number	Criteria of classification	Specific types	Specific explanations or examples
1	Extension	Safety intelligence	Intelligence related to safety (i.e., intelligence about accidents and disasters), which is SSI for safety management
		Security intelligence	Intelligence related to security (i.e., intelligence about terrorist attacks), which is SSI for security management
2	Area	Political SSI	SSI about political awareness, needs, content, activities, etc
		Territorial SSI	SSI about territory, territorial waters, territorial sky, key national infrastructure, etc
		Military SSI	SSI about armed forces, armament, military facilities, military secrets, military activities, etc
		Economic SSI	SSI about finance, fiscal issues and industry, and intelligence about economic and strategic resource crisis, etc
		Cultural SSI	SSI about languages, customs, values, lifestyles, etc
		Social SSI	SSI about society, traffic, life, manufacture, etc
		Scientific and Technological SSI	SSI about sci-fi achievements, personnel, products, facilities, etc
		Informational SSI	SSI about physics, network, mainframe (system), application, data, etc
		Ecological SSI	SSI about natural and ecological system, artificial ecosystem, intelligence about ecological crisis and vulnerability

(continued)

Table 2.1 (continued)

Number	Criteria of classification	Specific types	Specific explanations or examples
		Resource SSI	SSI about resources of water, energy, soil, mineral, etc
		Nuclear SSI	SSI about nuclear facilities, activities, materials, and radioactive matters, and intelligence about nuclear accidents
		Biological SSI	Intelligence about invasion of alien species and epidemic, and SSI about biological technology, laboratory, food, weapons and intelligence about biological terrorist attacks, etc
		…	…
3	Object	SSI of enterprises	SSI for SSM in enterprises, like SSI about SS hazards and risks, emergency capacity and so on in enterprises
		SSI of communities	SSI for SSM at communities, like SSI about sources of danger, public security, security incidents, harm, emergency, and so on at communities
		SSI of cities	SSI for SSM at cities, like SSI about sources of danger, public security, road safety, natural disasters, anti-terrorism, and so on at cities
		SSI of nations	SSI for SSM for nations, including SSI about territorial security, political security, economic security, and other fields
		…	…

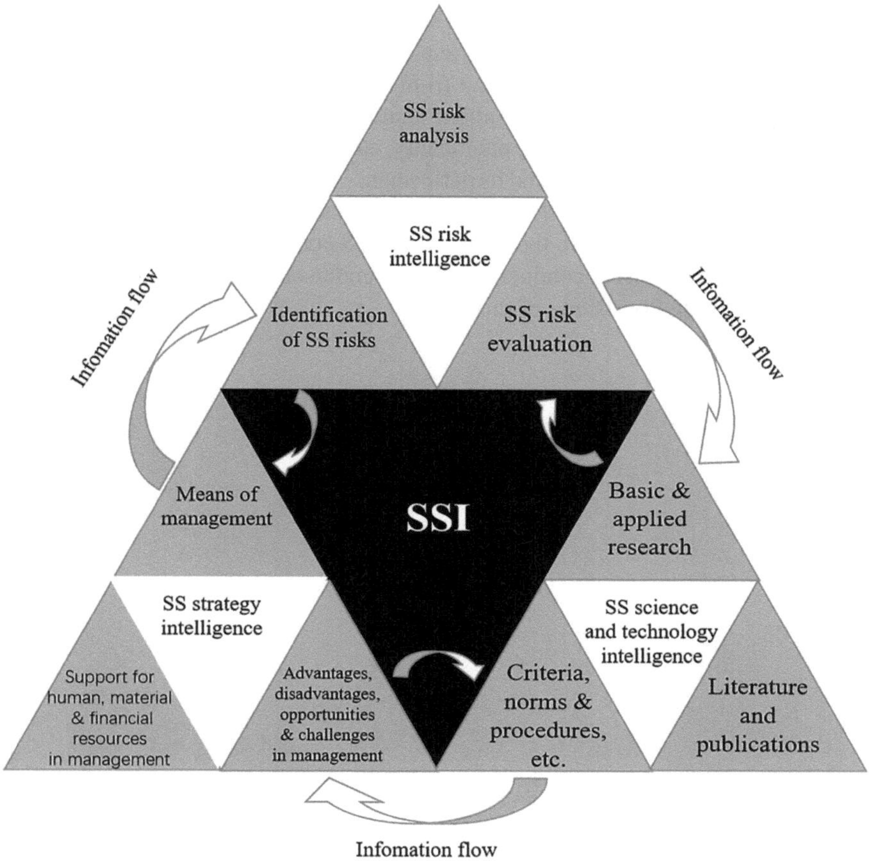

Fig. 2.3 Classification of SSI based on sources

third is based on the sources of SSI. The above classification of SSI helps clarify relevant research fields of SSI, define research scopes of SSI, and guide specific and in-depth SSI research and practice.

2.8 Chapter Summary

This chapter has observed and examined SSI from the height of SSI science in order to perceive and explore the essence, characteristics, and value of SSI for the promotion of SSI research and disciplinary construction. There are six major findings from the chapter: (1) From the theoretical perspective of SS science, the concept of SSI originates from an understanding newly acquired from SS science (especially SSM) in the realm of intelligence studies, that is, a fresh understanding of SSM from "SSi to SSI"; (2) The concept of SSI differs from the concepts of SSi and SS analysis, yet they are still closely related; (3) The SSI concept has generally evolved from "dispersion to unity" and from "specificity to universality"; (4) Seen from the perspective of SSM, the salient features of SSI include informativeness, transmission, utility (worthiness), practicality, strategy, predictability, purpose, and intelligence; (5) The role of SSI and its performance are mainly reflected in its reliable promotion and support for SSM; (6) SSI can be classified in various ways according to the links and levels of SSM, the categories of SS concepts and the sources of SSI. The classification of SSI is conducive to better understanding and studying specific types of SSI for the flourishing of this field.

Chapter 3
Theories of the Disciplinary Construction of Safety and Security Intelligence Science

3.1 Introduction

The theory of disciplinary construction aims to elucidate the developmental trajectory, core issues, foundational elements, system of sub-disciplines, and other critical aspects of an academic field, in other terms, a clear articulation of the disciplinary construction theory is not only the primary issue but also the cornerstone of the establishment and advancement of a discipline. Therefore, it is imperative to probe into the theory of disciplinary construction of SSI science, which constitutes the focal point of this chapter. The chapter discusses the origin, meta-concepts, and identity of SSI science, as well as the conceptual framework that guides the construction of SSI science. Then it addresses five fundamental inquiries: the definition and connotations of SSI, its intrinsic nature, the research object under scrutiny, the scope of research, and the overarching goals of the investigation. With these insights, the chapter proceeds to deliberate on the foundational principles and potential branches of SSI science. Ultimately, it culminates in the proposal of a methodology tailored for SSI science.

3.2 Basic Orientation of the Disciplinary Construction of SSI Science

As a new discipline, SSI science has only emerged within the past 2 years. Compared to its application, the development of SSI science as a field of study apparently lags behind. Its foundational theoretical framework remains underdeveloped, and fundamental theoretical research is limited. Currently, the development of SSI science as a discipline is in a new stage, facing challenges related to its origins, identity, and direction. These challenges hinder the construction of a sound theoretical framework of SSI. Therefore, it is crucial to analyze the reasons for these challenges and identify

B. Wang and C. Wu, *An Introduction to Safety and Security Intelligence Science*,
https://doi.org/10.1007/978-981-96-7261-5_3

solutions. This section addresses the key issues related to these challenges in order to facilitate its development as a rigorous field of study and promote its research and advancement.

3.2.1 Positioning of the Origin of the Discipline

The origin of SSI science remains undefined within the academic community. From a scientific perspective, SSI science emerges from the intersection of SS with other disciplines. This intersection can be conceptualized through two methodological approaches within SS science: "+SS" and "SS+" (Wang Bing and Wu Chao, 2019d). Their meanings and emphases are illustrated in Table 3.1, which shows that both approaches involve the integration of SS science with other disciplines or fields. However, they differ in their perspectives, leading proponents, and contributions to the development of SS science (Fig. 3.1). The relationship between " +SS," "SS+," and SS science is depicted in Fig. 3.2.

Methodologically, basic theoretical research in SS science transforms "+SS" inputs into "SS+" outputs, yet applied research transforms "SS+" inputs into "+SS"

Table 3.1 Basic meanings and emphases of " +SS" and "SS+ " (Wang Bing and Wu Chao, 2019d)

Name	Basic meaning	Emphasis
+SS	An approach of research and practice that takes other disciplines or fields as the main body, with SS or SS science playing a supporting role. In this model, SS science isn't the driving force but rather an adjunct, embodying the inherent dependency of SS and its applications. The emphasis is on initiating from other disciplines or fields, with the goal of enhancing and optimizing their development (progress featuring in SS). Here, SS science functions as an instrument. Evidently, it is not easy for this model to produce universally applicable theories, methodologies, and technologies of SS science	It emphasizes a trend-following development of SS science, in line with the growth and service needs of other disciplines or fields. It primarily builds on the established theories, methods, and technologies of other areas, employing principles of SS science to make minor adjustments or improvements, and safeguarding the secure advancement of other disciplines or fields
SS+	A mode of research and practice where SS science assumes the primary and propelling role. The focus shifts to the domain of SS science itself, leveraging theories and methods from other disciplines to foster its advancement. The crux of this approach is the complete integration of theories and methods of other fields with the SS science framework, converting the "+SS" inputs into "SS+" outputs. Clearly, the "SS+" approach is pivotal for SS science to evolve into an independent discipline and to establish universally relevant applicable theories, methodologies, and technologies of SS science	It underscores a counter-trend development of SS science, focusing on its own expansion and service. It involves directly adapting the theories, methods, and technologies from other disciplines to fit the SS science paradigm, aiming for a rapid and transformative growth in SS science

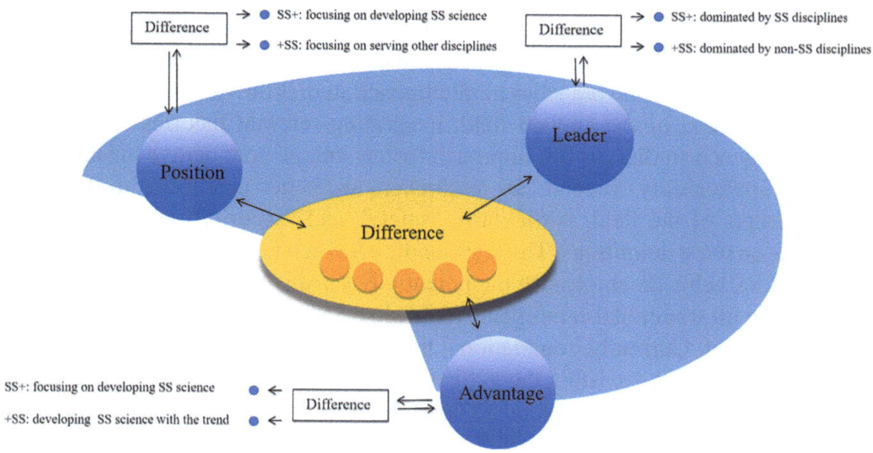

Fig. 3.1 Differences between "+SS" and "SS+"

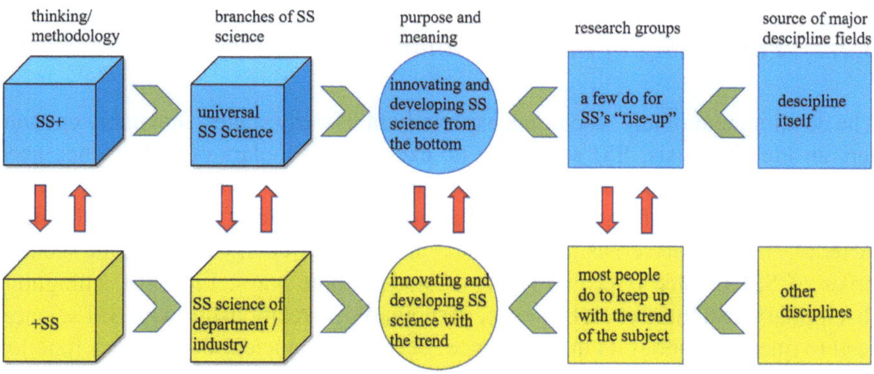

Fig. 3.2 Relationship between "+SS," "SS+," and SS science

outputs (Wang Bing and Wu Chao, 2019d). Seen from the perspectives of "+SS" and "SS+," the concept of SSI is born out of the integration of SS and intelligence (SS+ intelligence). In the same vein, SSI science is a product of the "SS+" approach. SSI science, led by SS science, applies the theories, methods, technologies, and principles of intelligence studies to the SS field, aiming to serve the development of the SS science. While SSI also represents a significant area of applied intelligence studies and a branch within the broader field, its development reciprocally contributes to the expansion and development of intelligence studies research. The core task of SSI researchers is to transform "+SS" inputs (theories, methods, technologies, and principles related to SS within intelligence studies) into "SS+" outputs (generally applicable theories, methods, technologies, and principles within SS science). In other words, SSI researchers adapt and refine concepts from intelligence studies for application within SS science.

Crucially, the development of SSI science should be grounded in SS science for three primary reasons: First, intelligence studies are inherently applied. Experience demonstrates that applied branches of intelligence studies thrive when it is developed within the context of the applied field, integrating relevant theories and methods from intelligence studies, as this approach fosters unique connotative meanings and enhances applicability. Second, SS researchers are better positioned to define the meta-concepts of the field, particularly "safety" and "security," and to achieve a consensus on these definitions. Third, grounding SSI science in the broader scientific context of SS helps de-emphasize the potentially sensitive or political connotations associated with traditional intelligence, allowing SSI to realize its full potential in SS management. Currently, foundational theoretical research in SSI science is still limited, as much of the existing theoretical basis is adapted from intelligence studies research, lacking a distinct SSI focus. Therefore, future research should prioritize the development of original theoretical contributions within SSI science, strengthening the "safe and security" aspect of its theoretical foundation.

3.2.2 Positioning of the Disciplinary Identity

The development of any discipline, especially in its early stages, inevitably encounters an identity crisis; SSI science is no exception. Scholars of SSI science need to recognize that as a young and developing discipline, such a crisis is unavoidable, and should confront this challenge and seek solutions to establish the field. Afterall, this identity crisis has historical and logical roots. Currently, the identity crisis of SSI science is primarily manifested in three ways: First, the ambiguous disciplinary attributes of SSI science, particularly its relationship with SSi science, lead to questions about its independence. Second, the unclear disciplinary function of SSI science creates uncertainty about its identity. Third, the lack of foundational theoretical research tends to weaken the discipline's theoretical basis. This study will focus on addressing the first two challenges, as the third (strengthening the theoretical foundation of SSI science) is a long-term endeavor requiring continuous and in-depth research. Since disciplinary attributes and functions are ultimately matters of disciplinary identity, accurate understanding and positioning of the disciplinary identity of SSI science is required for resolving the current identity crisis.

At first, the author of this book argued that SSI science has six basic disciplinary attributes: interdisciplinary, social science, cognitive science, management science, applied science, and a shared branch of both SS science and intelligence studies (Wang Bing and Wu Chao, 2019b). Yet subsequent academic exchanges revealed that it is insufficient to establish the independent status of SSI science by defining these attributes, as a significant point of confusion remains within the academic community regarding the relationship, and particularly the distinction, between SSI science and SSi science.

To address this confusion, it is crucial to clarify the relationship and distinction between SSI and SSi. From the perspective of system SS science (an advanced stage of SS science), SSI refers to SSi that affects system SSM, while SSi refers to the self-representation of a system's SS state and its changes (Wang Bing and Wu Chao, 2019a). Therefore, all SSI is SSi, but not all SSi is SSI (Wang Bing and Wu Chao, 2019a). This understanding highlights the difference between SSI and SSi. SSI science should focus specifically on SSI research and be distinct from SSi science in order to foster academic recognition of its independence. The differing research and practice paradigms of SSI and SSi have led to two paradigms within the "informatization" of SS science (especially the discipline of SSM): the SSi (science) paradigm and the SSI (science) paradigm. Table 3.2 provides a comparative analysis of the two paradigms.

Table 3.2 illustrates the major differences between the two paradigms. SSI emphasizes openness, sharing, and public service, while SSI focuses on providing targeted, "one-to-one" SSi services that are often shared but limited, or even confidential (It's important to note that the traditional view of SSI as "absolutely secret" should be reconsidered, as not all SSI is secretive, especially in peacetime when it is similar to SSi, hence the necessity of sharing SSI in certain aspects), emphasizing targeted, personalized use of SSi. This "one-to-one" service is personalized and tailored to ensure the target audience maximizes the use of SSi and enhances its effectiveness and security in an environment of information overload and chaos (Yang Yuanli, 2013). This approach not only meets the precise needs of the audience but also significantly increases the value density of SSi service. Therefore, the "one-to-one" service is an effective strategy for improving service quality and a practical requirement for achieving personalized, precise, and intelligent SSM. In short, SSI is specifically produced, unlike SSi (including SS knowledge) which is shareable, SSI is generated in response to the needs of specific individuals, times, SS issues, SS needs, SS scenarios, and SS matters. From this perspective, the SSI paradigm represents

Table 3.2 Comparison of the SSi (science) paradigm and the SSI (science) paradigm

Aspects of comparison	SSi (discipline) paradigm	SSI (science) paradigm
Orientation	Development of SS	SSM
Purpose	Solving the problem of absence of SSi in SSM (traditional understanding of SSM)	Solving the problem of absence of SSi in SSM (new understanding of SSM)
Audience	Satisfying SSi needs of the masses	Supporting SSM
Spread	Public sharing	Limited sharing, even confidential
Feature	Beneficial to all	Pertinent
	Mainly group-based	Mainly personalized
	Large quantity	Small quantity
	Expendable	Must be valuable
Focus	Well-organized	In-depth analysis
Goal	Effective supply of SSi	Accurate SSM and intelligent (smart) SSM

a refined and intelligent version of the SSi paradigm—an upgrade. Thus, the SSI paradigm = SSi paradigm + "one-to-one" SSi service strategy (Yang Yuanli, 2013).

Another key difference is that SSI exists only when it holds value (i.e., it is useful SSi), whereas SSi exists objectively regardless of its value. The volume of SSi is generally larger than that of SSI. In the era of big data, the SS field also faces an overwhelming amount of data that is often chaotic and disorganized. Obtaining valuable SSI (useful SSi) is increasingly challenging, making the "one-to-one" service of SSI (science) even more valuable. That SSI is distinguished from SSi is primarily because of its scarcity and higher value. Despite these differences, the two paradigms share a strong foundational connection: they both rely on SS data, and the processes of SSi and SSI (i.e., collection, processing, dissemination, and utilization) are roughly the same, ultimately serving as forms of service. Therefore, while acknowledging the differences between SSI and SSi, it's crucial to recognize the fundamental support SSi provides to SSI science and SSI operation.

Furthermore, Table 3.2 clarifies the disciplinary function of SSI: to provide SSI services for SSM, particularly SS decision-making. In summary, the traditional "informatization" paradigm of SS science, especially SSM science, is primarily the paradigm of SSi, targeting universal SS variables and SSi utilization services. A new understanding of SSM is yet to be formed for the SSI (science) paradigm to be proposed in the shift from SSi to SSI. SSI (science) targets specific SS variables and aims to provide SSI services for precise and intelligent SSM, particularly for SS decision-making. Obviously, the two paradigms are complementary, each with its own strengths, weaknesses, and focus. The question of which paradigm is "more important" is a false dilemma, as both are necessary and should coexist and develop together. SSI science that has once been diluted or has been ignored, submerged, or avoided by SSi science deserves emphasis and attention. Only then can both sciences define their respective development paths and thrive in mutual innovation and growth.

3.2.3 *Positioning from the Perspective of Discipline Construction*

The perspective from which SSI science is developed directly determines its research object, research categories, overall structure, scientific validity, and ultimately, its potential for a healthy and sustainable growth. Therefore, choosing the right perspective for disciplinary development is crucial. At present, there is a lack of consensus within academia regarding the appropriate perspective for building SSI science as a discipline. Thus, it is essential to establish the perspective of disciplinary construction of SSI science. A philosophically grounded discussion of disciplinary perspectives for SSI science is necessary for determining a scientifically appropriate perspective for all SSI sciences, which allows for the most comprehensive and insightful exploration of the characteristics, connotations, and scope of SSI science. Therefore, by adopting a philosophical viewpoint, and considering the connotations and development trends of modern intelligence studies and modern SS science, and emphasizing the applied and practical nature and value of SSI science, the disciplinary development SSI science should adhere to the "major SS" and the "major SSI" perspectives, while remaining firmly oriented toward SSM.

3.2.3.1 Concept of "Major SS"

Since the twenty-first century, a growing number of SS issues have exhibited the complex characteristics of an intricate giant system: increased complexity, openness, and interconnectedness. In recent years, both academic and practical circles in the SS field have recognized the arrival of the era of "major SS." The concept of "major SS" is an emerging philosophical viewpoint gradually formed in modern SS science research and practice. For this reason, this book adopts the "major SS" perspective as its guiding framework. Also known as "human SS," it refers to the comprehensive set of SS issues centered on human's well-being. It represents the most common and the broadest understanding of SS, encompassing all potentially unsafe factors that may pose threats or harm to humans, and emphasizes a highly comprehensive understanding of SS (Wang Bing and Wu Chao, 2019b). From the perspective of modern SS science, the connotation of "major SS" is primarily reflected in the following two aspects:

1. Seen from the domains or subjects involved in SS, the concept of "major SS" signifies a paradigmatic shift from a fragmented understanding of "partial SS" to a more holistic and comprehensive understanding of "holistic (comprehensive) SS" (Wang Bing and Wu Chao, 2019b). On April 15, 2014, during the inaugural session of the National Security Commission, the "Holistic Approach to National Security" was proposed by General Secretary Xi Jinping for the first time, with the goal of forging a cohesive national security framework that encompasses a spectrum of elements, including political, territorial, military, economic, cultural, social, technological, informational, ecological, resource, and nuclear security. In the wake of this declaration, Chinese scholars in the field of SS studies have reached a consensus on the imperative of adopting a holistic SS perspective. This is due to the inescapable interdependency and dynamic transformation of SS risks that transcend specific sectors, such as public and industrial safety, or various entities ranging from corporate to community, urban, societal, and national security. Authentic SS development is contingent upon the cultivation of an awareness and the conceptualization of "holistic (comprehensive) SS." The author posits that the essence of holistic SS lies in the all-encompassing advancement of a secure nation and society. The crux of the matter is captured by the term "comprehensive," demanding that there be no vulnerabilities in any aspect of SS development. The SS agenda must be directed toward remedying any deficiencies in comprehensive SS, with a relentless pursuit to ensure that no sector is left unattended and underdeveloped. This approach epitomizes a pivotal

trend in the evolution of global SS science in recent years (Wang Bing and Wu Chao, 2019b).

2. With regards to the scope of SS, "major SS" emphasizes a shift from viewing "safety" and "security" as separate entities to a more integrated understanding of the two. In other words, the scope of "major SS" encompasses both safety and security, signifying their integration. This point has already been discussed in detail.

In summary, "major SS" transcends the scope of specific fields, entities, or levels (including the levels of "safety" or "security"). Targeting universal security, it encompasses all fields, entities, and levels of security concerns. The definition of SS provided in Sect. 1.2.1 aligns with this "major SS" perspective. In short, this concept broadens the scope, fields, and entities encompassed by traditional security concepts, reflecting the general trend in SS science research and practice toward integration, fusion, and a universal approach to SS risks.

The concept of "major SS" also represents a mindset and framework for conducting modern SS research and practice amid the interweaving of contemporary SS issues, as well as the integrated development of SS science. It emphasizes the use of systematic perspectives and methods for understanding and addressing SS challenges and advocates starting from a holistic view and providing comprehensive and systematic prevention for all potential hazards. Modern SS science aims to tackle multifaceted and complex issues. Driven by critical SS challenges, strategic needs, and national and societal development goals, major SS has revolutionized our understanding of risk, leading to fundamental and widespread changes in the field, thus shaping the new era and the future of SS science. In the context of major SS, the modern discipline of SS possesses large "groups," including a "disciplinary group" (such as SS science and technology, national SS studies, information SS studies, cyberspace SS studies, nuclear SS studies, and emergency science), a "professional group" (including SS science and engineering, cyberspace SS, information security, national security, and emergency management), and "field groups"(like military security, economic security, cultural security, social security, technological security, information security, ecological security, resource security, and nuclear security). All SS-related concepts discussed in this book (including SS science, SSI, SSM, and SSI science) are consistent with the concept of major SS, as they are all universally applicable concepts.

Since SS is a subordinate concept of SSI, the SS within SSI should be understood in the same lineage as the "major SS." Therefore, when discussing the development of SSI science, ensuring the universality of its theoretical framework requires grounding the discussion in the concept of "major SS." For instance, Wang Bing and Wu Chao (2019b) address fundamental disciplinary issues of SSI science from the perspective of "major SS," which aligns neatly with the concept of "Holistic Approach of National Security" proposed by General Secretary Xi Jinping. It represents a new way of thinking about SS, encompassing a broad range of fields and industries. Correspondingly, the current research and service targets of SSI science should not be limited to narrow SS fields but expand further to encompass the broader field of "major SS."

Simultaneously, it is crucial to integrate SSI theories from various specialized subfields to form scientific, systematic, and unified overarching principles and theories for SSI science.

3.2.3.2 Concept of Major SSI

The disciplinary development of SSI science should be based on the "major SSI" concept as it corresponds to the major SS. The "major SSI" integrates the "broad intelligence" (Liu Zhihui, 1999; Liu Zhihui, 2000; Ma Dehui, 2011) with the "major SS." It belongs to a philosophical category, representing the concretization of system thinking within the field of intelligence. It emphasizes that intelligence studies should address intelligence problems across various fields, evolving from single-field intelligence systems to comprehensive social intelligence systems. It advocates for the development of "broad intelligence studies" (Liu Zhihui, 1999; Liu Zhihui, 2000; Ma Dehui, 2011). Considering the connotations of both concepts of "major SS" and "broad intelligence," it symbolizes an idea of "general service," "grand process" (extending from emergency intelligence focused on emergency management to SSI oriented toward the entire process of SSM), and "general method" (constructing a universal methodology for SSI science). Therefore, SSI, as understood within the "major SSI" concept, is a complex, extensive, broad, multifaceted, and integrated hybrid. Essentially, the "major SSI" refers to the all-encompassing field of SSI. It requires that SSI services adapt to the evolving needs of SS science and SSM practices, which necessitates moving beyond the traditional limitations of SSI, integrating with SSI practices in various security-related fields, addressing "major SS" and the entire SSM process, enhancing the SSM service function of SSI, and meeting the increasing demand for SSI in SSM practice.

From the "major SSI" perspective, the research and practice of SSI across various sub-fields tend to share common principles and similar methods. Research in SSI science guided by the concept should not only study SSI as traditionally separated fields (such as SS, emergency management, and national SS) but also examine SSI holistically across all SS fields, focusing particularly on the commonality of challenges across these fields. In other words, SSI science is not limited to a specific department or field but studies SSI as a universal phenomenon and its general laws oriented toward the universality of research. Regarding the disciplinary system of SSI science, a disciplinary system of major SSI science should be constructed, incorporating all SSI subsystems (such as emergency intelligence, public SSI, military SSI, scientific and technological SSI, and resource SSI) into a unified and organically interconnected "major SSI" system.

Upon thorough examination, the conventional methodology for advancing the science of safety (or security) intelligence studies is anchored in two pivotal aspects. Firstly, it should take the basic process inherent to SSI as the cornerstone of the logical framework for disciplinary evolution. Secondly, it ought to construct SSI science within this framework and orient itself toward practical application, resulting in a compartmentalization of the content of SSI operation. In relation to this methodology, the author endorses the former approach (recognizing its alignment with the intrinsic regularities of SSI science), and opposes the latter (while it mirrors the current state of SSI theoretical research and practical endeavors, it falls short of embracing the universality, scientific rigor, and developmental aspirations essential for a robust disciplinary growth). Thus, the author advocates for the disciplinary development of SSI science to be steered by the "major SSI" paradigm. This approach seeks to transcend the constraints of narrow, insular, and fragmented thinking and adheres to the principles of universality, inheritance, and progress in the construction of an academic discipline. By doing so, it facilitates a harmonized development strategy across diverse SSI systems, broadens the horizons of SSI science, fosters the emergence of a cohesive academic community, and nurtures the development of interdisciplinary SSI professionals.

3.2.3.3 Orientation Toward SSM

The heart of SSI endeavors is to fill the gaps in SSI within the SSM decision-making process. To truly capture the practical spirit of SSI and its role in supporting SSM, the growth of SSI as a discipline should naturally revolve around SSM. Ensuring that SSI research provides valuable services to SSM is both its primary task and ultimate goal (Wang Bing and Wu Chao, 2019b). Wang Bing and Wu Chao (2019c) have offered insights about why the development of SSI should be SSM-focused. The disciplinary system of SSI science should be built with consideration of the current digital age and the trend of informatization in SSM, and the unique traits and practical needs of SSM operations, which keeps SSI firmly rooted in practicality and problem-solving, steering clear of abstract theorizing and positioning it as a discipline that can genuinely inform and guide SSM practices.

Although SSI science is a branch of intelligence studies, there are notable distinctions between the two in terms of their meanings and extents, especially considering the unique characteristics of SSM and SSI endeavors. For instance, within the domain of SSM informatization, the concepts and models related to SS strategy and SSM operations are in a state of perpetual evolution, which in turn creates divergences in the disciplinary development trajectories of SSI and intelligence studies. Consequently, the maturation of SSI as an academic discipline—particularly its research agenda and the structure of its field—must be aligned with the objectives of SSM and with the backdrop of SSM informatization serving as a pivotal reference point. This alignment necessitates an in-depth analysis of the intrinsic development and imperatives of reform within SSI in the context of SSM informatization, with a focus on the facilitation of intelligence-led SSM (Wang Bing and Wu Chao, 2019 g). The ambition is to ensure that SSI efforts are not only in service of the development and progression of SSM but also contribute to the enhancement of SSM practices.

3.2.4 Conclusions

The complexities of the development of a discipline are indicative of the multiplicity of challenges and disputes inherent in its establishment and maturation. This section investigates three principal sources of "perplexity" in the quest to delineate the disciplinary development of SSI: the origins of the discipline, its core identity, and the most fitting vantage point for its construction. By examining these issues through the lens of disciplinary evolution, and considering the intrinsic qualities and attributes of SSI, this analysis has delineated a trajectory for dispelling the ambiguities surrounding SSI's academic development. This inquiry aims to inform the building of a sound framework for the advancement of SSI. To be precise, the disciplinary development of SSI is to be rooted firmly in its own domain, assimilating theoretical insights and methodological approaches from the broader field of intelligence studies. Oriented toward SSM, this endeavor is to form the concerns, theories, and methods of commonalities of SSI science by assembling various types of SSI, constructing a theoretical system that is not only scientifically robust but also cohesive and applicable. The author has recognized that, amidst the dynamic shifts occurring within SS science and intelligence studies, the development of SSI must be anchored in a vision that is both harmonious and expansive, seeking to chart a scientific developmental course for SSI among the uncertainties and aims for the strengthening of the disciplinary stature. In essence, it is a critical imperative for professionals of SSI science to accelerate the disciplinary development of SSI, cultivating it into a mature discipline.

3.3 Fundamental Issues about the Discipline of SSI Science

The field of SSI science has struck root in fertile soil, with a growing consensus among scholars regarding the significance, imperativeness, and viability of its establishment as an independent academic discipline (Wang Bing and Wu Chao, 2018a). The moment is opportune for the growth of SSI science. Since the cornerstone for affirming the emergence of any discipline and the impetus for its evolution is the articulation of its central inquiries (the definition of the discipline, the research subject, the research scope, etc.), the major endeavor in establishing SSI science as a scientific discipline is the elucidation of these foundational issues. As SSI science is still in the nascent phase of its exploration, the ambiguity surrounding its foundational questions has led to the absence of a foundational theoretical underpinning and a guiding framework, which in turn impedes the construction, scholarly inquiry, and progress of the discipline. Therefore, it is necessary to probe into these basic questions of SSI science, which are closely associated with the disciplinary development.

Informed by the discussion in Sect. 1.3, and enriched by the investigation into the development of the SS field (elaborated in the publication *New Branches of Safety and Security Science*, by Wu Chao and Wang Bing), this section ventures into theoretical contemplation. It addresses four primary inquiries of the SSI science from the perspective of the "major SS" (the definition and connotation, the nature, the research objects, and the breadth of its research content), with the objective of establishing a foundational framework for SSI science, thereby enhancing its construction and development.

3.3.1 *Definition and Connotation of the Discipline*

No consensus has been reached on the definition of SSI science. Wang Bing and Wu Chao (2018c) state that from a philosophical stance, when a science is entitled with the name of its research object, like "xx science" (e.g., "information science" or "SS science") or "xx studies" (like "cultural studies" or "intelligence studies"), the fundamental methodology for defining this science is "xx science (studies) is the scientific study of xx and its underlying principles of dynamics." Evidently, SSI science is named in direct reference to its research object SSI. According to the philosophical methodology of defining a certain discipline mentioned above, SSI science can be defined as "SSI science is the scientific inquiry into SSI and the principles governing its dynamics."

Despite the advantages of conciseness, inclusion, and accuracy of defining, the nuanced richness of SSI science is shielded; a more elaborate definition is warranted. SSI science represents a direct and interdisciplinary blending of intelligence studies with SS science. It stands as an offshoot of both domains and shares boundaries with other sub-fields within the realm of SS science (such as SSi studies and SSM studies). To be more specific, SSI science is an emerging, interdisciplinary applied science dedicated to tackling the deficiencies of SSI within the SSM process. It adopts SSI as its central research object and concentrates on the intrinsic nature, function, structure, genesis, dissemination, and application of SSI as its core research foci, with the ultimate goal of rendering effective SSI services to SSM endeavors. Against this backdrop, the essence of SSI science includes:

1. The underlying cause of SSM failures is often traceable to the deficiency of SSI, which serves as the foundational rationale for establishing SSI science. Thus, remedying the deficiency of SSI in the SSM process is not only the initial impetus but also the ultimate objective of SSI science. The inquiry, structuring, and advancement of SSI science must be directed toward addressing this issue.
2. The purview of SSI science is centered on SSI, which includes the static elements of SSI and the dynamic processes involved, and the entities or organizations that partake in SSI endeavors. SSI science extends its investigative and scholarly efforts to encompass not merely the essence of SSI but also the ancillary activities, technologies, methodologies, and the entities or organizations in SSI operations.
3. SSI science is a multidisciplinary field that incorporates theoretical frameworks, methodologies, and technical approaches from both intelligence studies and SS science. Additionally, it necessitates the collaborative insights from

allied disciplines, including philosophy (SS philosophy), management science (SSM science), information science (SSi science), sociology (SS sociology), and computer science. This interdisciplinary support enriches the body of knowledge within SSI science and enhances its comprehensiveness and effectiveness.

4. SSI science is oriented toward SSM. Its research goal is to empower SSM professionals to conduct SSI activities (such as collection, dissemination, analysis, and utilization) more effectively, thereby addressing the problem of the absence of SSI in the SSM process.
5. SSI science is a discipline that stresses practicality and is oriented toward application. It embodies the application of general intelligence theory in SS science. The core research object of SSI science and intelligence studies is the concept of "intelligence." So far, intelligence studies have evolved into a distinguished field with systematic theories and methodologies that are scientifically rigorous. The foundational theories, principles, and research methodologies from intelligence studies can be creatively adapted and applied within the scope of SSI science research. There is a direct lineage and strong interconnectedness between these two domains, highlighting the synergistic relationship between their theoretical underpinnings and applications.

Furthermore, it is imperative to elucidate the relationship between SSI science and SSi lest any misconstruction should occur. The former is concentrated on the dynamics of SSI (valuable or subjective SSi that has a bearing on SSM), and there is a significant overlap in the research content between SSI science and SSi. However, it is crucial to recognize that SSI and SSi are not synonymous. The relationship between them mirrors the nuanced distinction between "information" and "intelligence." From the vantage point of the theory of SS science, the emergence of the SSI concept represents an original interpretation of SS science (especially SSM) informed by the insights of intelligence studies—a paradigmatic shift from "SSi to SSI." Consequently, the research object and content of SSI science are characterized by a distinctive and irreplaceable attribute, which affirms its autonomy as a discipline. Moreover, contemporary SSI research underscores the application of modern information technologies, including SSi, to enhance the efficacy of SSI processes and systems. This approach enables researchers and practitioners to harness the full potential of modern information technologies, including SSi, to augment the efficiency of SSI generation, processing, storage, retrieval, communication, and utilization.

3.3.2 *Nature of the Discipline*

According to Ma Dehui and Su Yingjie (2013) and Liang Huiwen (2017), the intrinsic nature (or "disciplinary attribute") of SSI science addresses the fundamental question: "What kind of science is SSI science?" Based on the definition and connotation of SSI, the nature of this discipline can be analyzed from the following six perspectives:

3.3.2.1 SSI Science is an Interdisciplinary Science

SSI science originates as an innovative academic discipline, born at the intersection of two scholarly domains: intelligence studies and SS science. In other words, this field is shaped by the direct convergence and synthesis of the two disciplines. In line with the established mechanisms of interdisciplinary science (Wu Chao and Wang Bing, 2018b), the formation of SSI science is characterized by two approaches:

1. Certain major research topics within SS science (such as the absence of SSI in SSM) require the involvement of both intelligence studies and SS science, hence the emergence of SSI science within these combined research areas.
2. SSI science is formed in the process of applying application of theories and methods from intelligence studies in research and problem-solving within the field of SS science.

3.3.2.2 SSI Science is a Branch of Intelligence Studies and Security Science

In terms of academic classification, SSI science belongs to the two first-level disciplines—intelligence studies and security science. It stands as an offshoot of these two intimately connected disciplines, thus, intelligence studies and security science jointly generate SSI science. The connection between SSI science and its foundational disciplines is characterized by closeness, mutual reinforcement, interdependence, and a dialectical unity (Zhao Rongying, 2017).

The development of the foundational disciplines significantly promotes the development of the sub-discipline, and the development of the sub-discipline enriches the theories of its foundational disciplines (Zhao Rongying, 2017). Intelligence studies provide analogous frameworks and identify differentiating characteristics for the construction and development of SSI science, while security science offers areas of application for its development. Attention should also be given to strengthening research within the branches of intelligence studies and security science, which constitute the disciplinary background for the emergence of SSI science.

3.3.2.3 SSI Science is a Social Science

The classification of intelligence studies as a discipline has been a contentious issue, with two predominant perspectives emerging in scholarly discourse. The first perspective holds that intelligence research is underpinned by a confluence of theories, methods, and technologies drawn from both the social and natural sciences, thereby positioning it as an interdisciplinary science that spans these two realms (Ye Ying and Wu Yishan, 2012). The second perspective contends that the core research object of intelligence studies—encompassing intelligence, its activities, and processes—is inherently a social phenomenon. Given that intelligence studies concentrate on dissecting significant societal events, threats, and crises, with a focus on assessment, warning, advocacy, and planning, it is, by nature, a social science. This view aligns with the classification of intelligence studies as a social science in China (Ye Ying and Wu Yishan, 2012; Ma Dehui and Su Yingjie, 2013; Shulsky and Gary, 2002).

The academic categorization of intelligence studies remains a subject of debate, with two principal viewpoints gaining traction in the scholarly community. The first viewpoint suggests that intelligence research is supported by a synthesis of theoretical frameworks, methodologies, and technological advancements from across the social and natural sciences, thus establishing it as an interdisciplinary field that traverses the boundaries of these two domains (Ye Ying and Wu Yishan, 2012). The second viewpoint argues that the essence of intelligence studies (including the study of intelligence, its operational activities, and procedure) is fundamentally a prevailing social phenomenon. Since intelligence studies are primarily concerned with the analysis of critical societal occurrences, threats, and crises, emphasizing evaluation, alert systems, policy advocacy, and strategic planning, it is inherently classified as a social science. This perspective is consistent with the way intelligence studies is recognized as a social science within the academic landscape of China (Ye Ying and Wu Yishan, 2012; Ma Dehui and Su Yingjie, 2013; Shulsky and Gary, 2002).

3.3.2.4 SSI Science is a Cognitive Science

Qian Xuesen (1986) proposed that intelligence, as information rendered useful to humanity, inevitably intersects with human consciousness and cognition, thus, the field of intelligence should be regarded as part of cognitive science. Bao Changhuo (2007) suggested that intelligence encompasses the interpretation, judgment, and analysis of information, and is thereby a fruit of human contemplation. Devlin (2000) further expounded on this concept, observing that the crux of the intelligence process is a cognitive and thinking process, which transforms innumerable data into meaningful intelligence products or actionable strategies through the alchemy of creative thought.

3.3.2.5 SSI Science is a Management Science

SSI science exhibits pronounced characteristics of management science, which are primarily reflected in four aspects:

1. SSI exerts a direct influence on SSM and also serves it. The purpose of SSI science is to furnish SSM with effective SSI services, remedying the deficiencies of SSI inherent in the SSM process.
2. The competent management of SSi is a cornerstone of both SSI science and its practical applications. SSI necessitates oversight, embracing the investigation

and formulation of SSI policies as well as the development and governance of SSI systems.
3. SS data and information constitute the foundational resources for SSI endeavors, and the "intelligence process" to which they are subjected is classified within the sphere of knowledge management. Concurrently, the SSI process, which catalyzes the activation of SS data and information, is inherently a management procedure for the generation of SSI outputs.
4. The governance of SSI entities and personnel represents an essential management process.

In conclusion, SSI science must be oriented toward SSM, employing theoretical frameworks, methodologies, and means from the field of management science to efficiently administer SSi data, SSI, SSI entities, and personnel. The ultimate goal is to render effective SSI services in support of SSM operations.

3.3.2.6 SSI Science is an Applied Science

Within the contemporary framework of science and technology, Qian Xuesen (1986) aptly identified intelligence studies as an applied scientific discipline. Paralleling this, SSI science, too, is classified as an applied science, with emphasis on two principal aspects:

1. Philosophically, SSI science originates from the pragmatic endeavors of SSI operations, evolving toward theoretical development and systematic organization. It employs theoretical insights, methodologies, and technical approaches from intelligence studies, SS science, SSM research, SS statistics, and SSi studies, to address tangible SSI requirements within the SSM process.
2. In terms of the application and evolution of intelligence studies, SSI science epitomizes the extension of theories and methodologies of intelligence studies into the arena of SS science, with a particular focus on SSM. It constitutes a specialized application sphere of intelligence studies, imbued with characteristics unique to specific departments or industries. This field is shaped by the abstraction and synthesis of principles and laws selected from the practical experiences of SSI operation.

3.3.3 Research Objects of the Discipline

As the research object of a discipline is pivotal to the development of its theoretical framework and methods, delineating the research object of SSI science is essential for its recognition as a legitimate scientific discipline. In other words, identifying the research object is the most fundamental endeavor in the academic development of SSI science and represents the foundational point of departure for its progression. The contours of the research object of SSI science have come into focus through extensive theoretical exploration and practical engagement with SSI. In line with the definition and conceptual breadth of SSI science, its primary research object is the phenomenon of SSI itself. Within the purview of SS science, the basic research subject of SSI science includes both the essence of SSI and its manifestation, notably SSI activities and the institutions or organizations associated with SSI. Thus, SSI science is concerned with examining the underlying phenomena, intrinsic nature, functionality, structure, and governing principles of SSI, its associated activities, and the institutions or organizations involved in SSI.

3.3.3.1 Research Object of SSI Includes the Entity of SSI

SSI, as the overarching "meta-concept" in SSI science, is fundamentally a sub-concept of intelligence, more specifically, it is SS-related intelligence (Wang Bing and Wu Chao, 2019a). In the context of SSM, the pursuit of SSI is directed toward enhancing SSM, with a primary focus on SS prediction, decision-making, and execution. SSI encompasses all SSi (content) that bears implications for SSM (Wang Bing and Wu Chao, 2019a). Adhering to the principle of the information (SSi) chain (Wang Bing and Wu Chao, 2019a), the majority of SSi is directed "downward," occupying the lower links of the SSi chain, whereas SSI resides at a higher link, "looking up" toward SSM. Therefore, SSI is acquired from the analysis and processing of the SS data, SSi, and SS knowledge that are requisite for SSM. The categorization of SSI is elaborated in Sect. 2.7.

3.3.3.2 Research Object of SSI Science Includes SSI Activities

SSI activities refer to practical operations associated with the life cycle of SSI. Drawing from the paradigm of the intelligence process, the SSI process encompasses a spectrum of activities including the identification of SSI needs and planning, collection, processing (organization), analysis, production, dissemination, application, and feedback. It also intersects with broader managerial activities such as leadership, organization, coordination, control, and evaluation. The dynamic between SSI and SS managers, who are the consumers of SSI, constitutes an essential domain of inquiry. Given that SSI is tailored to support SSM, SSI activities must be congruent with SS strategy, policy, planning, and SSM activities to effectively inform and facilitate SSM endeavors. However, a certain degree of detachment is also needed in maintaining impartiality and fairness of decision-making. Furthermore, safeguarding the security and confidentiality of SSI is a critical SSI activity in itself, as it plays a vital role in SSM, particularly within the management of "security" aspects.

3.3.3.3 Research Object of SSI Includes SSI Institutions or Organizations

From a managerial standpoint, the execution of SSI operation entails the involvement of dedicated personnel and organizational frameworks, thereby necessitating the establishment of specialized SSI agencies or organizations. An SSI agency is an organization that participates in SSI activities (tasked with the organization and execution of the SSI process). Within the ambit of SSM, the SSI agency functions as a subsidiary unit of the SSM agency. Thus, the personnel of an SSM agency should encompass both SSI specialists and SSI end-users, namely SSM personnel. Inquiry into SSI agencies extends across a multitude of facets, including the examination of organizational structures, operational dynamics, agency governance (including personnel, establishment, roles, and management), human resource administration, personnel development, asset and equipment management, system management (including design, development, maintenance, and upgrading of systems), resource allocation and optimization, as well as the allocation of responsibilities and the harmonization of efforts between the SSI agency and the broader SSM agency.

3.3.4 Research Content of the Discipline

From a macroscopic perspective, the research content of a discipline includes three systems of a hierarchy: foundational research, which pertains to basic theoretical exploration; intermediate research, which is concerned with applied foundational theory; and downstream research, which focuses on the practical application of knowledge. The research content of SSI science is thus divided into three systems, as depicted in Fig. 3.3.

According to the analysis of the research object of SSI science and Fig. 3.3, it is evident that the specific research content associated with various research objects (SSI itself, SSI activities, and SSI agencies or organizations) can be scrutinized across the three levels of research: upstream, midstream and downstream. This framework establishes a "3-3" matrix of research content within SSI, which elaborates on the three principal systems illustrated in Fig. 3.3. For further details, please refer to Table 3.3.

The research content of SSI science is expanded and enriched, propelled by several key factors. These include the intrinsic evolution of SSI science, the progress in SSM informatization and its applications, the adoption of intelligence-driven SSM paradigms, the rise of new security challenges (risks), and the ongoing reform and development of SSM practices within the broader context of societal informatization. Particularly in the age of big data, these dynamics will significantly influence the development of SSI science, and the dynamic demands of SSI operation will serve as a catalyst for the ongoing enrichment and extension of SSI science.

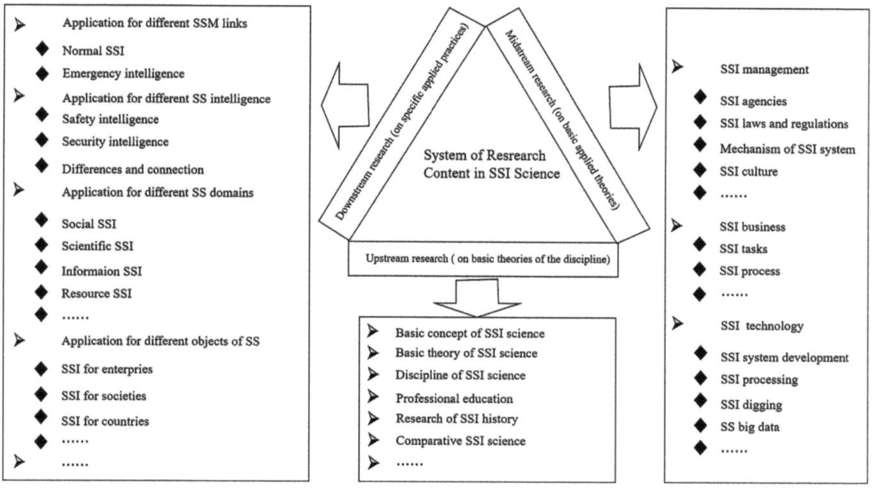

Fig. 3.3 Three systems of research content in SSI science

3.3.5 *Research Goals of the Discipline*

The establishment of a discipline is contingent upon three fundamental criteria: a distinctive and indispensable research object, an evolved or emerging theoretical framework, and the presence of established research methodology, preferably unique to the field. The pursuit of SSI science is aimed at meeting these criteria. The principal objectives of SSI science research are:

1. To establish a comprehensive theoretical framework and system: Current research in SSI is in the empirical or exploratory phase. Many scholars concentrate on partial or micro-level applied research, overlooking the need for a comprehensive and macro-level examination of SSI and its mechanism. The existing SSI theory is fragmented and lacks a cohesive, systematic, and in-depth theoretical framework. Therefore, the establishment of such a framework should be the foremost objective and undertaking of SSI research at present.
2. To balance theoretical and applied research: Research in SSI science should strike a balance between theoretical and applied dimensions. As a discipline with significant practical implications, SSI science presents itself with more features of applied research. Applied research provides new materials for theoretical exploration, which in turn fuels its advancement and informs practical applications. This reciprocal relationship fosters a full cycle, enhancing the quality of both theoretical and applied research within SSI science.
3. To drive innovative development: The ultimate aim of SSI is to make implicit SSI explicit, centralize dispersed SSI, convert voluminous SSI data into knowledge, maximize the value of SSI data information, and establish an effective SSI service system for SSM activities. This system should guide, support, and serve

Table 3.3 The "3-3" system of research content in SSI science

Levels of research	Research object		
	SSI itself	SSI activities	SSI agencies or organizations
Upstream research (on basic disciplinary theories)	SSI epistemology, basic terms and concepts of SSI science; connotation and development, nature, content and features, functions and utility of SSI	Internal laws, principles, subject and object, scientific and practical nature, core, development of SSI activities	Principles of construction and operation mechanism of SSI agencies, principles and methods of management of SSI agencies, theories of organization of SSI, theories of human resource management of SSI
Midstream research (on basic disciplinary theories)	Theories and methods of the spread, origin, management, organization, analysis, coding, of SSI	Theories and methods of operational principles, collection, processing and analysis, storage, utility, and intelligence-led SSM of SSI	Theories and methods of management system, working mechanism, facility and equipment management, service platforms, technological application of SSI
Downstream research (on basic disciplinary theories)	The downstream research in Fig. 3.3 involves various types of SSI, including application for different SSM links, application for different SS fields, and application for different SS objects	Searching (collection) processing, analysis, storage, retrieval, evaluation, transmission, and utilization of all kinds of SII	The downstream research in Fig. 3.3 involves various types of SSI and corresponding agencies, such as agencies of emergency intelligence, intelligence agencies of information SS, agencies of national SS, etc.

SSM endeavors, and achieve innovation and development within the field of SSI science through the dissemination and sharing of SSI.

4. To establish SSI science and conduct comprehensive research: The foundation of SSI science should be solidified, and a thorough investigation into its various aspects should be conducted.

3.3.6 Conclusions

In this section, the perspective of major SS is adopted to analyze the five fundamental issues that underpin SSI science: its definition and meaning, its essence, research objects of its study, the scope of its research, and its objectives. This inquiry serves to establish a robust theoretical bedrock for the development of SSI science as a discipline, offering a direction for its sound and sustainable evolution. It is important to acknowledge that this research marks only the inception of an exploration into the foundational issues of SSI science; it is not an exhaustive or final pronouncement. The discourse presented here is ongoing and remains an undertaking in progress, open to further refinement and enhancement. Feedback, constructive criticism, and suggestions for improvement are welcomed.

3.4 Disciplinary Foundation and Branches of SSI Science

3.4.1 Disciplinary Foundation

The formation and development of a discipline are predicated on a robust theoretical underpinning. Wang Bing and Wu Chao (2018a) have stated that from both philosophical and pragmatic standpoints and in light of the practice of SSM, the theoretical foundation of SSI science is rooted in the philosophy of intelligence management within the classical SSM paradigm. This paradigm covers the "problem-oriented," "risk-oriented," "statistics-oriented," and "intelligence-led" SSM models. In this section, the theoretical basis of SSI will be formed from a disciplinary standpoint.

In accordance with the definition and the interdisciplinary nature of SSI science, SSI science is a field where SS science intersects with intelligence studies. Its core theoretical framework is derived from the theories and principles inherent to both domains. However, akin to intelligence studies, SSI is characterized by its integrative and interdisciplinary approach. Beyond the spheres of SS science and intelligence studies, SSI research extends to embrace theories and methodologies from a variety of other disciplines. These include philosophy (specifically, the philosophy of SS), related social sciences [such as management (SSM), economics (SS economics), sociology (SS sociology), communication studies, archival science, and library science], and related natural sciences [including statistics (SS statistics), information science (SSi science), data science, and computer science]. Therefore, the disciplinary foundation of SSI is a confluence of theories and methods from these diverse disciplines, as illustrated in Fig. 3.4.

3.4.2 Branches of the Discipline

Theoretically, well-established disciplines are characterized by the presence of several sub-disciplines. For instance, both intelligence studies and SS science have evolved to include a range of specialized branches. While the discussion of SSI science's branches may seem premature given its nascent stage, theorizing, planning,

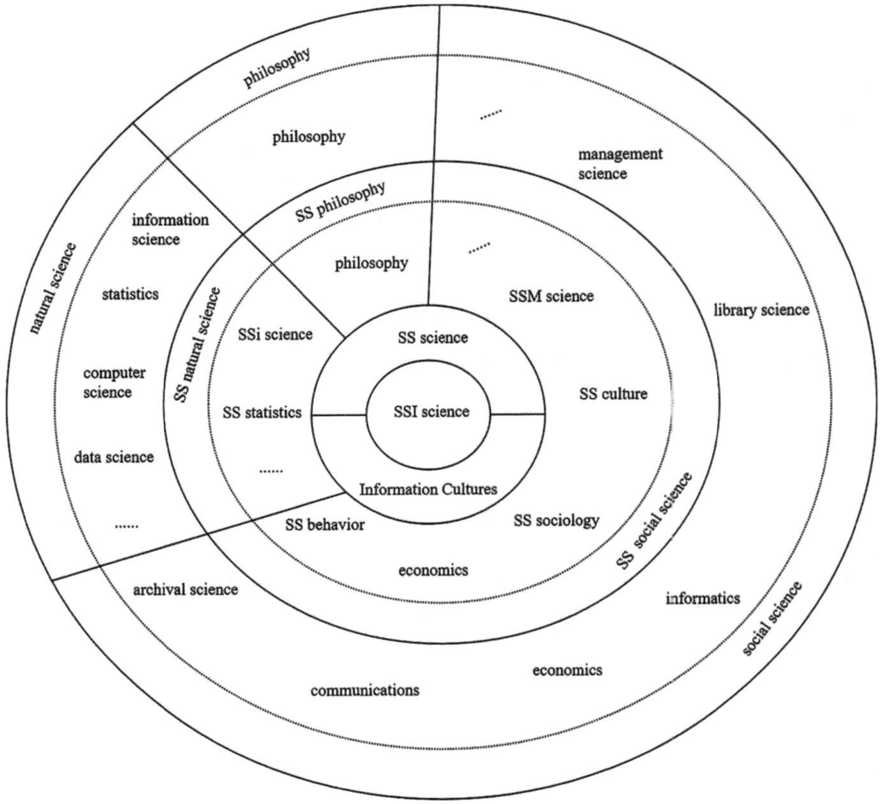

Fig. 3.4 Structure of the disciplinary foundation of SSI science

and anticipating its potential sub-disciplines is essential for steering its scientific, healthy, and sustainable development. From a doctrinal and theoretical standpoint, the branch structure of SSI science can be built in the following way:

1. From a "macro–micro" perspective, SSI science can be divided into macro- and micro-branches. Macro-SSI science concentrates on the efficient aggregation, analysis, and application of SSI resources within SSM practices. In contrast, micro-SSI science involves the specific components of the SSI process, encompassing collection, analysis, and utilization.
2. From a "theory–practice" standpoint, SSI science can be divided into theoretical and practical branches. Theoretical SSI science addresses an array of foundational theoretical issues, such as the core concepts of SSI, the essence and characteristics of SSI thought, its disciplinary nature, research domains, branch systems, and associated educational research, policy research, and the psychology of SSI users. Practical SSI science, on the other hand, is embodied by research on SSI activities (including studies of collection, processing, analysis, storage, retrieval, evaluation, dissemination, and utilization of SSI), SSI technology research (including

studies of the technologies of SSI collection, processing, analysis, transmission, storage, utilization, and monitoring), and research on SSI organization (including studies of management systems, operational mechanisms, asset and equipment management, development of service platforms, and human resource management).

3. From the perspective of SSI types, specific categories of SSI can be developed into corresponding branches of SSI science. For example, normal SSI science and emergency intelligence studies can emerge from different stages of SSM; safety intelligence studies and security intelligence studies can be developed from different extensions of SS; political SSI science, economic SSI science, social SSI science, scientific and technological SSI science, resource SSI science, and nuclear SSI science can be born from different fields of SS; and corporate SSI science, social SSI science, and national SSI science can be formed by different SS subjects.

In conclusion, SSI science is an emerging discipline. Current research, both domestically and internationally, is fragmented and lacks a unified focus, hindering the development of a comprehensive and systematic discipline. Therefore, it is suggested that scholars concentrate their efforts on specific areas within SSI science to achieve breakthroughs. This targeted approach will encourage the gradual emergence of distinct branches within SSI science, ultimately contributing to the establishment of a complete and robust disciplinary system.

3.4.2.1 Conclusions

This section has laid the groundwork for the establishment and development of SSI science, constructing a framework for its academic structure. SSI science is not merely a theoretical construct; its disciplinary development is anchored in a solid foundation. The multitude of potential branches within SSI science underscores its extensive reach and depth as an academic domain, indicating a new and rich field for further discussion and exploration.

3.5 Methodology of SSI Science

The distinctive research object and innovative research content of SSI science, particularly its unique positioning within the broader realms of SS science and intelligence studies, demand the development of the methodology of SSI science. Since there are entirely new research object and content for SSI science, SSI science cannot solely depend on conventional methodologies from SS science and intelligence studies; a new methodology suited for its research subject and content is needed.

From a theoretical and scientific standpoint, the methodology of a discipline serves as the guiding principle and toolkit for the establishment, development, and refinement of its theoretical framework, as methodology is required for every discipline (Liang Huiwen, 2017). Since SSI science is in its infancy, its methodology has not been clarified, thus its construction, research and practical application cannot be guided by effective guidelines and methods, which seriously hinders its disciplinary development. Therefore, the investigation and development of methodology for SSI science are of paramount importance.

3.5.1 An Overview of Methodology

The classical Chinese text *Mozi· The Will of Heaven* (《墨子· 天志》) explains the principles of identification through the metaphor of geometric shapes: "Shapes that align with my carpenter's square are deemed squares; those that do not are deemed non-squares. Thus, both squares and non-squares can be discerned. Why is this so? Because the criteria are clear." It further states: "Shapes that align with my compass are deemed circles; those that do not are deemed non-circles. Thus, both circles and non-circles can be discerned." These passages illustrate the renowned "square method" and "circle method," which refer to the techniques for measuring squares and circles. The term "method" (方法), initially denoting these specific techniques, has since expanded to include a broad range of approaches or means for accomplishing various tasks. In the context of academic discipline, research methods are divided into methodology and specific methods.

3.5.1.1 Basic Meaning of Methodology

Methodology is the science that studies methods (Wu Chao, 2016). Philosophically, it probes into the human endeavor of devising and employing a variety of methods, in pursuit of generalized knowledge about them. Methodology rises above the constraints of specific tools and empirical knowledge, venturing into the domain of abstract reasoning and theoretical frameworks. As a general theory of universal applicability, it has two principal connotations: one concerns the doctrine of research methods, or the conduct of research; the other is related to the doctrine of specific research methods and approaches within a given scientific field. Neither of these connotations take any specific matter as the research object or grapples with certain scientific issue concerning a specific research object; instead, they offer theoretical guidance for the research procedure about an object.

As a guide for universal application, methodology provides the general rules and principles that govern scientific research, influencing and shaping the selection and application of specific methods. Individuals who employ different methodologies, whether consciously or unconsciously, have distinct perspectives when addressing problems and will exhibit variations in the processes and methods they choose for research.

Methodology fulfills a directional role in scientific inquiry. New scientific discoveries hinge not only on the insights and experiments of scientists but also on the methodological framework that steers their work. Thus, methodology has emerged as a significant field of study for philosophers, scientists, and sociologists alike (Wu Chao, 2016).

Scientists tend to display distinctive habits, styles, and characteristics in their research methodologies, different schools of thought possess their own features and traditions, and various academic disciplines have distinctive research methods. The study of methodology aims to uncover the commonalities and patterns within these diverse practices across individuals, schools of thought, and disciplines. To employ a metaphor, methodology does not provide a bridge or boat for crossing a river; rather, it imparts the significance of such conduits and the general principle for constructing them. Such a principle provides intellectual cues, inspiration, and guidance, empowering researchers to devise methods suited to specific circumstances, avoid the pitfalls of the past, and accomplish tasks creatively and efficiently. This guidance constitutes the quintessential value of methodology.

3.5.1.2 Relationship Between Methodology and Method

Methodology and method are intricately linked yet fundamentally different concepts, forming a dialectical relation (Wu Chao, 2016). In terms of their interconnection, since methods provide methodology with raw and disconnected empirical materials, they cannot exhibit their efficacy without the guidance of methodology. However, since specific methods constitute the foundation and substance of methodology, the latter would fail in informing the utilization, summarization, and refinement of methods without the support from methods. Methodology is not merely an assembly of methods; rather, it is a systematized and organized collection of methods, directed by specific principles and perspectives. Method, in contrast, is the object of methodological scrutiny, and the raw material to be refined. It is individual and specific in methodology. The method that has not been systematized and organized under certain principles and viewpoints cannot be termed as methodology.

Disciplinary research methods are divided into methodology and methods. The methodology of a discipline represents a theoretical generalization and summarization of its specific research methods, embodying a systematized approach. Methodology is not mere accumulation of various methods, but rather the distillation and abstraction of them. The methodology of a discipline mirrors its investigative perspectives (approaches), principles, methods, means, and tools. Generally speaking, various disciplines exhibit unique methodological traits. Sherman Kent believes that methodological issues are the quintessential challenges encountered in the quest for truth, while methods are the specific paths, procedures, means, and tools employed to achieve a particular objective.

3.5.2 An Overview of the Methodology of SSI Science

3.5.2.1 Basic Meaning of the Methodology of SSI Science

Like any academic field, the methods and methodology of SSI science are crucial for the development and refinement of its theoretical framework. As previously discussed, methodology extends beyond a catalog of specific research methods; it represents a distillation and abstraction of these methods (Wu Chao, 2016). The methodology of SSI science includes the principles and theories that guide SSI research. It is a systematic approach to problem-solving, often involving discussions of phases, objectives, instruments, and methodological techniques (Bao Changhuo et al., 2018). Methodology directs specific methods, and these methods, in turn, inform methodology; methodology and method are interdependent (Wu Chao, 2016). Thus, the methodology of SSI science is not a collection of various methods but a structured framework informed by relevant content, with the study of methods offering significant theoretical guidance (Hu Shuangqi, 2003; Li Meiting and Wu Chao, 2015).

Drawing on the foundational concepts of SSI science and methodology, the methodology of SSI science is illustrated as follows: Philosophically grounded and building upon the methodologies of intelligence studies and SS science, the methodology of SSI science is primarily concerned with SSI science, encompassing a synthesis of research methods particular to the study of SSI science. In essence, it includes all methods and techniques used in SSI research to uncover the patterns governing its subjects (SSI itself, SSI activities, and SSI institutions). From the vantage point of SSI science, its methodology constitutes a system of characteristic methods employed to investigate and understand the essence, principles, and developmental trajectories of SSI science. It offers broad theoretical guidance for SSI research, assisting in the establishment of a unique and coherent theoretical framework and ensuring the discipline's relevance to contemporary issues.

As an evolving applied discipline with an inherently interdisciplinary character, the methodology of SSI science involves research methods from SS science and intelligence studies at the microscopic level, and from philosophy, natural science, and social science at the macroscopic level. Any method applicable to SSI research can be integrated into its methodological system. However, given its distinct nature as an independent discipline, SSI science requires a discerning approach when adopting methods from other fields. The suitability, benefits, and drawbacks of borrowed methods must be carefully weighed. Not all research methods from related disciplines are appropriate for SSI research, and not all borrowed methods are ideal. Moreover, as highlighted in the fundamental characteristics of methodology (Liang Huiwen, 2017), the strength of the methodology of SSI science lies not in a focus on individual research methods but in examining the structure, function, and interconnection among its array of research methods from a holistic perspective, harnessing the collective systemic power of this methodological ensemble.

3.5.2.2 Main Attributes of the Methodology of SSI Science

The objective and subjective dimensions of SSI science should be examined before employing the methodology of SSI science. Since the objectivity of the methodology of SSI science is derived from its historical underpinnings, the theoretical milestones of SSI science are anchored in the progress of information and intelligent technologies, the development of methodologies and tools, and the shift in cognitive paradigms. The subjectivity, on the other hand, arises from the researcher's proactive interaction with the research object. It is pivotal in the selection and utilization, adaptation and integration, and reorganization and synthesis of methods. The advantage of the methodology of SSI science is not about highlighting individual methods but about holistically examining the structure, function, and interplay among the group of methods so that the collective function of methods could be exercised.

3.5.2.3 Significance of the Methodology of SSI Science

It is significant to establish a sound methodology for the academic development and research of SSI science. The advancement of SSI science is contingent on continuous methodological breakthroughs and innovations. A well-defined methodology not only charts the course for the field's evolution but also defines research objectives and strategies. It is instrumental in constructing a multifaceted, multi-tiered, and comprehensive disciplinary framework, thereby promoting the field's expansion and competitive edge. Moreover, a sound methodology empowers researchers and elevates the caliber of SSI endeavors. The key significance is elaborated as follows:

1. The advancement of SSI science depends on methodological innovation. The methodology of SSI science serves as a scaffold for probing into and comprehending the essence and principles of the field, as well as the customary approaches within its genesis, evolution, and refinement. For SSI science and its sub-disciplines to make progress, methodological breakthroughs are essential, necessitating rigorous discourse and the fusion of a variety of theoretical insights and viewpoints. Therefore, the methodology of SSI science establishes the basis and promotes the development of the field.
2. The methodology of SSI science serves as a road map for its development. It elucidates the research perspectives, principles, methods, and procedures of the field, providing a clear methodological structure for SSI research. It facilitates the establishment of a multi-dimensional and holistic disciplinary structure, defining research directions and strategies. A robust methodology steers the formation of research focus, principles, methods, and procedures of SSI science, offering a blueprint for in-depth exploration.
3. The methodology of SSI science accelerates the pace and enhances the competitiveness of SSI science. It expedites the development of research objects within the field, contributing to the overall growth and specialization of the SSI science. This enhancement does not only accelerate the evolution of the field evolution

but also amplifies its competitive strength, making a substantial contribution to the SSI and SSM sectors.

4. The methodology of SSI science empowers researchers and elevates SSI operations. For instance, a reasonable methodology equips SSI researchers with sharper analytical skills, enhances the robustness and competitiveness of their research, and supplies a sustained impetus for SSI operation and the advancement of national and social security.

In conclusion, the methodology of SSI science is vital for its ongoing research and development, both at present and in the future. The development of SSI science should prioritize methodological advancement, confront potential challenges and nurture innovation, guiding the innovation and progress of SSI science.

3.5.2.4 Main Content of the Methodology of SSI Science

According to its meaning, there are six aspects of the methodology of SSI science (Liang Huiwen, 2017), as listed in Table 3.4.

Table 3.4 Main content of the methodology of SSI science

Number	Main Content	Objective of research
1	Discussing the research perspective and principles of SSI science	To propose the selection criteria for adopting various methods in studies of SSI science
2	Studying the features of functions of various scientific methods applied in the study of SSI science	To better harness the scientific and applicable properties of applying the methods that originate from different disciplines to the study of SSI science
3	Exploring the application of various methods in the studying of SSI science and analyzing their advantages, limitations, and scopes of application	To select proper research methods or improve methods of other disciplines for the study of SSI science according to the features and needs of SSI science
4	Verifying the features and effects of application of the adopted methods with case analysis	To judge the merits and demerits of research methods by evaluating their practicality
5	Discussing the comprehensive application of different methods in the study of SSI science	Comparing different methods for selecting the top one or propose a new comprehensive one through an integrated approach
6	Studying the logical structure and relationship among the methods of SSI science	To establish a system of the methodology of SSI science

3.5.3 Research Perspective of SSI Science

The research perspective of a discipline is an essential element in its methodological exploration, offering a starting point for a scientific inquiry. Without an empirically grounded research perspective, it might be impossible to initiate research, or even lead to an erroneous conclusion. The principal research perspectives for SSI science are perspectives of major SS, major SSI, SSM, cross-integration, the systemic (holistic) perspective, and the partial perspective. The first three have already been detailed in Sect. 3.2. This section concentrates on the perspective of cross-integration, the systemic (holistic) perspective, and the partial perspective. It is important to recognize that SSI research often necessitates the flexible application of one or more of these perspectives based on the specific demands of the research at hand.

1. The Perspective of Cross-Integration

Given its intrinsic nature and mechanism of formation, SSI science is inherently interdisciplinary, borrowing insights from intelligence studies and SS science. From the perspective of intelligence studies, SSI entails the application of intelligence theories and methods for examining and tackling SS challenges, making the perspective of intelligence studies indispensable for SSI research. From the standpoint of SS science, SSI signifies the application and adaptation of intelligence studies within the "SS+" paradigm. Recognizing this perspective is vital for enhancing the distinctive features and depth of SSI science within the domain of SS science.

In essence, the perspective of cross-integration should be adopted for doing SSI research, integrating insights from both intelligence studies and SS science. This involves perceiving and organizing SS-related information, applying theories and knowledge from SS science to identify and analyze intelligence, and evaluating the role and value of intelligence for SS. Such an approach equips managers with the capacity to make informed predictions, decisions, and actions regarding SS, thereby maximizing the utility of intelligence studies within the SS domain.

2. The Systemic (Holistic) Perspective

To achieve a comprehensive examination and precise comprehension of the dynamics of SSI and the principles that direct SSI operations within open and complex systems, researchers must adopt a systemic and holistic approach, which entails the investigation of SSI phenomena, underlying laws that govern them, and issues of SSI operation. This means studying the dynamics and operation of SSI under the overall "temporal-spatial" conditions.

3. The Partial Perspective

The partial perspective is concerned with studying specific research ideas of a certain aspect of SSI. In other words, it refers to the examination of a particular pattern or rule of SSI, or various SSI types, including SSI of different research objects (communities, enterprises, cities, and nations), different levels of SSI (strategic, tactical, and operational) and different aspects of SSI (nature, attributes, acquisition, analysis, and dissemination of SSI).

3.5.4 Research Principles of SSI Science

The principle acts as the benchmark and guideline for problem-solving. In scientific research, it is imperative to adhere to foundational principles. Methodologically speaking, and drawing from a comprehensive review of the literature, a minimum of seven principles should inform the research of SSI science: the principles of objectivity and systematicity, universality and particularity, theory and application, qualitative and quantitative inquiry, historical and contemporary relevance, individual and collective perspectives, and practice and assessment (Liang Huiwen, 2017). These principles are detailed in Table 3.5.

3.5.5 Research Paradigm of SSI Science

The research paradigm of SSI science acts as the point of departure for the inquiry of SSI science, unifying the research objects and methods of SSI research. Derived from the research content of SSI science and through a review of extant research outcomes, four sub-paradigms are identified as depicted in Fig. 3.5:

1. Sub-paradigm of SSI practitioners and users (the dimension of subject): SSI operation is characterized by its intellectual intensity, with individuals and organizations at the forefront, exerting a major influence on the SSI process. Research on the SSI workforce and users can be conducted from the dimension of the subject of SSI operation.
2. Sub-paradigm of SSI materials and management (the dimension of object): SSI materials are the object of SSI endeavors. SSI professionals participate in the manipulation and interpretation of SSI materials to generate SSI outputs that cater to the needs of SSI users. Research on the SSI materials can be carried out from the dimension of the object of SSI operation.
3. Sub-paradigm of SSI activities (the operational dimension): SSI activities refer to the array of activities conducted to secure the intelligence that is necessary for individuals or organizations to take SS-related actions. The activities include production, analysis, application, and management of intelligence, which are essential to the coordination, evolution, and exploitation of SSI resources. Research on the SSI activities can be done from the operational dimension of SSI.
4. Sub-paradigm of intelligence-led SSM (the dimension of application): The application of SSI concerns the convergence of SSI with SSM, with the objective of utilizing intelligence to direct the SSM process. This sub-paradigm explores the implementation of the "intelligence-led SSM" approach, which seeks to amplify the role of SSI within the SSM context. In other words, the application of SSI advocates for and employs the principles and methods of intelligence-led SSM.

Table 3.5 Research principles of SSI science

Number	Research principle	Connotation
1	Combining objectivity and systematicity	Objectivity requires that when studying the SSI of a certain system, it is necessary to understand the SSI needs, and characteristics and reality of SSI operations; Systematicity requires that the study of SSI science is conducted in the way of regarding all the elements of SSI and all the links of SSI operation as an integrated whole
2	Combining generality and specificity	Generality requires that the research of SSI science should establish a unified and general SSI strategic thought and theoretical method based on the macroscopic view of SS and of SSI. The specificity requires that theoretical systems should be built for the specific industries, fields or SSI of different levels in the study of SSI science
3	Combining theory and application	According to its disciplinary features, SSI science is a discipline integrating theories and application. The research should not only focus on the theoretical level but also apply theories to specific fields of SSI operation, utilizing the theory of SSI science to guide SSI practices
4	Combining qualitative research and quantitative research	The qualitative research of SSI science aims to observe and analyze its nature and apparent features of the research objects at the level of logical thinking for comprehending the focus and direction of the research and providing a theoretical basis for quantitative research. The quantitative research is using corresponding methods (such as probability statistics, mathematical modeling and risk assessment. etc.) to process and collect large SS data information, which can provide a data basis for SSI science and make the research results more objective and accurate
5	Combining history and reality	The research of SSI science is both restricted by the practical scope and knowledge structure of SSI researchers and promoted by modern science and technology and social progress. Such influences progress at different rates. The research should effectively link the history of SSI research with the reality of modern scientific and technological development, improving and developing the theory and methods of SSI science on the basis of research inheritance

(continued)

Table 3.5 (continued)

Number	Research principle	Connotation
6	Combining individualism and holism	Individualism focuses on the understanding of individuality, while holism focuses on explaining individual activities through the relationship of the whole. In the study of SSI science, it is necessary to study the individual behaviors of SSI users and SSI practitioners by using the method of individualism, and it is also necessary to apply the method of holism to comprehending the full cycle of SSI activities and its general laws as a whole
7	Combining empirical research and evaluative research	Pure empirical research is conducted from facts to facts, that is, to use fact A to explain fact B, but fact B does not explain or prove fact A; pure evaluative research is conducted from meaning to meaning, namely, meaning A is used to explain B, but meaning A itself is not well explained. In the research of SSI science, the combination of these two research is to put an end to the situation that the fact and value, and meaning of SSI are at opposing ends, enabling the fact to carry the meaning, and present the fact of SSI science as the research object since the meaning is determined; The meaning can be better understood, accepted and used by the outside world according to the fact of SSI operation and its factual occurrence

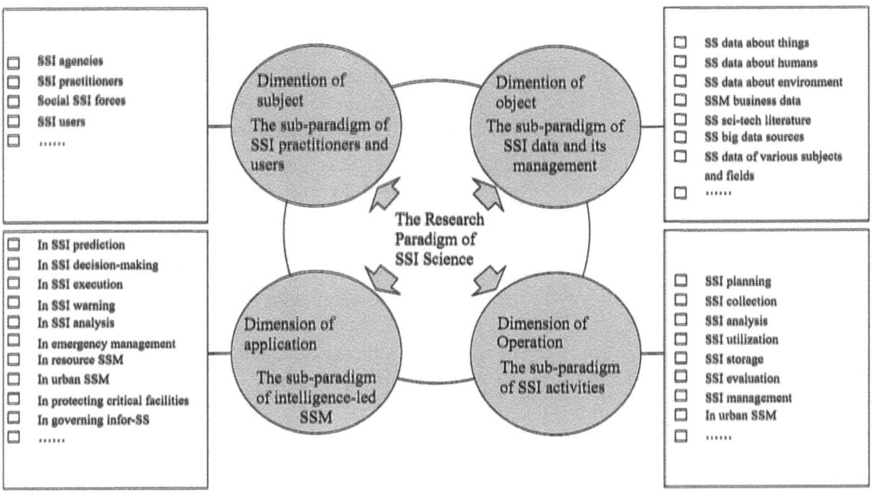

Fig. 3.5 Research paradigm of SSI science

From the dimension of SSI application, research can be conducted on the application of principles and methods of intelligence-led SSM across the fields of SSM.

As illustrated in Fig. 3.5, these four sub-paradigms are integral to SSI science. They are interconnected and mutually reinforcing, forming a complete research paradigm of SSI science. This paradigm emphasizes both the in-depth exploration along different dimensions and the interaction and integration between them, promoting the overall development of SSI science—both vertically and horizontally.

3.5.6 Methodological System of SSI Science

The methodology of SSI science encompasses an array of methods, ranging from methods for examining specific instances of SSI within the sphere of public SS at the micro-level to methods for investigating the overarching laws governing SSI agencies and their operational processes at the macro-level. The methods can be stratified into four levels, differentiated by their degree of generalization and the breadth of their applicability: philosophical methods, discipline-specific research methods, general research methods, and proprietary research methods. The integration of the four levels forms the methodological framework of SSI science.

1. Marxist Philosophical Method

The philosophical method (i.e., dialectical and historical materialism) is the most broadly applicable scientific method for all research endeavors and the overarching guiding method. It is not commonly employed to address specific issues within a given discipline but acts as the dominant principle that guides research. Although SSI science is an applied discipline with a focus on practical problem-solving, its research domain also extends to exploring foundational questions, such as the essence of SSI and the principles governing SSI activities. Consequently, philosophical methods are not only relevant but also directly influential in shaping SSI research.

2. Subject-Specific Research Methods

Given the emergence of SSI science at the intersection of intelligence studies and SS science, the research methods from these disciplines constitute the principal methods of SSI research.

(1) Research Methods of Intelligence Science: The methods within intelligence studies are crucial to SSI research and are not as prominently featured in other areas of SS science. Established paradigms such as the institutional paradigm (regarding intelligence institutions as social institutions) and the cognitive paradigm (highlighting the user's knowledge structure) are central to SSI research. Additionally, general research methods like mathematical and logical methods, and proprietary methods such as informetrics and citation analysis, are effectively employed to guide SSI research.

(2) Research Methods of SS Science: As an offshoot of both intelligence studies and SS science, SSI research also incorporates the methods of SS science. This includes methods like the "forward (initiating from SS) path—middle (initiating from SS risks) path—reverse (initiating from SS incidents) path" method (Wu Chao, 2016).

3. General Research Methods

General research methods are common among disciplines, representing broad methods applicable to various fields of study. They are characterized by their universal, interdisciplinary, and comprehensive nature. For instance, the comparative method is a general research method that plays a significant role in SSI science. By comparing the developmental patterns of SSI science with those of intelligence activities in other disciplines, the unique laws and characteristics of SSI science can be discerned for gaining a clearer understanding of its current state. Comparative analysis of different SSI phenomena also promotes mutual learning and the exchange of ideas. The general research methods used in SSI science can be divided into four groups (Liang Huiwen, 2017).

(1) Methods from interdisciplinary sciences: These include methods of mathematics, systems theory, information theory, control theory, synergy theory, catastrophe theory, and dissipative structure theory.
(2) Methods from empirical studies: These comprise literature reviews, social surveys, empirical summarization, simulations, observations, experiments, and case studies.
(3) Methods from noetic sciences: This category includes methods of abstraction, dialectical reasoning, induction and deduction, comparison and classification, analysis and synthesis, proof and refutation, the unity of logic and history, and the hypothesis method.
(4) Research methods from related disciplines: These draw upon methods from related disciplines, including information science, management science (SSM science), systems science (SS systems science), statistics (SS statistics), behavioral science (SS behavioral science), psychology (SS psychology), data science, intelligence science, and communication science.

Each of these categories includes a variety of specific research methods. The application of general research methods in SSI science ensures a more standardized and scientific research process, leading to findings that are more credible and readily accepted.

4. Proprietary Research Methods

Although some critics assert that SSI science lacks proprietary methods and is overly dependent on borrowed techniques, either directly or indirectly, others argue that it has its own distinct methodological system that is essential for the field. As a burgeoning discipline, SSI science is in the process of developing its own proprietary research methods. While many methods are indeed adapted from other disciplines, they are often modified and refined by SSI scientists, evolving into specialized tools that offer unique insights into the intrinsic laws and phenomena of SSI. Table 3.6 illustrates examples of such proprietary research methods. The evolution of a sound methodology is a dynamic process for any academic field, and SSI science is no different. Ongoing collaborative endeavors within the SSI community will be instrumental in establishing and perfecting a methodological system that is finely tuned to the specific demands of SSI research.

Table 3.6 Examples of some typical proprietary research methods of SSI

Research methods	Specific explanation
Method of SS event analysis	The occurrence of SS events can be analyzed and reasoned from the cause to the result by gradually deducing intelligence information. The start of every incident may potentially lead to a disastrous result. The logical relationship in the evolution of an event can all be reflected by intelligence information, with each step of the progression of the event being manifested in different forms of intelligence information. As useful SSI can be obtained by analyzing the causes, mechanisms, and prevention and control measures of SS events, one can alter the outcome of the event if one fully understands the pattern of the evolution of SSI and adopts preemptive measures
Method of SS big data analysis	Based on the modern network technology of big data information, reliable SS conclusions can be obtained by fully searching a vast amount of intelligence resources of SS big data, constructing a platform of SSI information data to retrieve, identify, filter and analyze massive SSI information or data, comparing and analyzing the different characteristics and attributes of SSI, and determining valuable and highly important SSI data
R-M method of SSI analysis and acquisition	The R-M method for SSI analysis and acquisition (Wang Bing and Wu Chao, 2019) suggests that SSI information acquisition and analysis should start from the perspective of SSM itself, and intelligence identification and research should be conducted from the two aspects of "risk module" and "management module." The R-M method covers methods of intelligence analysis for treating external and internal SS factors of organizations such as national, urban, social, and corporate entities. When using the R-M method for SSI analysis, the characteristics of internal and external SS factors of the organization should be considered on the basis of the actual SS situation, with a focus on the SS factors of the organization, to ensure the result of SSI analysis is systematic and reasonable
SS analysis method	Analysis is an essential process in transforming SSi data into SSI. Various SS analysis methods (such as event tree analysis, fault tree analysis, and preliminary hazard analysis, etc.) can all yield effective SSI

3.5.7 Conclusions

Disciplinary research is inherently guided by methodology, and SSI science is no exception. This section has probed into the methodology of SSI science to reinforce its theoretical underpinnings. It explicated the meaning, characteristics, significance, and substance of SSI science methodology through the lens of methodological principles. It also proposes research perspectives and principles of SSI science, and delineated a methodological framework that includes philosophical methods, discipline-specific research methods, general research methods, and proprietary research methods.

3.6 Chapter Summary

The establishment and advancement of SSI science as an academic discipline necessitate a solid theoretical bedrock. This chapter has concentrated on the theoretical aspects of SSI science's disciplinary construction, yielding three principal insights:

1. The disciplinary construction of SSI science should be anchored in SS science, with precise delineation of foundational concepts. It should actively assimilate theories and methods from the intelligence studies, underpinned by the notions of "major SS" and "major SSI." Concentrating on SSM, it should incorporate diverse forms of SSI, identify shared issues, theories, and methods, and construct a scientific, cohesive, and pragmatic theoretical system.
2. SSI science encompasses the essential elements to emerge as an autonomous discipline. Its core research inquiries, disciplinary underpinnings, and specialized branches could serve to lay a sound theoretical foundation and form a road map for future development, fostering a robust, scientific, and sustainable evolution of the discipline.
3. The methodology of SSI science is vital to its construction and development. It offers a strategic direction by elucidating research perspectives, principles, and a methodological system that includes philosophical methods, discipline-specific research methods, general research methods, and proprietary research methods.

Chapter 4
Theories of SSI Science

4.1 Introduction

The foundational theory (including methods) of a discipline provides the essential basis for research and practice within that field. This chapter introduces an array of key foundational theories of SSI. Firstly, as the perception, acquisition and analysis of SSI serve as the first step of SSM, this chapter presents a method for acquiring and analyzing SSI (the "Risk-Management (R-M) method") and a model for SSI perception and analysis, taking a comprehensive perspective of SSM and SSI and integrating SSI practices in SSM. Secondly, as the integration of SSI is an importance means of realizing the value of SSI and a necessary approach to acquiring and integrating multi-source SSI in the big data environment, the chapter proposes a model of SSI integration in the environment of big data. On this basis, it establishes a theoretical framework of SSI literacy and SSI system according to the abilities required of SSI personnel (SSI literacy) and the primary conditions for SSI operation (SSI system). Finally, building upon these foundations, the chapter introduces a model of intelligent SSI analysis, a theory of evaluating SSI services, a system model of intelligent SSI services, and a causal model of SIF, all aimed at addressing the challenges of SSI operation within SSM.

4.2 R-M Method for SSI Acquisition and Analysis

From an intelligence perspective, the SSI operation is a crucial component of SSM, serving as its foundation (Wang Bing and Wu Chao 2019f). As established in intelligence studies, the effective acquisition and analysis of SSI are fundamental to SSI operations (Ye Ying and Wu Yishan 2012). However, current SSI research is still in its early stages, lacking established methods for SSI acquisition and analysis, which hinders both research and practical application. Therefore, this section

B. Wang and C. Wu, *An Introduction to Safety and Security Intelligence Science*,
https://doi.org/10.1007/978-981-96-7261-5_4

proposes the R-M method for acquiring and analyzing SSI. This method, developed from a combined perspective of SSM and intelligence, and informed by SSI operations within SSM, aims to provide theoretical guidance and a reference framework for both research and practice in SSI acquisition and analysis, thereby advancing the field.

4.2.1 Approaches to the Acquisition and Analysis of SSI

Throughout the entire process and different aspects of SSM, SSI plays a critical role. The relationship between SSM and SSI is not unidirectional but mutually reinforcing (Fig. 4.1). Specifically, SSI supports and informs SSM, while SSM serves as the "carrier" for acquiring and analyzing SSI. The accuracy and rationality of this reciprocal relationship is illustrated from two aspects below:

1. From the perspective of management, Yovits and Kleyle's generalized system model of intelligence (Ye Ying and Wu Yishan 2012; Yovits and Kleyle 1993) posits that intelligence manifests through management and is perceived, acquired, and analyzed by intelligence users (managers). Similarly, within the context of SSM, SSI emerges from SSM processes, content, and activities. Therefore, the acquisition and analysis of SSI should be grounded in SSM itself.
2. The managerial perspective on intelligence defines SSI as any information that affects SSM (Chen Chao 2017). This definition reveals two key aspects of SSI: (1) SSI is an information product—the result of analysis and processing. (2) From the perspective of the generation and analysis of SSI, it is a process—a comprehensive activity involving the monitoring and analysis of factors influencing SSM and SSM behaviors. This aspect highlights that SSI involves the systematic and long-term tracking, collection, analysis, and processing of information that could impact SSM, culminating in the refinement of key intelligence directly relevant to SSM (such as SS threats, "strengths and weaknesses," and "opportunities and challenges," etc.). The refined intelligence empowers SS managers to conduct SSM and take strategies with adequate SSi. Therefore, the acquisition and analysis of SSI are inherently tied to SSM itself.

In summary, the fundamental approach to SSI acquisition and analysis within the context of SSM is that it should be intrinsically linked to SSM. This approach offers at least two major advantages: (1) Deficiencies in SSM can be identified and addressed promptly; (2) This approach facilitates the development of an SSI acquisition and analysis model centered on SSM goals. It creates an SSI network that interacts in real

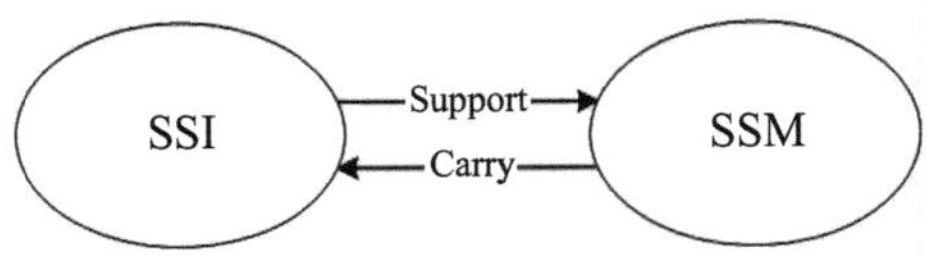

Fig. 4.1 Basic relationship between SSM and SSI

time with SSM content and processes, ensuring that SSi collecting and analyzing are more focused and targeted, ultimately leading to more effective, real-time, accurate, and efficient SSI acquisition and analysis.

Modern SSM science posits that "SSM" is in fact a shorthand for "SS risk management"—the intersection of "SS risks" (hereafter referred to as "risks") and "management" (Wang Bing and Wu Chao 2019f). SSM involves controlling risks by utilizing management theories, methods, and tools. This understanding allows one to decompose SSM into two basic modules: the module of risk (focuses on the content and process of risk identification, analysis, and evaluation, and is more concerned with SS science) and the module of management (centers on SSM behaviors and strategies for controlling risks, and is more concerned with management science). From an SSM perspective, the sources of SSI can be categorized according to these two modules. This leads to two distinct approaches for SSI acquisition and analysis: SSI of the risk module (which can be seen as SS risk information) and SSI of the management module (which can be seen as management information).

4.2.2 *R-M Method for the Acquisition and Analysis of SSI*

Based on the approach to SSI acquisition and analysis from the perspective of SSM, the R-M method is proposed as illustrated in Fig. 4.2. Obviously, this method is built upon the two distinct approaches to acquiring and analyzing SSI (represented by "I" in the figure) within the SSM framework: "the Risk Module" (represented by "R") and "the management module" (represented by "M"). Hence, the method is named the "R-M" method.

As illustrated in Fig. 4.2, the R-M method emphasizes that SSI acquisition and analysis should be SSM-oriented and grounded in SSM itself. The core principle of this method is twofold: (1) The SSM module (SM in Fig. 4.2) is divided into two modules: the Risk Module (R) and the Management Module (M); (2) SSI is acquired and analyzed sequentially, first for the Risk Module (I_R) and then for the Management

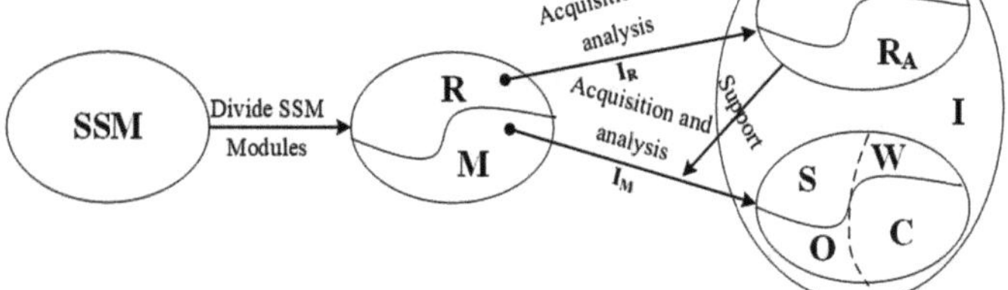

SSM—Safety and Security Management; R—Risk Module; M—Management Module; I_R—SSI of the Risk Module; I_M—SSI of the Management Module; R_U—Unacceptable Risks;
T—SS Threats; R_A—Acceptable Risks; S—SSM Strength; W—SSM Weakness; O—SSM Opportunity; C—SSM Challenge; I—SSI

Fig. 4.2 Basic framework of the R-M method for SSI acquisition and analysis

Module (I_M). Two further points are to be clarified for the R-M method: As SSM focuses on risks, I_R serves as a crucial foundation for acquiring and analyzing I_M; (2) The R-M method is a comprehensive approach encompassing both internal and external SS factors within organizations (e.g., countries, societies, cities, enterprises, etc.). Therefore, applying the R-M method requires considering both internal and external factors to ensure the systematic nature of SSI acquisition and analysis. In practice, the relative emphasis on SS factors of both internal and external should be adjusted according to specific contexts.

From the perspective of SS science, the key SSI within I_R comprises two aspects: Acceptable Risk (R_A) and Unacceptable Risk (R_U) (Wu Chao et al. 2018). A prevailing view in SS science is that SS involves managing risks within acceptable limits (Wu Chao et al. 2018). Therefore, R_U can also be seen as an SS threat. While similar in meaning to "SS hazard," the term "SS threat" is preferred here due to its wider international recognition. From the perspective of management science, the key SSI within I_M is related to SSM characteristics, covering two main categories: SSI is related to current SSM characteristics and SSI to future SSM characteristics. Specifically, I_M can be further divided into four aspects: SSI is related to current SSM characteristics including SSM (S) strengths like unique SS technologies, extensive SSM experience, an excellent SS culture, sufficient investment in safety and security, and high-quality SSM personnel and SSM weaknesses (W), such as a lack of advanced SS technology, methods, facilities, and equipment, limited SSM experience, a negative SS culture, and insufficient investment in safety and security; while SSI is related to future SSM characteristics including SSM opportunities (O) like significant potential for SS development, consistent attention to SS from an organization, the government or the nation, increasing demand for SS development, continuous improvement of the environment for SS development, and elimination and upgrading of outdated processes, technologies, and equipment, and challenges (C) for SSM, like continuous expansion of an organization's operational scale, emergence of complex and interconnected SS risks, continuous emergence of new SS risks, a mismatch between an organization's SSM capabilities and its development, an imperfect SSM system mechanism, and the uneven distribution of SS resources.

In summary, the key SSM-oriented SSI includes six aspects, represented by the set expression as $I = \{(R_A, R_U)\text{U}[(S,W)\text{U}(O,C)]\}$.

4.2.3 Implementing the R-M Method for SSI Acquisition and Analysis

Based on the essential idea of establishing the R-M method for SSI acquisition and analysis, the implementation involves two aspects: I_R acquisition and analysis, and I_M acquisition and analysis. The two methods for each below:

1. The R-I_R Method for I_R Acquisition and Analysis

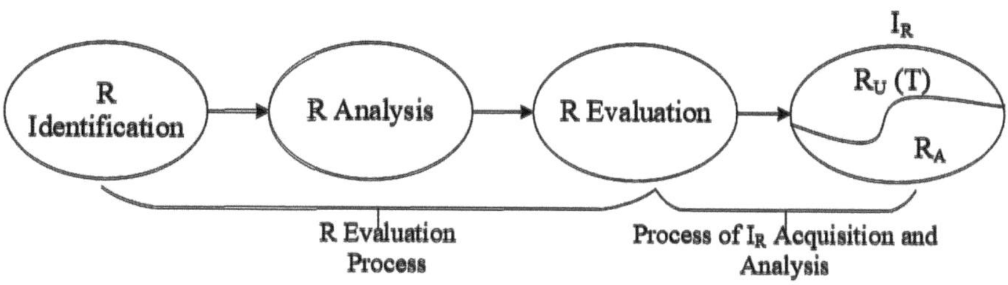

Fig. 4.3 Basic framework of the R-I_R method for I_R acquisition and analysis

From an SS science perspective, risk assessment forms the foundation for I_R acquisition and analysis. The process of I_R acquisition and analysis can be divided into two logically sequential processes: the "risk assessment process" and the "I_R acquisition and analysis process" (illustrated in Fig. 4.3). This method is thus termed the "R-I_R method." The risk assessment process, based on established risk assessment methodologies (Xu Zhisheng and Jiang Xuepeng 2016), is comprised of three steps: (1) Risk Identification: Identifying risks faced by various aspects of the organization (e.g., business units, major business activities, and critical business processes); (2) Risk Analysis: Analyzing and describing the identified risks (e.g., characteristics, probability of occurrence, conditions for occurrence, and potential consequences of risks); (3) Risk Evaluation: Measuring the magnitude of the risk value using risk assessment methods, and classifying risks accordingly. Upon the completion of the risk assessment process, R_A (acceptable risk) and R_U (unacceptable risk) can be identified according to the risk evaluation results (i.e., results of risk classification), thereby obtaining I_R. Apparently, I_R provides crucial SSI for risk control (i.e., SSM).

2. The SWOC Method for I_M Acquisition and Analysis

As indicated by the composition of I_M, the acquisition and analysis of I_M revolve around four elements: Strengths (S), Weaknesses (W), Opportunities (O), and Challenges (C). Hence, this method is termed the "SWOC Method." It should be noted that the R-M method emphasizes that I_M acquisition and analysis need to be targeted and centered around I_R, using I_R as its foundation and support. This method shares a similar conceptual approach with the SWOT method which is applied in strategic management and library and information science, yet the meaning of each factor in SWOC differs significantly. Here is a brief introduction to the SWOC method for I_M acquisition and analysis.

The SWOC method for I_M acquisition and analysis targets and centers around I_R, taking I_R as its foundation. This method, both present-focused and future-oriented, involves obtaining key SSI about the primary characteristics of organizational SSM with a comprehensive analysis of SSM. This includes SSI on current SSM characteristics (strengths and weaknesses) and future SSM characteristics (opportunities and challenges). The guiding principle is to maximize SSM strengths and opportunities while minimizing weaknesses and challenges. The theoretical framework encourages endeavors of exploiting strengths, mitigating weaknesses, seizing opportunities, and proactively addressing challenges to formulate organizational SS development strategies. The SWOC method can be visualized using a 2 × 2 matrix, as shown in Table 4.1.

By integrating and matching SSM strengths, weaknesses, opportunities, and challenges, the SWOC method generates four types of organizational SS development strategies: the SO (Strengths-Opportunities) strategy, the WO (Weaknesses-Opportunities) strategy, the ST (Strengths-Threats) strategy, and the WT (Weaknesses-Threats) strategy (Fig. 4.4). Referencing the SWOT method (Wang Zhijin and Ge Linlin 2013; Lan Xiaoxia 2016), these four SS strategies can be briefly explained as follows:

(1) The SO (Strengths-Opportunities) Strategy: This is an enhancement strategy focused on leveraging SSM strengths to capitalize on SSM opportunities, thus enhancing the ability of organizational SSM. It represents an ideal SSM state where an organization maintains its efforts and further enhances its SSM capabilities.
(2) The WO (Weaknesses-Opportunities) Strategy: This is a turnaround strategy aimed at overcoming SSM weaknesses by seizing SSM opportunities, potentially transforming weaknesses into strengths. Internal weaknesses may hinder the utilization of opportunities, so the organization must address its weaknesses to effectively capitalize on the available opportunities.
(3) The SC Strategy: This is a defensive strategy focused on utilizing SSM strengths to mitigate SSM threats. These threats can hinder the effective use of

Table 4.1 SWOC acquisition and analysis matrix

Prospect of SSM	Current Situation of SSM	
	Advantage of SSM (*S*)	Disadvantage of SSM (*W*)
Opportunity of SSM (*O*)	SO (Utilizing *O* to exert *S*)	WO (Overcoming *W* for seize *O*)
Challenge of SSM (*C*)	SC (Responding to *C* by using *S*)	WC (Reducing *W* to respond to *C*)

Fig. 4.4 Four Quadrant Coordinate System for the SWOC Method of I_M Acquisition and Analysis

strengths and even weaken them. The organization should proactively leverage its strengths to address and minimize the negative impact of these threats.

(4) The WC Strategy: This is an avoidance strategy focused on minimizing SSM weaknesses while addressing SSM threats. It is taken when the organization faces a precarious situation where weaknesses and threats combine to pose a serious challenge to its survival. In extreme cases, the organization (such as an enterprise) may consider halting development, suspending operations for rectification, or even closure to minimize losses. This strategy prioritizes damage control and survival.

The core of SSM lies in SSM strategies (content) and SSM behaviors (process). Modern SSM science summarizes SSM strategies as the "4E+C" strategy: SS Engineering (E_1), SS Education (E_2), SS Enforcement (E_3), SS Economics (E_4), and SS Culture (Cs). SSM behaviors is summarized as "FDA" behavior: SS Forecast (F), SS Decision (D), and SS Action (A) (Wang Bing and Wu Chao 2017b). I_M acquisition and analysis are primarily based on these strategies and behaviors. To enhance the operability of the SWOC method and ensure more detailed and comprehensive I_M acquisition and analysis, two dimensions are added apart from the dimension of characteristics: the dimension of SSM strategy and the dimension of SSM behavior. These combined dimensions ("SWOC-4E+C" and "SWOC-FDA") facilitate I_M acquisition and analysis. Building upon Table 4.1, two second-order matrices for I_M acquisition and analysis are formed (Tables 4.2 and 4.3). These tables enable systematic and detailed acquisition and analysis of I_M.

Table 4.2 Acquisition and analysis matrix SWOC-4E+C

Strategy of SSM	Feature of SSM			
	Advantage (S)	Disadvantage (W)	Opportunity (O)	Challenge (C)
SS project (E_1)	E_1S	E_1W	E_1O	E_1C
SS education (E_2)	E_2S	E_2W	E_2O	E_2C
SS enforcement (E_3)	E_3S	E_3W	E_3O	E_3C
SS economy (E_4)	E_4S	E_4W	E_4O	E_4C
SS culture (Cs)	C_SS	C_SW	C_SO	C_SC

Table 4.3 Acquisition and analysis matrix SWOC-FDA

SSM behavior	Feature of SSM			
	Advantage (S)	Disadvantage (W)	Opportunity (O)	Challenge (C)
SS prediction (F)	FS	FW	FO	FC
SS decision-making (D)	DS	DW	DO	FC
SS execution (A)	AS	AW	AO	AC

4.2.4 *Quantification of the R-M Method for SSI Acquisition and Analysis*

Apart from the qualitative analysis of the R-M method in the previous sections, the R-M method can also be analyzed quantitatively, which enables the acquired and analyzed SSI to better inform the selection and formulation of SS strategies. Regarding the formulation of an SS strategy, the acceptable risk (R_A) within I_R has minimal impact, while the unacceptable risk (R_U), or SS threat (T), can significantly influence the formulation, posing a substantial SS challenge (in other words, the "C" in I_M has included R_A). Therefore, the SSI that primarily supports the formulation of an SS strategy is I_M, with I_R serving as its foundation and support. Referencing the quantitative SWOT method (Lan Xiaoxia 2016), the logical framework of the quantitative R-M method is presented in Fig. 4.5. Step 1 belongs to the qualitative aspect of the R-M method. Steps 2 through 6 are further elaborated below, again referencing the quantitative SWOT method (Lan Xiaoxia 2016).

1. Quantitatively assessing the impact strength and occurrence probability of SSM strengths, weaknesses, opportunities, and challenges on SSM, and normalizing the collected data.
2. Calculating the total impact of SSM strengths, weaknesses, opportunities, and challenges on the SS strategy. The theory of strategic management (Wang Changjiang 2013) posits that strategy inherently involves planning and guidance for future actions, emphasizing foresight and predictability. Therefore, when selecting and formulating SS strategies, the considered strengths, weaknesses, opportunities, and challenges should reflect future trends. Their impact on the SS strategy is the mathematical expectation derived from their respective impact strengths and occurrence probabilities. The calculation formula is as follows:

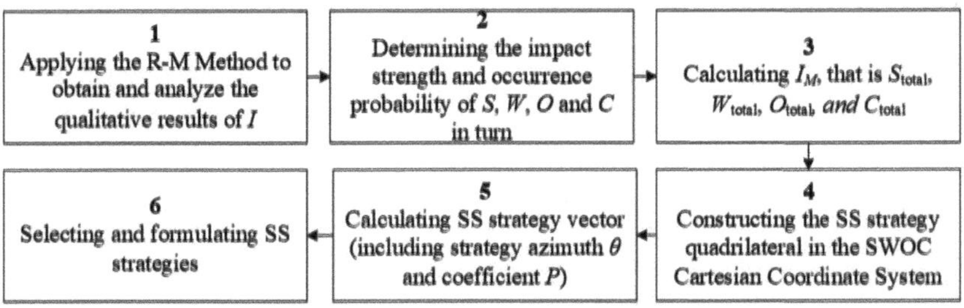

(S_{total}, W_{total}, O_{total}, and C_{total} represent the total impact of SSM advantages, disadvantages, opportunities, and challenges on SS strategies; the meanings of the other letters are the same as in Figure 4-2.)

Fig. 4.5 Logical framework of the quantitative R-M method

$$\begin{cases} S_{\text{total}} = \sum_{a=1}^{m} S_a \times s_a \quad (a = 1, 2, \ldots, m) \\ W_{\text{total}} = \sum_{b=1}^{n} W_b \times w_b \quad (b = 1, 2, \ldots, n) \\ O_{\text{total}} = \sum_{d=1}^{j} O_d \times o_d \quad (d = 1, 2, \ldots, j) \\ C_{\text{total}} = \sum_{h=1}^{k} C_h \times c_h \quad (d = 1, 2, \ldots, k) \end{cases} \tag{4.1}$$

where S_{total}, W_{total}, O_{total} and C_{total} represent the total impact of SSM strengths, weaknesses, opportunities, and challenges on the SS strategy respectively; S_a, W_b, $O_{d,}$, and C_h represent the impact strength of the ath strength, bth weakness, dth opportunity, and hth challenge respectively; s_a, w_b, o_d, and c_h represent the occurrence probabilities of the ath strength, bth weakness, dth opportunity, and hth challenge respectively; m, n, j, and h represent the total number (piece) of strengths, weaknesses, opportunities, and challenges respectively.

3. Constructing the SS Strategy Quadrilateral: Within a Cartesian coordinate system, use S, W, O, and C as semi-axes to create a four-quadrant coordinate system. Plot the calculated values of S_{total}, W_{total}, O_{total}, and C_{total} on their respective axes, and connect these points with line segments to form the SS strategy quadrilateral (Fig. 4.6).
4. According to mathematical knowledge, the relevant SSI information for SS strategy selection and formulation is reflected by two variables: the SS strategy azimuth (θ) and the SS strategy coefficient (ρ). In other words, the SS strategy position can be represented by the vector (θ, ρ), where θ represents the direction and ρ represents the magnitude. θ is calculated as the arctangent of the centroid of the SS strategy quadrilateral. The formula for ρ is:

$$\rho = \frac{S_{\text{total}} \times Q_{\text{total}}}{S_{\text{total}} \times Q_{\text{total}} + W_{\text{total}} \times C_{\text{total}}} \tag{4.2}$$

5. Selecting and Formulating SS Strategies: Plotting the calculated SS strategy vector (θ, ρ) on the SS strategy spectrum diagram (Fig. 4.7). For instance, based

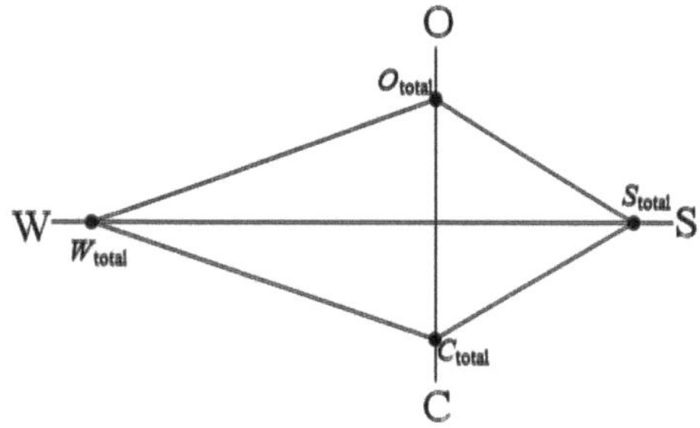

Fig. 4.6 SS strategy quadrilateral

Fig. 4.7 SS strategy spectrum diagram

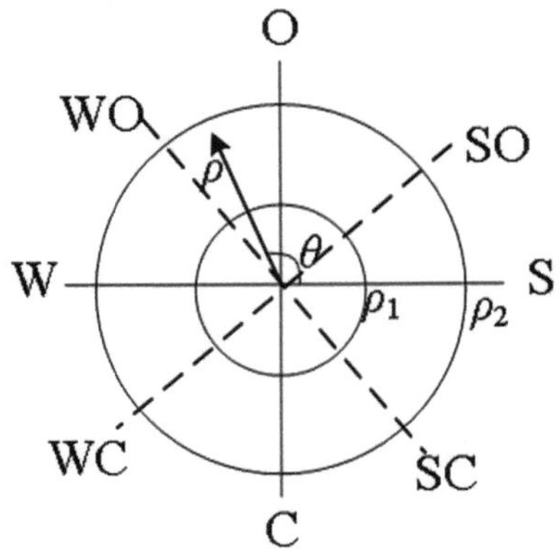

on the vector in Fig. 4.7, a turnaround SS strategy should be selected and formulated, indicating that the opportunities for SSM outweigh the disadvantages. Furthermore, referencing the quantitative SWOT method, two threshold values for ρ (ρ_1 and ρ_2 in Fig. 4.7) can be defined. If $0 < \rho < \rho_1$, it corresponds to a "conservative and stable SS strategy mode." If $\rho_1 < \rho < \rho$, it corresponds to an "aggressive SS strategy mode." The SS strategy vector (θ, ρ) in Fig. 4.7 corresponds to the latter.

4.2.5 Conclusions

This section first examined SSI acquisition and analysis from an SSM perspective. It then proposed the R-M method for SSI acquisition and analysis based on this perspective. The specific implementation of the R-M method was analyzed, followed by a discussion of its quantification. The section demonstrates that the R-M method emphasizes grounding SSI acquisition and analysis in SSM itself. By acquiring and analyzing SSI through the "risk module" and "management module," the R-M method facilitates comprehensive SSI acquisition and analysis while effectively supporting SS strategy selection and formulation. While this section thoroughly discussed the proposal, implementation, and quantification of the R-M method at a theoretical level (i.e., constructing a theoretical framework), it has not probed into the specific elements and modules of the R-M method, nor did it address empirical research, which are to be explored by future research.

4.3 Perception and Analysis Model of SSI

4.3.1 *Basic Elements of SSI Perception and Analysis*

Drawing upon the theory of intelligence perception and analysis, and considering the fundamental needs, together with resources, manpower, methods, and environment of SSI operation, the five major elements of SSI perception and analysis are categorized as: Needs of SS decision-making (Why), SSI sources (What), personnel of SSI perception and analysis (Who), the ecological environment of SSI perception and analysis (Where), and the process of SSI perception and analysis (How). These are referred to as the "4W1H" elements. Figure 4.8 illustrates the relationship between these elements.

1. Needs of SS Decision-Making (Why)

The needs are the primary driver of intelligence (Wu Qiong et al. 2018). From the perspective of macroscopic intelligence, the needs of SS decision-making serve as the compass for all SSI operations, representing a fundamental requirement. Perceiving and analyzing these needs involves two aspects: The first is explicit needs. SSI personnel communicate with all organizational levels to gather SSI needs from SSM and scientific research personnel. Through consultation and efforts of analysis with experts and decision-makers, they clarify the content and timeline of these needs. The second is implicit needs. Traditional SSI needs arise from decision-makers' awareness of SSM deficiencies. However, subjective biases can lead to overlooking hidden flaws of organizational resource allocation. Perceiving and analyzing implicit needs relies on SSI personnel's understanding of system SS and decision-makers. By leveraging their foresight regarding complex system SS issues and strategic considerations, SSI personnel reduce uncertainty and unforeseen challenges stemming from unknown problems. This fulfills SSI's role as the "eyes, ears and the vanguard."

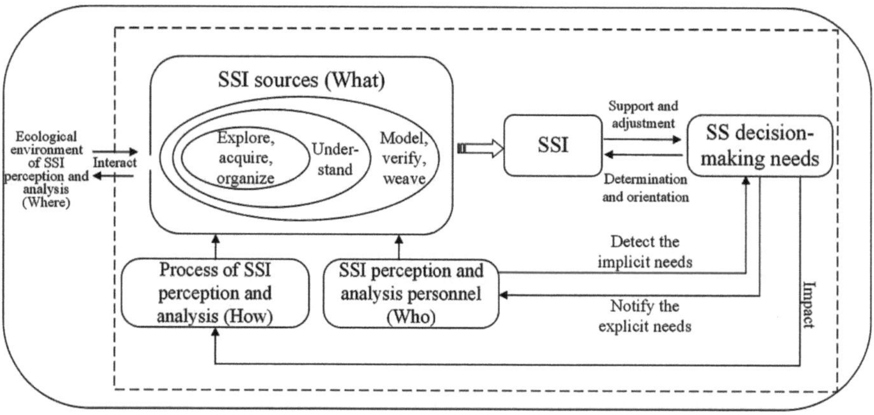

Fig. 4.8 Diagram of Element Relationship "4W1H" of SSI Perception and Analysis

2. SSI Sources (What)

SSI perception and analysis targets diverse resources, including literature, SS evaluation documents from production management, logs, various SS standards, global political, economic, cultural, and SS-related laws and regulations, and scientific technologies. Perceiving and analyzing these SSI sources requires a comprehensive, multifaceted approach to uncover potential obfuscations, reconcile heterogeneous data, discern the context of SSI generation, and develop a preliminary understanding of SSI content. This analysis also involves assessing the extent to which the SSI source has been edited, verifying its authenticity and validity, and identifying key information like location, time, individuals involved, and potential impact. Challenges such as data heterogeneity and content imbalance within these sources pose significant challenges for traditional SSI operation. Therefore, leveraging modern tools of information analysis is becoming increasingly important (Wu Qiong et al. 2018). For example, ontology tools can reveal implicit connections among different fields of knowledge, while techniques of co-occurrence association analysis can cluster and visualize SSI source content (Chen Meihua and Wang Yanfei 2018), ultimately reducing the manual workload associated with SSI analysis.

3. Personnel of SSI Perception and Analysis (Who)

While the demand for SSI provides the driving force, the personnel engaged in SSI perception and analysis are the essential gears that propel the process. Two key abilities are required of the personnel: The first is ability of intelligence perception and processing, which involves data mining, data analysis, model development and application, semantic processing, intelligence description, and exploitation. These are fundamental professional qualities for all intelligence personnel; the second is the basic understanding of SS science and the ability to comprehend the system SS of the organization to which it belongs. Since SSI perception and analysis are always performed in tandem with SSM, perceiving the knowledge of the SS field helps enhance the understanding of decision-makers' needs, reducing errors occurring in organizational collaboration. Furthermore, with the prevalence of computers and artificial intelligence, human–computer interaction will undoubtedly shape the future organizational structure of SSI perception and analysis. Data-driven SSI discovery within the context of big data will become a usual practice.

4. Process of SSI Perception and Analysis (How)

The process of SSI perception and analysis involves two key aspects: overall structure and methodology.

First, the process begins by identifying SSI needs, which then guides the search for relevant SSI sources based on the requirements of SS decision-making. The trajectory of operation is determined by the needs of SSI for searching SSI sources as determined by SS decisions, and different analysis procedures are adopted in accordance with the content, forms and extents required by the SSI. This can be summarized as a workflow model: SSI Needs → SSI Sources → SSI Transmission Process, which determines the model of operational process. Then, traditional methods of SSI perception and analysis include social survey, classification, comprehensive evaluation, induction, and cluster analysis (Xiao Lianjie et al. 2019). Recent technological advancements have enabled the integration of neuroscience, machine learning, and other advanced techniques with SSI analysis. These advancements facilitate data fusion, predictive modeling, and deep learning, overcoming the limitations of manual interpretation and enhancing the ability to process massive amounts of information.

5. Ecological Environment of SSI Perception and Analysis (Where)

The ecological environment for SSI perception and analysis consists of several key elements: the structure of organizational management within which SSI operations are conducted, the prevailing SS culture, resource allocation, assigned SSI tasks, and the quality of SSI personnel. From the perspective of complex systems, building a sound ecological environment for SSI perception and analysis hinges on harmonious coexistence and sustainable development. This environment influences every stage of the SSI process, directly impacting the sensitivity of SSI issues, the accuracy and depth of SSI processing, and the overall quality of SSI products. Therefore, a self-adaptive and continuously improving ecosystem needs to be established with support from several aspects: effective governance of SSI data, innovation in resources and tools for SSI perception and analysis, enhancement of the quality of personnel, and strengthening of organizational communication.

4.3.2 Construction of the SSI Perception and Analysis Model

Constructing a robust model of SS theories requires a clear understanding of its purpose and significance, ensuring a systematic and structured approach (Huang Lang et al. 2016). The mechanism of SSI transformation defines the essence of the SSI perception and analysis model. Needs of SS decision-making, SSI sources, personnel involved in SSI perception and analysis, the process of SSI perception and analysis, and the ecological environment for SSI perception and analysis form the general framework of this model. From the above, the core ideas of constructing an SSI perception and analysis model can be summarized:

1. The workflow of SSI perception and analysis must prioritize SSI transformation. The linear transformation of SS data → SSi → SS knowledge → SSI serves as the general workflow, illustrating the working mechanism of SSI perception and analysis and highlighting diverse methods of SSI transformation.
2. The model of SSI perception and analysis must be applied to real SSM operations, promoting collaboration between modern technologies with SSI operations.
3. The model should be constructed from a systemic and ecological perspective, ensuring a reasonable mechanism of operation for obtaining accurate and intact SSI products.

To refine the model and facilitate further evaluation and optimization, the following aspects related to SSI perception and analysis are considered: completion of a task, granularity of a process, and mechanism of operation.

1. Completion of a task: Three goals of SSI perception and analysis are to be fulfilled upon completion of the task of transforming SSI: reducing uncertainties, proactive foresight (as indicated by a Chinese saying "waking up early to look far"), and predicting future trends. The degree to which these goals are met reflects the model's theoretical effectiveness. By considering the balance between SSI supply and demand, one can estimate the direction and content of SSI perception, the depth and results of SSI analysis, and ultimately evaluate the model's overall performance.
2. Granularity of a process: The model's description of the SSI perception and analysis workflow should be detailed and concise. Each step should be clearly explained and specifically instructed without unnecessary complexity.
3. Mechanism of operation: A reasonable mechanism of operation is essential for a model to function systemically with self-adaptive abilities. As previously discussed, SSI perception and analysis requires a healthy ecological environment, supported by a well-defined mechanism. This includes timely communication of SSI needs, broad-spectrum perception of SSI sources, and bidirectional feedback throughout the process. The mechanism should continuously optimize the operation of SSI perception and analysis and ensure its effectiveness.

Based on these principles, the model of SSI perception and analysis (Fig. 4.9) has been constructed. The model incorporates a systemics and ecological view as its foundational concept, utilizes the SSI transformation chain as its core, employs various elements and their relationships as its framework, and follows the workflow of SSI perception and analysis as its guiding principle.

4.3.3 Interpretation of the Connotation of SSI Perception and Analysis Model

As depicted in Fig. 4.9, this model divides the SSI perception and analysis workflow into six stages: perception of SSI needs, perception of SSI sources and data, SSI perception processing and storage, establishment of SSI indicators, SSI analysis and mining, and SSI evaluation and improvement. The meaning of each stage and the model's mechanism are elucidated in the following content.

4.3.3.1 Perception of SSI Needs

SSI needs serves as a driver in all SSI operations. Primary participants in this stage include SSI personnel, decision-makers, experts and researchers. It is crucial to acknowledge that in the landscape of modern intelligence, a traditional SSI operation

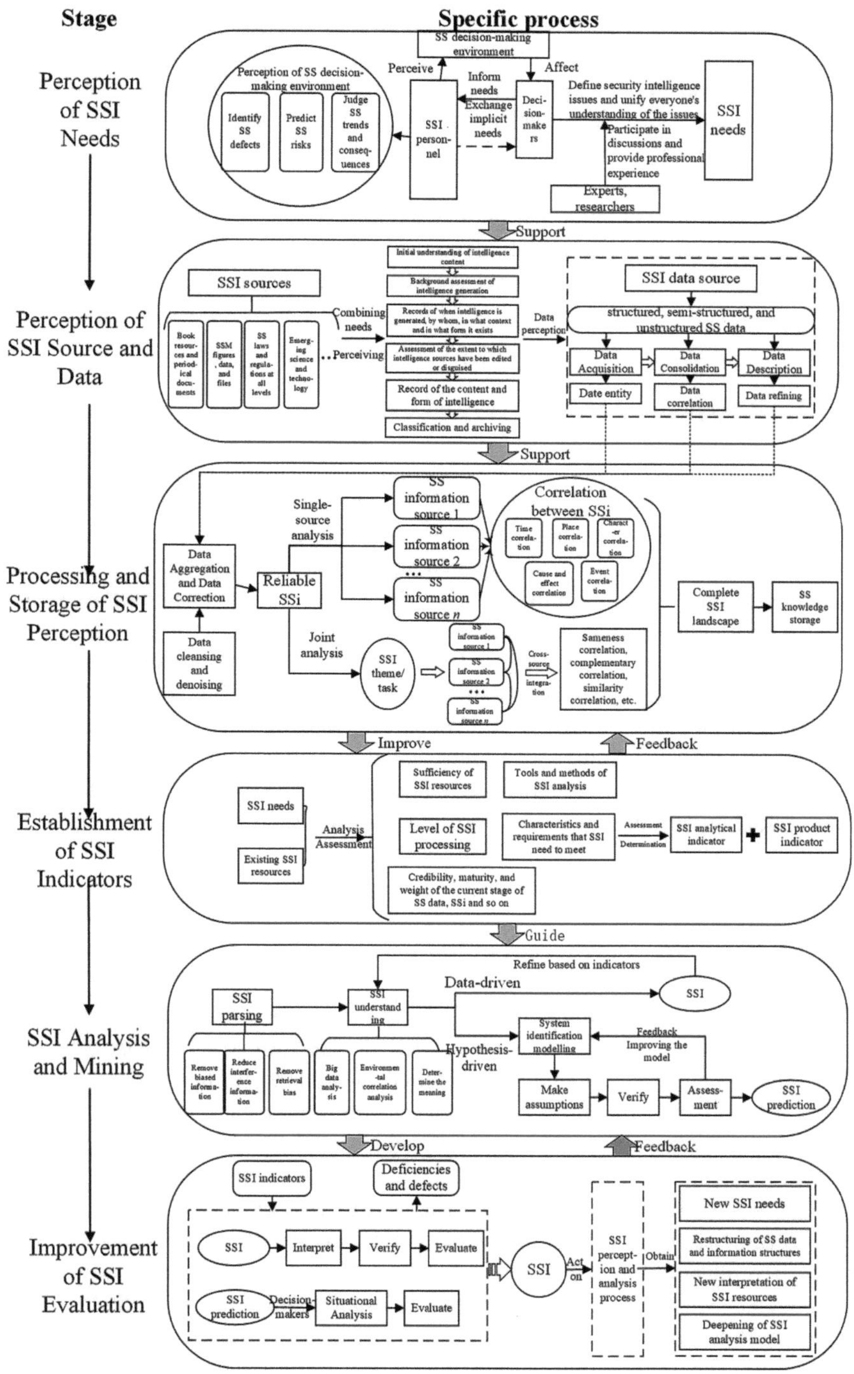

Fig. 4.9 SSI perception and analysis model

may provide limited system SSi, as it is liable to overlook hidden SS risks and SS scenarios for a possible future. Therefore, a proactive, SSI personnel-led mechanism for perceiving both explicit and implicit needs is essential. Beyond addressing explicit task requirements, SSI personnel must analyze the environment for SS decision-making. This includes identifying SS defects, predicting SS risks, assessing SS trends and consequences, gathering relevant policies, regulations, and technical information, and analyzing potential SSI gaps based on existing SSI results and identified problems. By proactively anticipating future needs and recognizing decision-makers' potential requirements, SSI personnel can address the shortcomings of traditional SSI approaches.

4.3.3.2 Perception of SSI Sources and Data

SSI sources are diverse and complex. Searching for relevant sources based on SSI needs relies heavily on the perceptive abilities of SSI personnel. This involves using search engines, SSM systems and specialized databases, and employing targeted keywords, topics or horizon scanning techniques. To mitigate potential search biases stemming from cognitive limitations, advanced perception technologies such as the IARPA's (Intelligence Advanced Research Projects Activity) "SHARP," can be employed. By leveraging insights from cognitive science, human behavior, and neuroscience, SHARP enhances reasoning and problem-solving abilities during SSI collection, broadening the scope of SSI sources (IARPA 2020). Since the forms of these sources vary—including text, numerical data, and images, they cannot be standardized under a unified format. Data perception involves integrating, describing, and duplicating structured, semi-structured, and unstructured SS data, and mining correlations to transform data into SSi.

4.3.3.3 Processing and Storage of SSI Perception

This stage focuses on refining and storing the data processed in the previous stage. Initially, data cleansing and noise reduction are performed to ensure data credibility, then data model calculation is applied in order to further integrate the data for deriving reliable SSi. Information processing in this stage involves mining the connections between different pieces of information. There are single-source analysis and joint analysis in terms of objects. Single-source analysis identifies the relationship within a single information source, such as time, location, task, cause-and-effect, and event correlations, with the results pertaining specifically to the events within that source. To enhance connections between various information sources, joint analysis is conducted, which involves integrating different sources based on SSI tasks and mining various correlations, namely, correlations of being identical, complementary and similar. This process not only helps uncover event patterns and rules but also strengthens the global interconnectedness of SSi, mitigating potential biases from localized SSi. A repository of SS knowledge is thus formed with a comprehensive SSI landscape.

4.3.3.4 Establishment of SSI Indicators

SSI indicators are crucial for quantifying the effectiveness of SSI perception and analysis and evaluating SSI products. By integrating SSI needs and existing SSI resources (e.g., data channels, technologies), the requirements for subsequent SSI analysis and product development are determined. Analyzing the sufficiency of SSI resources, the tools and methods employed, and the current level of SSI processing allows for defining achievable goals for SSI analysis. These goals encompass the hierarchy and diversity of analytical methods, the extent of data fusion, global analysis capabilities, and predictive analysis potential. The credibility, maturity, and weighting of current SSI resources reflect the effectiveness of operation in the previous phases, indicating the starting point and potential challenges for subsequent SSI analysis. The characteristics and requirements of SSI needs dictate the appropriate analysis tools and intelligence exploitation methods, shaping the final SSI product indicators. High-quality SSI products should support SSM efforts, reduce system SS uncertainty, alert decision-makers to potential SS risks, and forecast the future state of system SS. In summary, establishing SSI indicators not only facilitates quantitative evaluation but also provides a mechanism for ensuring the scientific development, continuous feedback, and ongoing improvement of SSI perception and analysis.

4.3.3.5 SSI Analysis and Mining

In general, SSI analysis comprises four key steps: SSI parsing, SSI understanding, SSI analysis, and feedback and improvement. SSI parsing involves using SS data, SSi, and SS knowledge as analytical materials, removing biased information and minimizing interference and errors which are caused by retrieval biases. It is important to note that error elimination at this stage complements, rather than conflicts with, similar efforts in the stages of SSI perception, processing, and storage. Performing cleansing at each stage minimizes errors throughout the entire process.

In terms of SSI analysis and mining, as the name suggests, it utilizes data tools for data analysis and environmental correlation analysis to determine the meaning and significance of SSI materials. Two primary approaches drive SSI analysis in terms of means of production: data-driven SSI production and hypothesis-driven SSI prediction. Data-driven SSI analysis utilizes big data analysis techniques, such as machine learning and intelligent algorithms, employing mathematical and quantitative methods for sampling, measurement, and problem testing. Hypothesis-driven SSI prediction relies on SSI analysts' perception of the future situation, building system identification models for various future SS scenarios. By inputting current SSI materials and continuously testing and refining the model, predictions are generated for different scenarios, providing decision-makers with a basis for proactive judgment. When system instability is high, these scenario-based predictions can help SS managers prevent SS incidents and reduce personnel and economic losses. SSI analysis should integrate both data-driven and hypothesis-driven approaches, utilizing a range of analytical methods, from data processing (e.g., principal component analysis) and prediction (e.g., time series prediction) to relationship analysis (e.g., co-word analysis) and evaluation analysis (e.g., fuzzy comprehensive evaluation), to conduct thorough SSI analysis. Table 4.4 lists common analysis types based on the functional attributes of SSI analytic methods (Fleisher and Bensoussan 2003; Qiu Junping and Wang Yufen 2008; George and William 1997; Jiang Xinyu and Wang Bodi 2014; Wang Jingyi et al. 2020; Tan Xiao and Wu Chensheng 2019). Finally, based on SSI analysis and product indicators, the analysis model is consistently optimized with evaluating efforts, so as to enhance the completeness and effectiveness of SSI products.

4.3.3.6 SSI Evaluation and Improvement

The initial review of SSI products and forecasts during the SSI analysis and mining stage can be considered as an internal audit conducted by SSI personnel. However, the ultimate quality and effectiveness of SSI must be validated by experts, decision-makers, and SSI personnel to ensure alignment with organizational SS needs and avoid being confined by the SSI team's thinking. Using the previously established SSI indicators, SSI is interpreted, verified, and evaluated. Decision-makers analyze the future SS landscape to assess whether the SSI forecast provides insights into potential SS incidents. Finally, any deficiencies or defects in the SSI products are identified and relayed back to SSI personnel for modification and optimization, ensuring alignment with current SSI needs. Importantly, the resulting SSI product, now a new SSI resource, feeds back into the SSI perception and analysis process. This injection of fresh information stimulates new SSI needs, reshapes the structure of SS data and SSi, and provides a new lens for interpreting SS resources. This allows SSI personnel to refine existing models based on the insights gained from mature product concepts. Thus, SSI perception and analysis is a multi-layered, cyclical process of feedback and continuous optimization.

4.3.3.7 Mechanism of SSI Perception and Analysis

A breakdown of the stage of SSI perception and analysis reveals that the need for SSI is the foundation and core of the entire process. The direct transformation of SSI flows from source identification to data processing, transformation, and analysis, guided by SSI perception and analysis. The establishment of SSI indicators serves as a crucial guide for this operation, providing essential criteria for evaluating and inspecting SSI at all levels. The final stage of evaluation and feedback acts as a control gate for SSI output and a critical link for reintegrating SSI into the process. The feedback mechanism involves extensive interaction between sub-processes. For

Table 4.4 Common SSI analysis methods

Types of method	Description of analysis	Specific methods
Analysis of early warning	Identifying and assessing factors that may impact system SS, thereby providing early warnings to decision-makers and allowing for proactive prevention. This method primarily addresses sudden SS event and operates within the initial identification and subsequent early warning phases of emergency management	Weak signal analysis, outlier detection, time series analysis, event correlation analysis, social network analysis, machine learning
Evaluative analysis	Constructing indicators to evaluate and authenticate aspects such as organizational SS, behavioral SS, SS events, as well as competitiveness and influence	Analysis of principal component, fuzzy comprehensive evaluation method, analytic hierarchy process
Prediction analysis	Based on historical records and intelligence resources, using statistical, mathematical, and inferential methods to predict future trends	Bayesian analysis, trend extrapolation, gray prediction method, artificial neural networks, Delphi method, regression analysis
Track analysis	Targeting specific subjects for broad-spectrum, source-based searches, and continuously recording and analyzing changes in the subjects	knowledge gene method, path evolution analysis, thematic evolution analysis, technology road map analysis
Correlation analysis	Uncovering hidden associations between subjects, including temporal, spatial, causal, and co-occurrence relationships to identify correlations within SSI	co-word analysis, gray relational analysis, association rule analysis, co-citation analysis

example, during SSI processing and storage, SSI resources inform the establishment of SSI indicators, while the indicators, in turn, can highlight resource gaps and provide feedback to the preceding stage. The feedback mechanism also operates within individual processes, such as the continuous optimization of the analysis process during SSI analysis and mining. In fact, even defining the SSI needs involves ongoing communication and feedback among various personnel. The final integration of the SSI product as a resource back into the workflow represents another form of feedback. The operational mode of SSI perception and analysis is task-oriented and intelligence-assisted. This is evident in the explicit acceptance of SSI needs, the implicit exploration of needs driven by SSI personnel and the technology of environmental perception, and the human–machine collaboration that combines the strengths of cutting-edge technology and SSI professionals in areas like needs perception and data analysis.

In summary, the SSI perception and analysis model incorporates the theory of intelligence transformation and modern scientific and technological thinking. It details SSI transformation mechanism in perception and analysis, and the feedback and optimization at various levels, focusing on the practical aspects of SSI operation. This section analyzes the mechanism of SSI perception and analysis through a theoretical model. Due to space constraints, a detailed discussion of model application testing is not conducted here. However, it's crucial to emphasize that the effectiveness of SSI perception and analysis depends on the interplay of work resources, personnel expertise, SSI transformation channels, feedback mechanisms, and other factors (which can also be understood as the ecological environment of SSI perception and analysis). Among them, resources of SSI operation include the information channels available to the SSI departments, resource allocation of methods, tools, data systems, hardware, and so on. Channels of SSI transformation refer to the capabilities and tools for various SSI transformation pathways (data-to-intelligence, information-to-data, intelligence-to-data, and information-to-intelligence). Therefore, in addition to theoretical research, application within SSI operations and capacity building is essential for testing and optimizing this model.

4.3.4 *Conclusions*

This section analyzed the elements of SSI perception and analysis, constructing and interpreting a model of this process based on its workflow. The analysis reveals that this model is grounded in SSI production, establishing a working system for SSI perception and analysis from both systemic and ecological perspectives. Integrating practical SSM considerations and modern technology, a task-oriented, intelligence-assisted working mechanism is proposed for building a modern, human–machine collaborative model of SSI perception and analysis. The establishment of the model contributes to the theoretical framework of SSI, providing theoretical and methodological support for practical SSM operations. Future research will involve further investigation and empirical studies to refine and enhance the SSI perception and analysis model.

4.4 Model of SSI Fusion

In the big data era, SS science places greater emphasis than ever before on the value of SSI. Within the framework of a "holistic approach to national security" (Ba Zhichao et al. 2018), SS big data has permeated nearly all SS fields, becoming a key focus of SS research. The focus of SSM informatization has shifted from "SS data of an SS event" to "SS data of all sources" (Wang Bing and Wu Chao 2018b), driving the evolution toward intelligent SSM. In practice, new information technologies like IoT, GIS, cloud computing, and AI are now widely deployed in the SS field, creating integrated SSM platforms characterized by intelligence, interconnectedness and accuracy. This backdrop further amplifies the significance and value of SSI. However, the growth spurt of SS data in this era does not automatically translate into high-quality SS data resources for SSI operations. Instead, it creates challenges of "data overload" and "sparse value." It has become a pressing challenge in both SSI research and practice to extract truly valuable SSI by fusing massive volumes of raw, multi-source SS data, and subsequently transforming that SSI into a foundation for SSM.

The problem of fusing multi-source data is commonly seen in the field of intelligence in the era of big data (Hua Bailin 2013). Research which addresses this issue exists in areas like competitive intelligence, public intelligence, and military intelligence (Sun Lin and Wang Yanzhang 2017; Peng Zhihui 2017a; Xu Huihui and Xu Rongrong 2014). However, within the SSI field, research on SSI fusion remains limited to discussions of crisis intelligence fusion in emergency management (Yuan Li and Yao Leyuan 2014; Wang Jingru and Song Shaocheng 2016). The existing theoretical and applied results fall short of meeting the needs of SSI fusion under the "holistic approach to national security." Furthermore, as a nascent discipline, SSI science lacks a fully developed theoretical framework, with related research remaining in its exploratory phase. Therefore, this section analyzes the concept of SSI fusion and proposes an SSI fusion system for the big data environment. It aims to enrich SSI science research and provide theoretical and methodological guidance for SSI fusion in the big data era.

4.4.1 An Overview of SSI Fusion

1. Concept of SSI Fusion

Intelligence fusion is a method of refining and combining raw intelligence to generate higher level intelligence (Cai Shilin 2019), and a crucial means for intelligence researchers to extract data value. SSI fusion is thus a product of applying principles, methods, and techniques of intelligence fusion to SSI analysis. Based on this understanding, the concept of SSI fusion is analyzed as follows:

(1) SSI fusion is a process of SSI analysis and processing oriented toward organizational SSM. Following the SSi chain of "SS data → SSi → SS knowledge → SSI," SSI fusion performs layered screening, integration, correlation, and serialization of various types and multi-directional SS data from both inside and outside the organization. It extracts valuable components from the SS data, ultimately outputting SSI products in a new logical format, helping SS decision-makers understand the systemic SS state and its dynamics, so as to provide a basis for SS enforcement, prediction, and decision-making.
(2) SSI fusion is guided by the SSI needs of organizational SSM. On the one hand, the plan of SSI source data collection is based on these needs, specifying the scope and objectives of raw SS data collection, thus defining the target of SSI fusion. On the other hand, the content of SSI fusion is determined by SSI needs. Different SSM practices, such as hazard identification, danger control and mitigation, SS resource allocation, and SS environment optimization, have varying requirements for the type, structure, and timeliness of SSI. SSI fusion considers these diverse needs based on the characteristics of organizational SSM to determine the specific methods, content, and goals of the fusion process.
(3) The ultimate goal of SSI fusion is to provide SSI for organizational SSM. Generating SSI requires SSI professionals to extract insights from a large volume of SSi with existing SS knowledge and experience. Therefore, SSI fusion must be grounded in human cognitive processes related to SS. This involves normalizing, validating, and contextualizing SS data in real time to transform it into SSi, then, based on existing SS knowledge or experience, SSI is derived from SSi. In other means, new SS knowledge is firstly generated through SSi correlation analysis and SS knowledge fusion, then SSI is obtained through the ongoing fusion of SS knowledge and dynamic SSi which are being continuously accumulated.
(4) Technologies and methods of information fusion are applied as means of SSI fusion. Considering the characteristics and state of raw SS data resources, SSI fusion involves setting reasonable objectives and selecting appropriate fusion methods. Both traditional techniques of information fusion and emerging comprehensive technologies of information fusion can be employed to achieve SSI fusion. Examples include methods of signal processing and estimation theory, statistical inference, decision theory, geometry, and artificial intelligence (Han Chongzhao et al. 2006).

In summary, SSI fusion is a multi-level process of SSI processing and analysis aimed at generating SSI products. Guided by the SSI needs of organizational SSM and following the SSi chain, it employs technologies and methods of multi-source information fusion to perform raw SS data cleansing, SSi correlation, fusion of SS knowledge, and SSI integration.

2. Significance and characteristics of SSI fusion in the big data environment

Cleansing and integrating SS big data resources rapidly and accurately, extracting valuable components and acquiring SSI are the core tasks as well as the most significant components of SSI fusion in the big data environment. The significance and characteristics of SSI fusion in this context can be fully understood by considering the state of an organization's SSI resources and its SSI needs.

The concept of SS big data analysis emphasizes full-sample data analysis (Ouyang Qiumei et al. 2016), advocating the use of all-source SS data to understand the overall context and development trends of SS issues. However, SS organizations often have complex structures, with resources like personnel, materials, and finances intertwined and dispersed. This leads to a fragmented distribution of SS big data resources across different domains and structures. Furthermore, full-sample data analysis implies a higher tolerance for data imperfections, because raw SS data often contain significant noise and interference, making them difficult to use directly for SSM. The raw quality of SSI resources resulting from the surge in SS big data application necessitates SSI fusion for cleansing and integration. This process eliminates irrelevant data from the raw SS data and incorporates multi-source heterogeneous SS data into a unified analytical framework, providing a solid SS data foundation for SSI analysis.

SSM in the big data environment requires comprehensive, logically clear, meaningful and timely SSI. SSI fusion utilizes diverse channels, multiple collection methods, and various sensors to acquire SS data, significantly expanding the temporal and spatial scope of SSI sources. In particular, the combined SS data obtained from multiple sensors can effectively reduce ambiguity, improve credibility, and enhance the spatial resolution of SS data (Wang et al. 2018). Through correlation analysis between SSi and SS knowledge, the relationships between various elements within the SS system can be explored in a more comprehensive manner, which helps clarify the trends of SS state changes. This facilitates the generation of an SS knowledge association network, ensuring the timeliness of SSI and providing robust SSI support for SSM.

4.4.2 An Overall System of SSI Fusion in the Big Data Environment

SSI fusion in the big data context must be based on the SSI needs and the current state of SSI resources within organizational SSM. It requires the integrated application of SS science and technology, intelligence fusion technology, and big data analysis to progressively elevate low-level SS data to high-level SSI. Building upon the existing SSI system structure (Wang Bing, Wu Chao 2019 h), an overall system is built for SSI fusion in the big data environment, as illustrated in Fig. 4.10.

From the perspective of realizing SSI fusion, the system can be described by five types of "fusion," as detailed below.

1. Fusion of SSI Needs and the SSI System

The fusion of SSI needs and the SSI system is reflected in the following three aspects:

(1) The SSI collection fusion system adheres to the plan of SSI source data collection, which encompasses the identification of SSI sources, the collection of SSI source

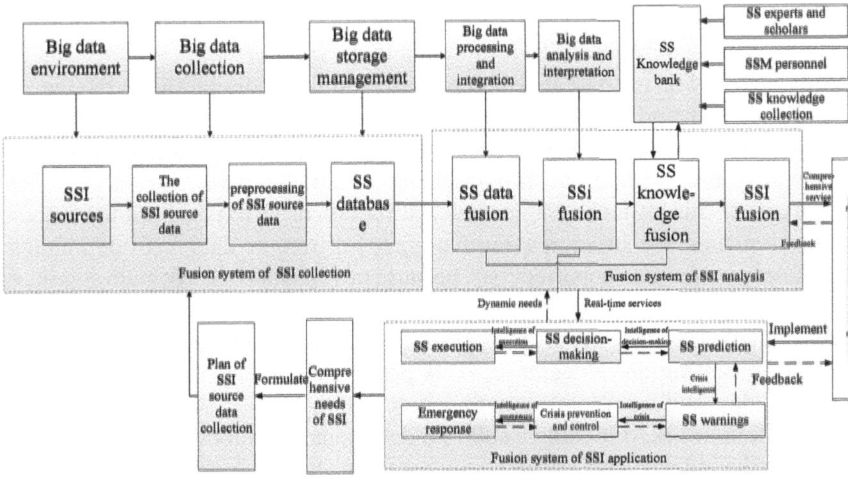

Fig. 4.10 Overall system of SSI fusion in the big data environment

data, and the preprocessing of SSI source data. This plan is based on SSI needs and practical constraints; (2) The effectiveness of SSI application is reported to SSI users (organization members or the SSM division) through a tiered feedback mechanism. Users conduct comprehensive effectiveness evaluations and provide feedback in conjunction with new SSI needs, prompting SSI professionals to optimize the SSI fusion process and methods; (3) Due to the real-time and dynamic nature of the organization's SS state and SS environment, the SSI achievement system cannot always encompass all real-time SSI. The SSI analysis fusion system addresses this issue by dynamically providing real-time SSI services to meet immediate needs at all levels. For example, crisis prevention and control personnel can request crisis data at any time and receive real-time SS data on changes in a state of crisis to implement timely preventative and control measures. Similarly, emergency management personnel can access the latest SS data for a given event to quickly initiate emergency responses and execute rescue operations.

2. Fusion of SSI Analysis and Big Data Technology

The fusion of SSI analysis and big data technology includes: The construction of a comprehensive network of SSI sources as facilitated by the big data environment expands the scale of SSI source data collection and eliminates blind spots in SS data; The technology of big data collection is crucial for gathering SSI source data; Experiences from big data applications in other fields provide valuable insights for selecting appropriate collection technologies, providing rich and valuable experience for the collection of SSI source data; The technology of big data storage and management is essential for building and maintaining SS databases; The technology of big data processing provides solutions for SS data cleansing, integration, and reduction; The technology of big data analysis provides a technical platform for generating SSi by adding contextual interpretation and performing correlation analysis on SS data.

3. Fusion of Multi-Level SSI Analyses Under the SSi Chain

Fusion of multi-level SSI analyses under the SSi chain involves: Cleansing, integrating, and consolidating SS data to eliminate conflicts, redundancies, and heterogeneities within the SSI source data. This process incorporates SS data into a unified analytical framework; Collaboratively using multi-source and multi-format SS data to create a comprehensive and objective contextual representation of the target SS state; Leveraging existing SS knowledge and experience to conduct comprehensive, comparative, and correlational analyses of SSi. This process discovers SSI and generates new SS knowledge through the fusion of SSi and existing SS knowledge; Integrating SSI fragments and performing correlation and serialization to generate a cohesive SSI achievement system.

4. Fusion of SSI and SSM

As an embodiment of SSI productions, this fusion is primarily manifested by content of the system of SSI application fusion. It focuses on using SSI in both normal SSM and contingency management (anomalous SSM). This system delivers SSI products to SSI users (organization members or SSM agencies) through various formats, including special SS reports, SS risk analyses, and reports on internal SS state (Wang Bing and Wu Chao 2019 h). Based on the SSI provided, users grasp the organization's SS state and development trends, enabling them to execute SS decisions and SSM actions. For acceptable risks, SSI users implement normal SSM measures, making SS decisions based on the results of SS prediction (intelligence on SS decision-making) and taking SS actions in accordance with these decisions (intelligence on SS execution). This process aims to reduce, eliminate, or control organizational SS risks and maintain or improve the organization's SS state. For unacceptable risks and accidents, SSI users implement measures of contingency management, which involves issuing SS warnings and implementing crisis prevention and control based on SS prediction results (crisis intelligence) to avoid or forcibly eliminate unacceptable risks. If prevention and control fail or an accident occurs (emergency intelligence), the procedure of contingency response is activated, initiating the stage of contingency response.

5. Fusion of Human Wisdom and Machine Computation (SSI Fusion)

On the one hand, the existing SS knowledge and wisdom, is essential for extracting SSI from raw SSi and generating new security insights. A base of SS knowledge, built upon the combined expertise of SS experts, scholars, and the practical experience of SSM personnel, along with publicly available SS research and internal organizational reports, provides the foundation for this transformation. On the other hand, this human-driven process is complemented by machine computation, which produces refined SSI (SSI fusion). This refined SSI is then utilized by SSI users to gain SS awareness, make predictions, and form judgments regarding the organization's SS state. These users, acting as decision-makers and implementers of SSM measures. Their cognition and utilization of SSI are the ultimate test of its effectiveness.

4.4.3 *Core Subsystem of SSI Fusion in a Big Data Environment*

The primary goal of SSI fusion is the generation of SSI. This process is facilitated by three core systems: the system of the fusion of SSI collection, the system of the fusion of SSI analysis, and the system of the fusion of SSI application. These subsystems work in concert to "produce" SSI, forming the core of the overall SSI fusion system. The following sections will detail the models, content, and functions of each of these subsystems.

1. System of the fusion of SSI collection

This system serves as the terminal for inputting raw SS data in the SSI fusion process. Driven by the organization's SSM needs and the plan of SSI source data collection, this subsystem identifies relevant SSI sources and collects the necessary data. The scope and speed of data collection directly impact the quality and efficiency of the resulting SSI. Therefore, this subsystem forms the foundation of the entire SSI fusion system and plays a critical role in its success. The subsystem model of the fusion of SSI collection in the big data environment is shown in Fig. 4.11.

Three are four aspects concerning the content of this systems: (1) Determining the SSI source according to the plan of SSI source data collection. SSI sources generally include internal sources within the organization (such as SSM data, SS documents, SS reports, etc.) and external sources (such as literature databases, various websites, national or industry SS laws and regulations, etc.); (2) Facilitating data sharing and business collaboration within the organization in the big data environment, creating a robust SS big data resource environment to support the collection of SSI source data; (3) Selecting and utilizing appropriate data collection techniques to accurately and efficiently gather raw SS data as required by the plan; (4) Decrypting, verifying, and registering the collected raw SS data. This data can be stored temporarily in

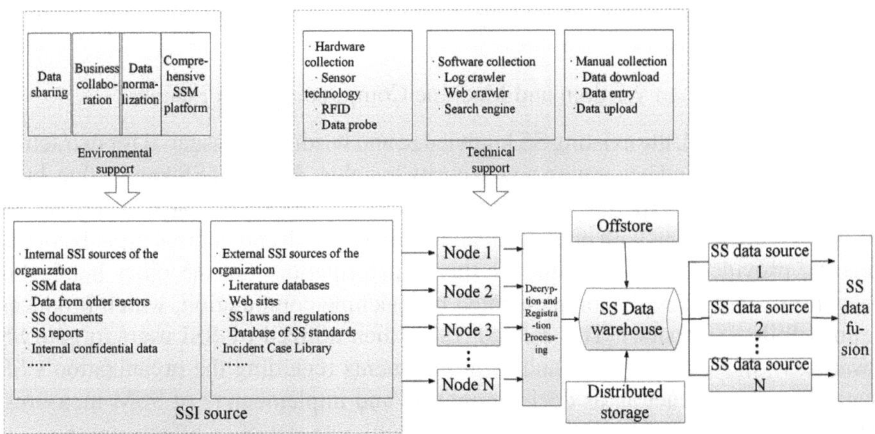

Fig. 4.11 System model of the fusion of SSI collection in the big data environment

the SS database via the data bus, and then stored online or offline for backup using HDFS (Hadoop Distributed File System), based on the data source. This process yields preliminarily classified and full-sample raw SS data, laying the foundation for subsequent SSI fusion.

2. System of the fusion of SSI analysis

This system is the core of the overall system of SSI fusion—the "production workshop" of SSI. In the big data environment, the fusion of SSI analysis leverages existing SS knowledge and experience with the technological support of big data processing and analysis. It follows a hierarchical fusion process: fusion of SS data, fusion of SSi, fusion of SS knowledge, and finally, fusion of SSI. The multi-source SS data is gradually transformed into SSI in the process. The system model of the fusion of SSI analysis in the big data environment is shown in Fig. 4.12.

The fusion of SSI analysis in the big data environment is a multi-level, multi-directional process. Its main content and functions include: (1) Adding value to SS data by improving its quality. This involves reducing the volume of data from multiple SS data sources, eliminating irrelevant or weakly related data categories, and cleansing and serializing SS data according to predefined SS data cleansing rules; (2) Contextualizing SS data and enabling real-time SSi services. This requires

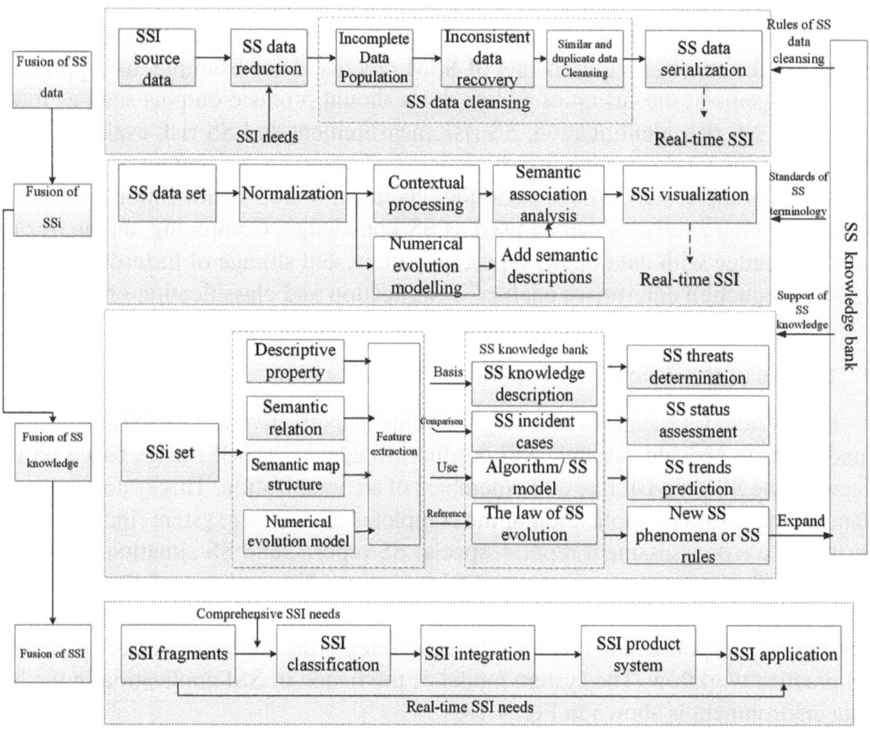

Fig. 4.12 System model of the fusion of SSI analysis in the big data environment

normalizing names and descriptions within the SS data set on the basis of the standards of SS terminology. SS data is then transformed into SSi, and SSi is associated through contextual processing, semantic enrichment, and semantic linking. Visualization techniques are employed to enable dynamic SSi display as the last step; (3) Generating SSI and discovering new SS knowledge. Utilizing the semantic descriptions and numerical evolution models of various SSi as features, SS knowledge is fused to identify SS threats, assess the SS state, and predict SS trends (i.e., obtain SSI fragments). The SS knowledge base is thus expanded with new SS phenomena or SS rules inferred from relationships between SSi and SS knowledge; (4) Combining SSI fragments to generate a cohesive SSI achievement system. SSI fragments are classified and integrated to produce SSI products (such as thematic SS reports and SS situation analyses) for SSI applications for meeting comprehensive SSI needs.

In the process of implementation, the fusion of SSI analysis supports the application of SSI in a targeted manner based on the content and characteristics of SSM practices. This is reflected in: (1) The degree of analysis fusion is chosen based on the requirement of timeliness of SSM practices. For instance, during an emergency rescue, only basic cleansing and integration are needed for essential SS data like event location, type, number of victims, and resource availability to become valuable SSI, meeting the immediate needs of rescue personnel. Conversely, SSM practices like SS education and system management which prioritize comprehensive, long-term benefits necessitate full utilization of SS knowledge, requiring fusion of at the levels of SS knowledge and SSI; (2) The goals and content of analysis fusion are determined by the specific requirements of SSM practices. For example, to support SS risk management, the fusion of SSI analysis should produce outputs such as intelligence of SS risk identification, SS risk measurement and SS risk evaluation; (3) The objects of SS knowledge fusion are selected based on the content of SSM practices. For example, standards of hazardous chemical source identification and known hazardous characteristics can be used as SS knowledge. Comparing and analyzing this knowledge with data on the types, quantities, and storage of hazardous chemicals at production enterprises enables identification and classification of hazardous chemical sources within those enterprises.

3. System of the fusion of SSI application

This system is crucial for realizing the value and effectiveness of SSI. Its primary function is to provide dynamic SSI products to appropriate SSI users based on the needs of the SSM division or other members of an organization. This system not only handles the fusion of application of the complete SSI product system (including SSI results like risk assessment reports, special SS reports, and SS situation analyses), but also efficiently delivers real-time SSI fragments. The real-time delivery is especially important for time-sensitive SS practices like crisis prevention and control, contingency response, and risk management, integrating SSI results directly into the SS practice workflow. The system model of the fusion of SSI application in the big data environment is shown in Fig. 4.13.

The content of the system includes:

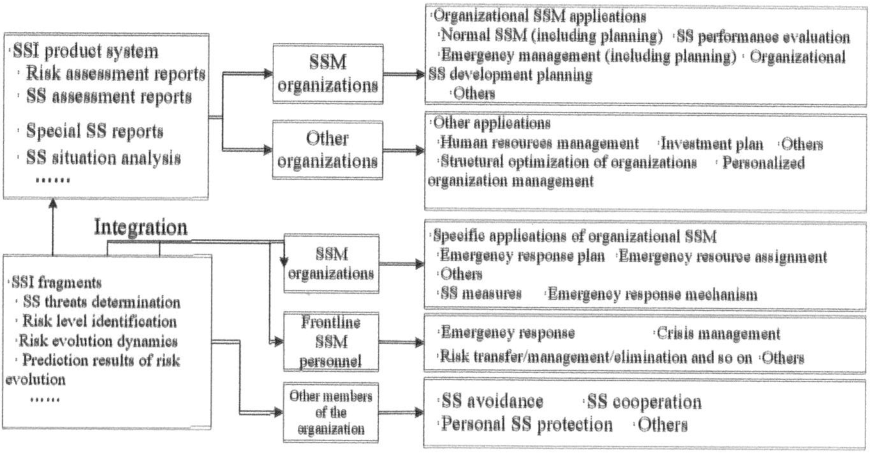

Fig. 4.13 System model of the fusion of SSI application in big data environment

(1) The integration between the system of SSI products and SSM practices of an organization primarily involves applying SSI products to comprehensive, long-term SSM practices; (2) The integration between the system of SSI products and other departments of an organization refers to providing SSI to those departments to help them understand the organization's SS state accurately, so as to promote optimization of the SS environment and improve SSM efficiency; (3) The integration between SSI fragments and SSM practices of an organization refers to utilizing real-time SSI fragments strongly correlated with hazards, hidden dangers, and risks to enable the SSM division of the organization to respond swiftly to crises and take preventative measures; (4) Real-time SSI fragments are utilized to enhance the efficiency of front-line SSM personnel, enabling quick identification of SS threats, understanding trends of risk development, and timely implementation of SS measures; (5) Real-time SSI fragments are used to improve the SS awareness of other members of the organization, helping them understand organizational risks and SS threats. This encourages active participation in SS avoidance, personal SS protection, and cooperation with SSM efforts.

4.4.4 Conclusions

SSI analysis is crucial for realizing the value of SS data, and SSI fusion is a key method for extracting that value. This section focuses on organizational SSM. It clarifies the basic connotation of SSI fusion and analyzes its significance and characteristics in the big data era, then considers the current state of SS data resources in the big data environment and needs of organizational SSM for intelligence to construct a comprehensive system of SSI fusion, and constructs and analyzes the core branches of this system. The section demonstrates that SSI fusion is an SSI analytic method that enhances the value density of SS data and extracts its value from multiple aspects and levels. The system of SSI fusion in the big data environment can be summarized as five types of "fusion": Fusion of SSI needs and SSI systems; fusion of SSI analysis and big data technology; fusion of SSI analysis; fusion of SSI application and fusion of human and machine. The section aims to theoretically realize SSI fusion in organizational SSM and construct an overall model of its functions, providing theoretical guidance for practical application. Further refinement of the research is needed, to explore the application of SSI fusion by examining the patterns and characteristics of various organizational SSM practices. Future research should also consider how methods of information fusion can be applied to realize SSI fusion in organizational SSM and how the technology of big data analysis can be specifically combined with SSI fusion.

4.5 Foundational Theories of SSI Literacy

This section aims to introduce the concept of SSI literacy, which is built upon the concept of SSi literacy introduced by Wang Bing and Wu Chao (2018e), which highlights its importance in SSM within the big data era, and the understanding of intelligence literacy as a higher-order form of literacy built upon information literacy (Qin Dianqi and Zhang Yuwei 2015). SSI literacy is posited as a valuable research area at the intersection of SS science and intelligence studies. SSI, the core element of SSI theory, represents the professional knowledge required by SS personnel engaged in SSI operations. It is expanded upon traditional SSi literacy, encompassing the qualities and ethical considerations necessary to effectively manage and utilize SSI for SSM purposes within the SSI field.

This section establishes the concept of SSI literacy from the theoretical level, clarifying fundamental theoretical issues to advance SSI theory and establish a new research direction in SSI literacy. This contributes to enriching the research content of intelligence literacy, fostering interdisciplinary development in research on intelligence literacy, and laying a solid theoretical foundation for SSI literacy research and education.

4.5.1 An Overview of SSI Literacy

1. Defining SSI literacy

The concept of SSI literacy derives from applying the broader concept of intelligence literacy to the specific domain of SS science. Conceptually, SSI literacy represents a synthesis of SS literacy (Wang Bing and Wu Chao 2018e) and intelligence literacy, and can be considered as a higher-order form of literacy built upon

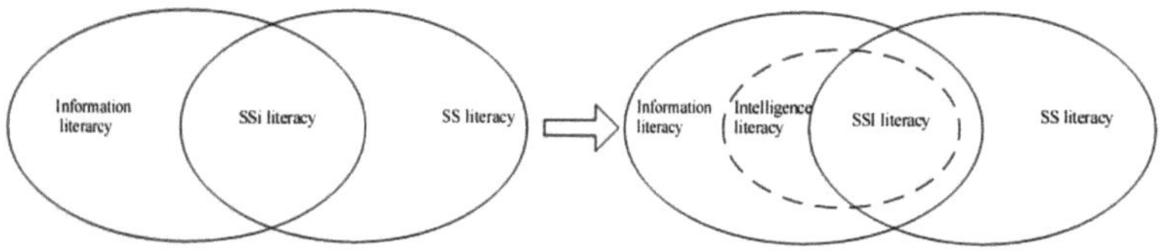

Fig. 4.14 Logic diagram of the definition of SSI literacy

the foundation of SSi literacy (Wang Bing and Wu Chao 2018e) (Fig. 4.14). Qin Dianqi and Zhang Yuwei (2015) define intelligence literacy as the knowledge, skills, and cultural cultivation related to intelligence acquisition and utilization, necessary for intelligence professionals to adapt to the demands and evolving circumstances of intelligence activities (e.g., the big data environment). They position intelligence literacy as a more critical, higher-order form of literacy built upon information literacy. Wang Bing and Wu Chao (2018e), after clarifying SS literacy and information literacy (Qin Dianqi 2012; Qin Dianqi and Zhang Yuwei 2017; Bernheim 1992), define SSi literacy as the set of abilities which enable SSi users to recognize SSi needs, acquire, evaluate, and utilize necessary SSi to make informed SS decisions.

Based on the definitions of SSi literacy and intelligence literacy provided by the above mentioned scholars, SSI literacy can be defined as follows: SS personnel (SSI users) must possess the knowledge, abilities, and cultural understanding related to SSI and its acquisition and utilization to effectively adapt to SS activities (such as SS prediction, SS decision-making, and SS execution in SSM) and the changing scenarios of those activities (such as the environment of complex and massive big data). The logical relationship between these concepts is illustrated in Fig. 4.14. Given the broadened definition and scope of SS within the context of integrated SS research, SSI literacy refers to the ability of SSI users in any industry or field (e.g., production, engineering, national security, social services, resource management, culture) to collect, analyze, and utilize intelligence related to SS issues.

2. Differences between SSI literacy and SSi literacy

Understanding the definition and connotation of SSI literacy requires distinguishing it from SSi literacy. This begins with understanding the difference between SSI and SSi. Japanese scholar Kenichi Shiraiwa defines intelligence as "a meaningful expression carefully processed from materials or data that meet a certain purpose after thorough evaluation, identification, and selection." Therefore, SSI is the SSi that has been further filtered, selected, processed, and potentially utilized. Its essence remains a form of information. SSi contains SSI, yet not all SSi is SSI. The "information chain" illustrates the source and evolution of intelligence: Fact → Data → Information → Knowledge → Intelligence (smart) (Zhang Jianian and Ma Feicheng 2018). The key difference is that SSI is more purposeful, service-oriented, and focused. It is highly condensed SSi with distinct characteristics of "intelligence." As previously mentioned, SSI is the information that affects SSM. This definition positions SSI as a higher level concept within SSi, with its value and utility derived from its influence on SSM, aligning with the increasingly generalized concept of SS.

Understanding the distinction between SSI and SSi clarifies that SSI literacy is the ability to refine SSI from SSi to in order to serve SSM. SSI literacy builds upon and surpasses SSi literacy, representing a higher level manifestation. Figure 4.14 illustrates this, showing that SSi literacy forms a part of SSI literacy, but SSI literacy stands for broader requirements.

The same term can have different meanings in different contexts. SSI literacy differs from "intelligence literacy" or "information literacy" in business, library and information science, and other sectors because the scope of SS science is vast. The "SS" in "SSI literacy" is universal, spanning various SS fields, including those within other sectors like business. However, the "intelligence" in SSI literacy is specific to SSI, focusing on SS issues rather than development or other concerns. Non-traditional SS issues permeate nearly all aspects of society. SSI literacy can therefore be understood as the ability and professional literacy of SS personnel to process diverse SSi resources, condense and extract SSI to serve various SS activities, and effectively address SS problems.

4.5.2 Basic Elements and System Construction of SSI Literacy

1. Basic Elements of SSI Literacy

Based on the constituent elements of intelligence literacy (Qin Dianqi and Zhang Yuwei 2015; Tan Bo 2017) and the basic elements of SSi literacy, the SSI literacy of SS personnel can be considered to cover the following six basic elements:

(1) Possessing Adequate SSI Theories

Practices is guided by theories, i.e., theories serve as the foundation for practical activities. The theoretical knowledge of SSI includes: (1) Understanding the connotations and basic theories of concepts like SSI and SSi, and fundamental theories of SS science; (2) Mastering the content and structure of relevant SSI, including the truthfulness and effectiveness of its core, value and influence, and the suitability and appropriateness of its carrier or format.

(2) Possessing Proper SSI Awareness

SSI awareness is a crucial element of SSI literacy. It refers to the attention, keen perception, and deep insight maintained toward SSI, as well as an inherent awareness of and need for SSI. This includes awareness of SS situations and trends, leading to more purposeful SSI collection. SSI awareness is decisive in SS activities like SSM. For example, attention paid to sensitive events and critical SS issues and the tracking of them directly impact strategic decision-making and the achievement of organizational SS goals throughout the life cycle of an SSI activity. As an element of SSI literacy, SSI awareness aims to leverage the subjective initiative of SS practitioners. In the current era of SSi management, SSI awareness plays a crucial role of early warning.

(3) Possessing Abilities of SSI Acquisition

Abilities of SSI Acquisition refers to: (1) Identifying SSI resources (sources), which covers various SSi resources across different fields and industries, such as SS big data and real-time SSi; (2) Skillfully utilizing intelligence technologies, understanding the characteristics, applicability, and use of SSI acquisition tools; (3) Employing effective SSI retrieval skills, and selecting optimal research methods and retrieval systems to acquire necessary SSI; (4) Demonstrating the ability to collect SSI, and understanding the channels for acquiring SSI and the characteristics of SSI sources (such as SSI networks).

(4) Possessing Abilities of SSI Analysis

Abilities of SSI analysis includes: (1) The ability to generalize SSI, that is, to classify, organize, and induct various types of SSI from diverse sources and with varying content; (2) The ability to integrate SSI, that is, to master basic information technologies like data format conversion and data mining to filter, analyze, and summarize acquired SSI; (3) The ability to transform SSI, or effectively absorb and appropriately modify acquired SSI.

(5) Possessing Abilities of SSI Utilization

Abilities of SSI utilization encompass the capacities of SS personnel (SSI users) to address SS challenges effectively with the acquired and analyzed SSI. This includes both the capacity to use SSI for problem-solving and to share SSI with other users, facilitating collaborative solutions to SS behavior issues. Obviously, SSI utilization is an ultimate key for realizing the full value of SSI—a manifestation of SSI literacy. Therefore, a robust foundation in SSI theory, awareness, acquisition, and analysis skills is essential for effective SSI utilization.

(6) Adhering to SSI Ethical Principles

SSI ethics, a crucial cultural element of SSI literacy, is shaped by the prevailing SS culture and its influence. The established intelligence culture further informs and strengthens these ethical principles. Integrating both SS ethics (Liu Xing 2007) and intelligence ethics (Qin Dianqi and Zhang Yuwei 2015), SSI ethics dictates that all SSI activities must adhere to relevant SS and intelligence policies, laws, and regulations. Key ethical considerations include: (1) Ensuring the legality and appropriateness of all SSI activities; (2) Assessing the potential adverse effects of disseminating SSI and considering the implications for the recipients; (3) Maintaining a strong sense of SS and privacy regarding SSI, protecting personal SSI spaces, carriers, and resources from threats, infringements, and misinformation, and ensuring privacy and SSs during transmission.

2. Theoretical Model of the SSI Literacy System

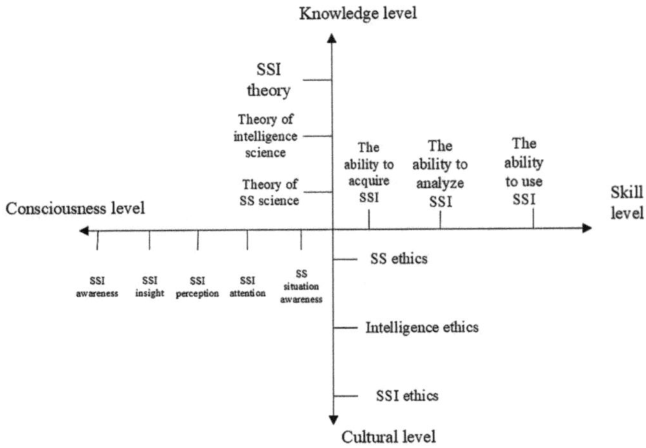

Fig. 4.15 Theoretical model of SSI literacy system

Based on the six fundamental elements of SSI literacy, a theoretical model of the SSI literacy system can be constructed (Fig. 4.15). This model comprises four levels: consciousness, culture, knowledge, and skill. The level of consciousness covers SSI awareness and other related items and runs through all stages of SSI activities; the level of culture includes SSI ethics and other related content, which is fusion of SS ethics and intelligence ethics, representing a unique value orientation and judgment system shaped by cultural influence and dissemination; the level of knowledge covers theoretical knowledge of SSI, intelligence studies, and SS science; the level of skill includes SSI acquisition, analysis, and utilization skills. Each element is comprised of several sub-elements (Fig. 4.16), part of which are listed here. It is believed that SS science researchers and intelligence scholars will participate in discoveries and discussions on more sub-elements through collaboration.

1. SSI theory forms the bedrock of the SSI literacy system. A solid theoretical understanding in conducive to fostering SSI awareness, which in turn drives the acquisition of SSI. Therefore, both SSI awareness and the ability to acquire SSI are built upon the foundation of the SSI theory. Since intelligence is defined as "a meaningful expression carefully processed from materials or data that meet a certain purpose, through evaluation, discrimination, and selection," the ability to analyze acquired SSI is essential for discerning and filtering valuable information. Just as the Chinese saying goes, "even a clever woman cannot cook without rice," the ability to acquire SSI is a prerequisite for possessing the ability to analyze it, and the latter is also a key element of the SSI literacy system. Highly analyzed SSI loses its value if not applied, for it ceases to be actionable intelligence and cannot contribute to SSM. The key distinction between SSI and SSi lies in SSI's stronger and more explicit focus on purpose and application. The ability to utilize SSI is therefore a critical component of SSI literacy, which enables the true value of SSI to be realized through application.

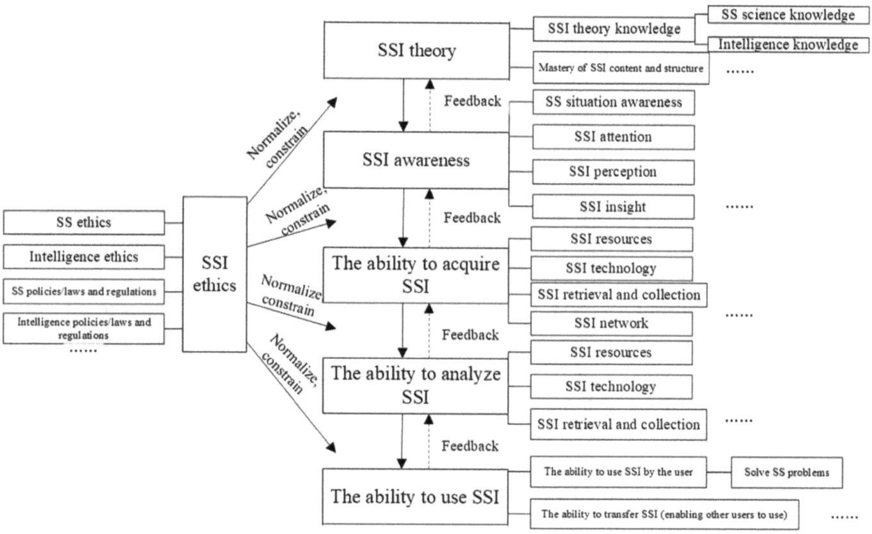

Fig. 4.16 System model of SSI literacy elements

2. In utilizing SSI, the effectiveness of SSI application can be reported back to the previous stage, contributing to the enhancement of the ability of SSI analysis. Similarly, feedback loops extend to earlier stages, creating a continuous improvement cycle. The bidirectional arrows connecting the five elements—SSI theory, SSI awareness, skills of SSI acquisition, skills of SSI analysis, and skills of SSI utilization—represent this iterative refinement and progressive enhancement. These five elements together form the core of the SSI literacy system.
3. The five elements—SSI theory, awareness, acquisition, analysis, and utilization—engage in a continuous cycle of mutual enhancement. SSI ethics plays a crucial governing role in this process, providing a framework for responsible practices. While these elements work together to improve the overall SSI literacy, it is important to adhere to ethical principles. Only through this "ethical lens" can these six fundamental elements form a robust and effective SSI literacy system.

4.5.3 *SSI Literacy Education*

Enhancing and cultivating SSI literacy is most effectively achieved through dedicated education. This section briefly outlines the necessity and basic strategies for implementing SSI literacy education.

1. Necessity of SSI Literacy Education

(1) SSI represents a user's understanding of and response to changes in the external environment, serving as the foundation and guide for developing strategies and

SS measures (Zhang Xiaojun 2017). Research on SS strategies hinges on the ability to collect, analyze, and utilize SSI to formulate strategic thinking and objectives, yet a deficiency in SSI literacy directly hinders the process of formulation. In the current landscape, marked by both traditional and non-traditional SS issues, various domains, including homeland security, public safety, production security, emergency management, and counter-terrorism, urgently require the safeguards, support, and services provided through effective SSI operations (Zhang Jianian and Ma Feicheng 2018). Therefore, SSI literacy is an essential quality for all types of SS activities.

(2) The development of SSI literacy education is crucial: Advancements such as cloud computing, big data, IoT, mobile internet, and AI have profoundly impacted all aspects of the social and ecological structure. While these technologies facilitate SSI operations, they also significantly disrupt existing SSI operation methods, processes, systems and mechanisms, which means that SSI users are required of a higher standard of SSI literacy. To keep pace with these advancements, SSI literacy education is both indispensable and imperative.

(3) Deficiencies in traditional SS education can be supplemented by SSI literacy education: SSi literacy education has been recognized for its ability to highlight the essence of SS education (Wang Bing and Wu Chao 2018e). As a higher level manifestation built upon SSi literacy, SSI literacy education can further address the gaps and shortcomings of traditional approaches. SSI itself doesn't directly solve SS problems; rather, it provides the "knowledge" necessary for solutions (Wang Bing and Wu Chao 2018e). SSI literacy reflects the ability to apply this knowledge effectively. While traditional security education often focuses on "giving a man a fish," SSI literacy education emphasizes "teaching a man to fish," representing a valuable evolution of the traditional idea.

2. Basic Strategies of SSI Literacy Education

SSI literacy education should prioritize the following five macro-level strategies:

(1) Formally Integrating SSI Literacy Education into SS Education

As a critical component of professional SS education, SSI literacy education should be formally incorporated into the training curriculum for SS professionals, and a tiered and differentiated mechanism of SSI literacy education is needed for conducting tailored and targeted education for different levels of learners. For example, following the approach outlined in Tan Bo (2017), an initial assessment of the target group's SSI literacy can inform a graded and tiered educational approach, which progresses from basic to advanced levels.

(2) Strengthening SSI Literacy Research

Research on SSI literacy can provide the theoretical basis and guidance for SSI literacy education. Despite the rapid growth of SS science, the crucial area of "SSI" at the high end of the SSi chain remains relatively unexplored by SS science researchers and practitioners. While the community of intelligence studies recognizes

SSI research as a critical area, professional limitations can hinder the integration of perspectives from both SS science and intelligence studies. *Intelligence Journal*, with its dedicated "Information and Security" column, increasingly publishes interdisciplinary papers by SS science researchers, bridging the gap between these two fields. Furthermore, the inclusion of SS science scholars in recent conferences hosted by the intelligence studies community signifies growing interdisciplinary exchanges. Strengthened collaboration between SS science scholars and intelligence scholars is crucial for the systematic and comprehensive development of SSI and SSI literacy theories. This collaboration will enrich SSI literacy education by integrating the unique characteristics and advantages of both fields.

(3) Appealing for Policy Support

The holistic approach to national security and the establishment of the Emergency Management Department signal increased demands on SSI operation. While the term "intelligence" has historically been associated with traditional SS fields like espionage and counter-terrorism, its role in management is equally crucial. The Emergency Management Department's need for intelligence in non-traditional SS areas is now more significant than ever, and these areas cannot be viewed in isolation under a holistic SS approach. The introduction of the Intelligence Law has laid the groundwork for SSI development. However, for SSI literacy education to flourish, SS regulatory departments of a government need to formulate detailed promotional policies, as this top-down approach will encourage organizations to actively implement SSI literacy education. In the meantime, collaboration between SS and non-SS regulatory departments of a government in developing these policies is essential for the effective implementation and growth of SSI literacy education.

(4) Promoting SSI Literacy Education with Platforms

Certain institutions and platforms possess inherent advantages in the collection, organization, and dissemination of SSI resources. For instance, a good many institutions are currently conducting research on big data about SS in their exploration, which gives them an edge in SSI collection, organization, and service compared to many SS researchers with universities. These institutions can collaborate with similar entities to promote the concept of SSI literacy education. Efforts of jointly formulating and implementing policies can support SSI literacy education as effective SSI management practices. Creating sections for SS data, information, and intelligence on websites like National Data enhances transparency, openness, and visualization of intelligence. Providing data storage, management, and sharing systems for SSI users, along with interactive platforms, can facilitate SSI literacy education. Integrating these platforms and institutions' intelligence, data management systems, and intelligence service projects into training programs will further promote SSI literacy development.

(5) Publicizing SSI Literacy for Education

While the program of publicizing SSI literacy has not officially started in China, films and television programs like *Homeland*, *24 Hours* and *Person of Interest* have contributed to raising public awareness of SSI concepts. Although espionage-themed films and television dramas have been produced in China, they often depict historical periods predating modern information technology. It is expected that films and television programs reflecting the current situation are produced and serve as a valuable tool for publicizing SSI literacy in China.

4.5.4 Conclusions

This section first defined SSI literacy, differentiating it from SSi literacy, based on established definitions of SSi and intelligence literacy. It then elucidated the fundamental elements of SSI literacy, constructing a theoretical model of the SSI literacy system and explaining its connotations. Finally, it advocated for SSI literacy education as the primary means of promoting SSI literacy, outlining its necessity and four core strategies. The exploration has revealed SSI literacy as a critical intersection of library and information science and SS science, serving as a key point for SS literacy promotion and intervention in SS behaviors. The six foundational elements of SSI literacy—SSI theory, SSI awareness, abilities of SSI acquisition, abilities of SSI analysis, abilities of SSI utilization, and SSI ethics—must work in concert to establish a sound SSI literacy system. The conceptualization of SSI literacy and the associated research has laid a solid theoretical groundwork for future research and application. While this section provides a preliminary theoretical analysis of SSI literacy, identifying entry points of fundamental issues and research is needed for exploring areas like as developing scientific evaluation indicators and systems for SSI literacy, as the future research on SSI literacy requires.

4.6 A System Framework of SSI

In practice, similar to intelligence practices in other fields (e.g., competitive intelligence), the SSI system provides the organizational framework and physical foundation for conducting SSI operations. Establishing an SSI system (SSIS) is crucial for organizational SSM. However, as an emerging field at the intersection of intelligence studies and SS science, research on SSI remains in its exploratory stage. Thus, the concept of SSIS is also in its early stage of development, lacking a clearly defined definition and framework within the academic community. There is limited research. Scholars like Fan Bo and Li Jinhong (2011) focus on planning methods for emergency intelligence systems within the specific context of anomalous SSM (emergency management). The scarcity of foundational research hinders both theoretical understanding and practical application of SSI systems. Therefore, this section aims to provide a theoretical definition and framework for the SSIS, grounded in the realities of SSI operation within SSM, hoping to provide guidance and reference for future research and development of SSI systems.

4.6.1 Definition and Connotation of SSIS

The concept of SSIS emerged from the practice of SSM, particularly within the context of SSM informatization. Given its nascent stage of development, a precise definition of the SSI system remains elusive within the academic community. This section attempts to provide a definition by analyzing the key connotations of the SSI system. Conceptually, the SSIS is a subordinate concept of the broader intelligence system, representing the application of the intelligence system concept within the specific domain of SS science, particularly SSM. Therefore, the SSI system is a specialized type of intelligence system, possessing both general characteristics and functions of a typical intelligence studies while incorporating the unique features of SSM, especially its intelligence operation. It is inherently an SSM system. Based on the general features and functions of intelligence systems (Zhao Rongying 2012) and the unique characteristics of SSM (Wang Bing and Wu Chao 2018b), six key connotations of the SSI system can be identified:

(1) SSIS as an Intelligence System

Intelligence systems can be categorized based on the specific areas of application (e.g., competitive intelligence systems, scientific and technological intelligence systems). The SSIS is a specialized type of intelligence system applied to the field of SSM. Therefore, it remains as an intelligence system, albeit one specifically designed for the collection, analysis, and utilization of SSI. Consequently, the design, development, and implementation of SSI systems must be grounded in the theories and methodologies of intelligence systems.

(2) SSIS as an SSM System

The study of SSI is inherently geared toward supporting and serving SSM. The SSIS, as a core component of SSI research, shares this same objective. Therefore, from an SSM perspective, the SSI system functions as an SSM subsystem. More specifically, considering the content, processes, and behaviors within SSM, the SSIS can be viewed as contributing to four key SSM subsystems: (1) The SSIS is an SS prediction (early warning) support system. The SSI system facilitates proactive, accurate, scientific, and precise SS predictions, enabling preventative measures; (2) The SSIS is a support system of SS decision-making. It provides support for making rapid, scientific, reliable, effective, and efficient SS decisions; (3) The SSIS is a support system of SS execution. It aids in the timely, effective, and thorough implementation of SS decision plans; (4) The SSIS is an SS learning system. It provides SS managers with continuous exposure to new SS concepts and advanced SSM methods, while also facilitating the learning of SSM experiences and lessons. In all, SSM theories and methodologies constitute an essential theoretical foundation for the design, development, and utilization of SSI systems.

(3) SSIS as an SSi System

The SSI, while a high-level node in the SSi chain, remains fundamentally a form of SSi. Therefore, the SSIS can be understood as an information system designed to develop and utilize SSi resources both within and outside the organization. More specifically, it collects, stores, processes, and analyzes security data, SSi, and SSI that reflect an organization's SS state and its changes. This process generates valuable and refined SSI for SSM purposes, which is then disseminated (supplied) to relevant SSM agencies and personnel through specific channels and formats.

(4) SSIS as a Human–Computer Interaction System: The SSIS comprises both human and machine components. Human components include SSI personnel and SSM personnel at all levels, while machine components encompass computers, communication networks, infrastructure, hardware, software, data, and SS protection equipment. The purpose of the system is to support SSM operations, which are ultimately carried out by humans. Therefore, humans are the primary users of the SSIS, and effective human–computer interaction is essential for its successful utilization. The development of an SSIS must carefully define the roles and responsibilities of both human and machine components, leveraging the strengths of each to optimize the overall system. In essence, being a human–computer interaction system, the SSIS relies on machines as tools and infrastructure while prioritizing human intellect and decision-making.

(5) SSIS as an Open and Complex System: The SSIS is inherently open, continuously exchanging materials, energy, and information with its external environment. It takes in raw SSi data, outputs processed SSI products, and adapts by incorporating new technologies and resources. Furthermore, the system's complexity arises from several factors: the diverse human and machine components, the intricate non-linear relationships and interactions between these components, the diverse and varied sources of SSi, the complex procedures of SSI processes, and the inherent complexities of SSM operation itself. These interwoven complexities create an intricate operational mechanism within the SSI system. Therefore, the SSI system is best understood as an open and complex system.

(6) SSIS: An Auxiliary Support System for Intelligent SSM

From an intelligence perspective, SSM can be viewed as a progressive upgrade and layering process along a "SSi chain." This chain is often conceptualized as a progression from raw facts to data, information, knowledge, intelligence, and ultimately, wisdom. SS scholars have described similar SSi chain elements, progressing from upstream physical attributes to downstream cognitive attributes. Following this principle, the SSI system acts as an auxiliary support system that facilitates the realization of intelligent (or smart) SSM.

In summary, the SSI system is not merely a combination of an intelligence system and an SSM system. Instead, it represents an intelligence system specifically designed to enhance an organization's SSM capabilities by effectively developing and utilizing SSi resources within the context of organizational SSM. It is a new type of SSM and intelligence system that integrates and coordinates organically with the organization's existing SSM and intelligence systems. Based on this analysis, the definition of intelligence systems, and the nature of SSM (particularly SSI operation),

The SSIS is defined as follows: It is an intelligence system designed to directly enhance an organization's SSM capabilities, with the extended goal of achieving intelligent SSM; it primarily relies on human intellectual labor, utilizing information networks (including infrastructure and equipment) as a means to collect, organize, store, process, analyze, and research SSi reflecting the organization's SS state and its changes; it disseminates the analysis and research results (i.e., SSI) through appropriate formats and channels to support SSM endeavors.

4.6.2 An Overall Model of the SSIS

A clear understanding of the basic framework of the SSI system is essential for its effective implementation. Based on the definition and connotations discussed earlier, the SSI system is built around the following key requirements and objectives: within a given organization, and under the joint leadership of the organization's SSI and SSM agencies, the system supports SSM behaviors (including SS prediction, decision-making, and subsystems of execution). It utilizes information networks as the primary means, supplemented by organizational and interpersonal networks. The system's core elements are based on the system of SSI operational procedure, with functional layers comprising SSI collection, analysis, and service subsystems. This framework can be summarized as "one organization, one intelligence source, two agencies, two major systems, and three major networks." This model is represented in Fig. 4.17, where a detailed explanation of the core components is made:

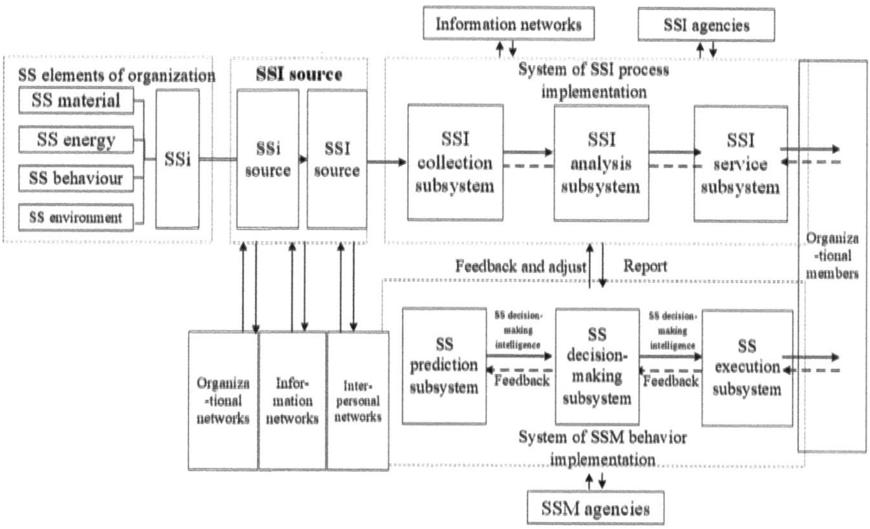

Fig. 4.17 An overall model of SSIS

1. One Organization

To ensure the effectiveness and practical application of an SSIS, its design and development tend to be tailored to the specific SSM needs (including SSI needs) of the organization it serves. This includes considering the organization's SSM performance (including SSI operation) and the environment. In other words, as SSI systems inherently vary between organizations, the design and development of an SSIS should be organization-specific.

2. One Intelligence Source

The single intelligence source refers to the source of SSi, which provides "raw materials" (i.e., SSi) for the SSIS. The SSI is essentially "activated" SSi, meaning it is derived directly from SSi. Therefore, the SSI source is fundamentally an SSi source. Furthermore, Wang Bing and Wu Chao (2018b) highlight the integrative function of SSi, which consolidates SS elements within an organization—including SS materials, SS energy, SSi itself, SS behavior, and the SS environment—into a form represented by SSi (as depicted in Fig. 4.17). This representation provides a comprehensive overview of the organization's overall SS state and its dynamic changes.

3. Two Divisions

The establishment and operation of an SSIS require the involvement of two key organizational entities: the SSI division and the SSM division. The SSI division designs, develops, and maintains the system, primarily responsible for operating the system of SSI operational procedure. The SSM division, on the other hand, is the system's primary user and feedback provider, responsible for operating the system of SSM behaviors. Therefore, the successful establishment and operation of an SSIS depend on the coordinated efforts of both the SSI and SSM divisions. They share responsibility for the system's development, operation (including systems and plans of formulating, supervising, and managing) and jointly participate in the core affairs of the entire SSI system business. Effective collaboration between these two divisions is essential for building a robust SSI system and ensuring its continued functionality.

4. Two Major Systems

The establishment and operation of an SSIS are oriented toward SSM. So the effectiveness of an SSI system hinges on the seamless interaction between two major systems: the system of SSI operational procedure and the system of SSM behaviors. This interaction forms the core of the SSI system. The system of SSM behaviors resembles a general SSM system, which is not detailed due to space limitations. Here the focus will be on the system of SSI operational procedure, which comprises three key subsystems: collection, analysis, and service. Drawing upon existing literature (Wang Bing and Wu Chao 2019a; Zhao Rongying 2012), the functions of these subsystems are: (1) Subsystem of SSI Collection: This subsystem gathers and organizes relevant SSi based on predefined SSI needs. It performs initial screening of the collected SSi and carries out preliminary SSI tasks, such as documentation, record-keeping, and regular archiving; (2) Subsystem of SSI Analysis:

This subsystem utilizes appropriate analytical methods and tools to process the SSi collected from the subsystem of collection, generating valuable SSI products; (3) Subsystem of SSI Service: This subsystem tailors and packages SSI products using various methods and formats for delivery to SSI users (i.e., SSM agencies). It also provides user-friendly browsing and querying services for SSM agencies and other authorized organizational members.

5. Three Major Networks

An effective SSI operation relies on three interconnected networks: the information network, the organizational network, and the interpersonal network. These networks provide essential infrastructure and platforms for SSI activities (Zhao Rongying 2012). (1) Information Network: Built upon an organization's information system (including the SSi system), this network utilizes the intranet as its primary platform. It includes specific modules for SSI collection, analysis, and service delivery; (2) Organizational Network: This forms the foundational infrastructure and safeguard for the SSI system. The SSi network relies on the organizational structure and its members for implementation. A core agency dedicated to SSI operation—the SSI agency—should be established within the organizational network. (3) Interpersonal Network: A well-rounded interpersonal network, encompassing both internal and external connections (e.g., organizational members, SS oversight departments of an government, SS experts, industry associations, consulting agencies, professional conferences, exhibitions, literature databases), is essential for effective SSI collection, analysis, and service delivery. It serves as both a crucial resource for gathering information and a primary channel for disseminating insights and enhancing the impact of SSI.

In addition, Fig. 4.17 reveals five key characteristics of the SSIS:

(1) Integration of SSI operation and SSM; (2) Interconnected information, organizational, and interpersonal networks; (3) Human–machine collaboration; (4) Integration of SSI and organizational informatization platforms (especially SS informatization platforms); (5) Combined SSI safeguards and SSI support.

4.6.3 *Subsystem Model of the System of SSI Operational Procedure*

The system of SSI operational procedure is the core of the SSIS, distinguishing it from other intelligence systems. This section details the structure and functions of its three subsystems: Collection, Analysis, and Service.

1. Subsystem of SSI Collection

This is a crucial component of the SSIS, serving as its input system and the foundation of all SSI operations. Its efficiency and performance directly impact the overall effectiveness of the SSIS. This subsystem operates on the basis of the main

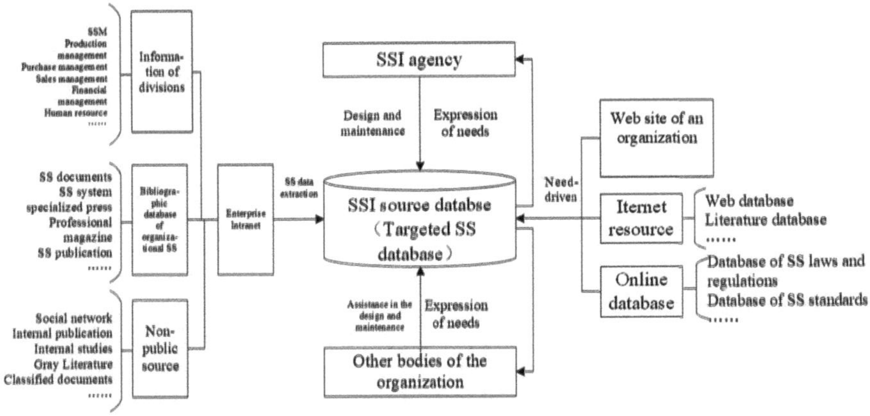

Fig. 4.18 Subsystem model of SSI collection

tasks and defined SSI needs stipulated by the broader SSIS. The basic framework of the SSI Collection Subsystem is illustrated in Fig. 4.18.

The subsystem of SSI collection performs six primary functions: (1) Mastering various methods and channels for acquiring SSi through diverse media; (2) Developing and executing SSI collection plans based on an organization's internal SSI needs, ensuring accurate and timely completion of SSi collection tasks; (3) Tracking and providing timely feedback on SS materials, energy, behaviors, organizational SS state and its dynamics, SS environment, and organizational SSM activities; (4) Staying informed about relevant national and local security policies, laws, regulations, and standards, and keeps abreast of international SS research and practices; (5) Maintaining effective communication and collaboration with relevant consulting departments, information departments, and intelligence agencies; (6) Organizing and supervising SS research and inspection activities.

2. Subsystem of SSI Analysis

This is the core of the SSIS, acting as its "manufacturing workshop" where raw SSi is transformed into valuable SSI. This subsystem relies heavily on human intellect, employing a "black box" approach to integrate, reorganize, and intellectualize SSi. SSI analysts utilize appropriate analytical methods and techniques which combines manual analysis with machine-assisted processing to systematically and hierarchically organize the SSi collected in the subsystem of collection. This process transforms raw SSi into actionable SSI, providing useful intelligence for SSM activities. The basic framework of the subsystem of SSI analysis is depicted in Fig. 4.19.

This subsystem performs six primary functions: (1) Developing SSI research plans aligned with an organization's overall SSM planning; (2) Identifying, screening, archives, and saving SSi according to established protocols; (3) Determining the optimal SSI analysis plan to ensure accurate and timely completion of analysis tasks with consideration of the specific characteristics of the organization's internal SSI needs and factors such as time, cost, complexity, urgency; (4) Providing tailored

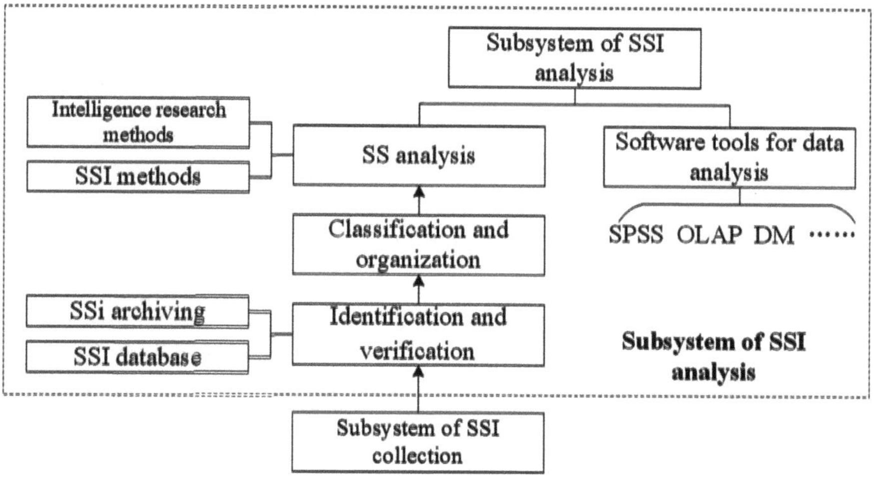

Fig. 4.19 Subsystem model of SSI analysis

reports, including SS situation analysis, SS investigation and inspection reports, and SS environment analysis, to support organizational SSM activities; (5) Delivering SS prediction (early warning), decision-making support, and execution support services to the organization; (6) Identifying new opportunities and evidence for improving organizational SSM effectiveness.

3. Subsystem of SSI Service

This subsystem is the output system of the SSIS, responsible for dynamically providing SSI products and services to an organization's SSM division and relevant members based on their specific needs. It delivers a range of SSI outputs, including specialized SS reports, risk analyses, environmental analyses, regular SSI briefings (daily/weekly/monthly), updates on internal SS state, and access to SS databases. To maximize the timeliness of SSI, the subsystem of SSI service ought to emphasize efficient and rapid delivery, and organizations' intranet platforms are conducive to enhancing the efficiency of SSI services. The basic framework of the subsystem of SSI service is illustrated in Fig. 4.20.

This subsystem has four primary functions:

(1) Serving the SSI needs of an organization's SSM division and related departments, while also providing SSI information to other authorized users within the organization; (2) Disseminating SSI through various formats and channels, including written reports, electronic documents, conversations, SS meetings, and SS training sessions; (3) Collecting feedback on SSI services and relaying it to the Collection and Analysis Subsystems, enabling adjustments and supplementary collection/ and analysis activities to better meet user needs; (4) Providing training and support to organizational members on how to effectively utilize SSI services.

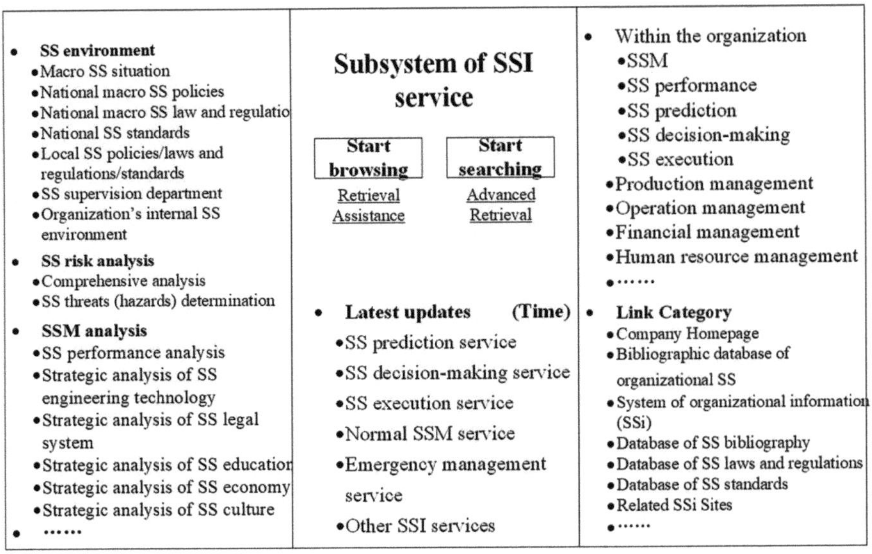

Fig. 4.20 Subsystem model of SSI analysis

4.6.4 Conclusions

The SSIS provides the organizational infrastructure and operational foundation for conducting SSI operations. This section focused on the theoretical underpinnings of the SSIS, specifically within the context of SSM, and draws upon practical intelligence operations. It analyzed the concept of the SSIS, proposes its definition and presents a theoretical framework for this system. First, by examining the general characteristics and functions of intelligence systems, along with the specific characteristics of SSM, the connotation of SSIS and its definition were analyzed. Second, based on this definition and conceptual analysis, a comprehensive theoretical model of the SSIS was constructed. Finally, a theoretical subsystem model of the system of SSI operational procedure was developed and analyzed. The analysis reveals the SSIS as a new SSM and intelligence system that operates in close coordination with existing SSM and intelligence systems. Its fundamental framework can be summarized as "one organization, one intelligence source, two divisions, two major systems, and three networks." This research provides theoretical guidance, a reference framework, and a comprehensive blueprint for both research and application related to SSI systems.

4.7 Intelligent Analysis Model of SSI

With the rapid development of technologies of mobile internet and information, data is being updated in real time and in vast quantities. When faced with the enormity of multi-source, heterogeneous data, AI stands out with its outstanding analytic capabilities. It amplifies people's capacity to gather, dissect, and generate new insights and revolutionizes paradigms of intelligence analysis (Tang Xiaobo 2019; Xi Caiyun et al. 2018). Integrating AI into the SSI field becomes essential in this age, where analytical thinking and technical methods of AI are utilized for managing and mining SS data in order to forge new SSI analysis modes. In other words, it is an important goal to conduct intelligent SSI analysis when SSI analysis is concerned.

At present, research on SSI analysis is predominantly focused on analytical methods (methodologies), often sidelining the substantive integration of AI technology in SSI analysis. The academic community has yet to probe into intelligent SSI analysis in depth. However, some research within the academic sphere has already touched upon intelligent analysis of intelligence, which holds significant potential for advancing intelligent SSI analysis. The research can be classified into two main areas. One is the area of theoretical exploration, where Xi Caiyun et al. (2018), Wu Chensheng et al. (2015), and Gan Yi et al. (2018) have analyzed the advantages of applying artificial intelligence in the field of intelligence operations. Scholars like Hua Bailin and Li Guangjian (2017), Feng Qiuyan and Zhu Xuefang (2019), and Ding Xiaowei and Su Xinning (2019) have proposed frameworks for systems of intelligent analysis of intelligence, establishing a new system for intelligence analysis workflows. The other is the area of practical application. For instance, Ding Xiaowei and Su Xinning (2019) elaborated on a new financial intelligence analysis architecture based on trustworthy AI. Zeng Qinghua and Chen Chengxin (2018) constructed a "human-centered combination of human and machine" analysis system of counter-terrorism intelligence. Tang Xiaobo et al. (2019) utilized AI technology to acquire intelligence and build an intelligent analysis model for corporate competitive intelligence. Zeng Wen et al. (2018) conducted research on intelligent analysis of intelligence that extended beyond literature, adopting a data engineering perspective.

In light of this, this section investigates the role of AI in SSI analysis, clarifies the basic model of intelligent SSI analysis, and integrates AI technology with SSI analytical methods and processes, with the aim of constructing a comprehensive model of intelligent SSI analysis and subsystem models, so as to provide theoretical and methodological guidance for research and practical operation related to intelligent SSI analysis.

4.7.1 The Role of AI in SSI

Artificial Intelligence, which stimulates human intelligence processes through machines or software, is also known as "Machine Intelligence" (He Zhiming 2018). It is a multidisciplinary field that develops and studies intelligent systems (machines) to simulate, extend, and expand human intelligence (He Zhiming 2018; Nie Hua 2019).

Features of SS events such as suddenness, randomness, and destructiveness decide that there is need for real-time, dynamic and visible SSI. As traditional analysis relies heavily on structured SS data, which lead to a narrow scope of SSI, it may present challenges in accurately and efficiently forecasting SS events in a timely manner, resulting in a passive stance in SSM. In the era of big data, AI technology enables real-time, targeted tracking and collection of SS data, alongside comprehensive mining and analysis. This allows for the early containment of many SS events, for instance, Guan Lei et al. (2016) proposed a perception platform of SS situations that integrates data collection, processing, analysis, risk discovery, monitoring, and early warning, which perceives SS risks and provides early warnings with AI technology.

When it comes to analytical methods, AI employs more scientific approaches, leading to a more precise analytical structure (Xi Caiyun et al. 2018). It excels at handling large-scale and multi-dimensional data, offering analytical capabilities that surpass human analysis, especially when dealing with uncertain research objects. This advantage not only frees up cognitive resources but also helps to mitigate biases that can arise from human factors (Xi Caiyun et al. 2018). SS managers face varying intelligence needs at different stages of SS events, and AI technology plays a crucial role in effectively preventing and controlling these incidents. By enabling real-time tracking, targeted data collection, and in-depth data mining, AI provides dynamic and visualized SSI. This support aids SS managers in executing intelligent SSM strategies and gives them the upper hand in preventing and controlling SS events.

Regarding the foundation of data, intelligent SSI analysis relies on a substantial amount of SS data. From a disciplinary perspective, SS data is the fundamental research resource for SSI, as big data environments inevitably require a data-intensive research paradigm for SSI studies (Wang Bing and Wu Chao 2020a). The total value of SSI covered by massive SS data is so vast that a greater amount of valuable SSI can be uncovered. Big data uses computer systems to design intelligent algorithms to solve missions that are difficult for humans, aiding AI in data collection, mining, and analysis (Guan Lei et al. 2016). For example, Wang Yuefen et al. (2019) proposed that a big data-driven intelligence research knowledge base can calculate, analyze, manage, and store massive multi-source data of a specific field, thereby supporting intelligent analysis of intelligence.

It is important to note that AI is "artificially realized human intelligence," meaning that when analyzing SSI with AI, the intelligence of humans (SS and intelligence professionals) cannot be overlooked. Therefore, "AI + Human in the Loop" is the fundamental approach for data-driven, intelligent SSI analysis systems (Gan Yi et al. 2018). The intuition, experience, and theoretical knowledge of SS experts provide intellectual support for SS data analysis, adding value to the intelligence analysis results obtained by machines in a conventional thinking model, realizing human–machine collaborative SSI analysis, and thus producing valuable SSI products.

4.7.2 *Basic Mode of Intelligent SSI Analysis*

At the core of the intelligent SSI analysis system, the quality of SSI products directly determines the scientific accuracy of SSM. To maximize the leading role of SSI products in SSM and their early warning function for SS events, and to achieve the "eleven-ization" of SSM (Wang Bing and Wu Chao 2020c), a basic mode of intelligent SSI analysis is proposed. This mode is based on a "target-centered" SSi sharing network and combines the "SSI + " analysis pattern with the " + SSI" analysis pattern.

1. "Target-Centered" SSi Sharing Network

The traditional SSi analysis follows a singular and linear approach, characterized by long analysis cycles and poor timeliness. Each subject of analysis works independently, lacking lateral communication. This "chimney-style" intelligence analysis mode makes it challenging to accurately predict SS trends or acquire SSI prior to the occurrence of SS events, often leading to severe consequences. For instance, the unanalyzed symptomatic information in the "chimney-style" intelligence system used by American agencies before the intelligence reform led to the Pearl Harbor event. The "9/11" event and the "Intelligence Gate" further exposed issues of unhealthy competition, erroneous intelligence, and resource closure in large intelligence agencies. Based on these events, Clark (2010), an American intelligence scholar, proposed a "target-centered" networked process model of intelligence analysis, which focuses on targeted intelligence analysis, incorporating all intelligence analysis subjects into the process and achieving interactive collaboration among intelligence participants. In the SS field, Zhai Zhikai et al. (2017) established a target-centered intelligence analysis network that allows for prompt sharing and communication of intelligence collectors, analysts, and decision-making users in the face of emergencies.

SSI refers to the information that directly impacts SSM. By definition, SSI is produced for the conducting SSM and serves as a direct resource for SSM (Wang Bing and Wu Chao 2020e). The ultimate objective of SSI is to facilitate and direct SSM efforts. Wang Bing and Wu Chao (2019f) have concluded the fundamental strategy for acquiring and analyzing SSI: these processes should be inherently linked to SSM itself. Consequently, the collection and analysis of SSI must be centered on the overarching objectives of SSM. It's essential to employ technical and managerial approaches to integrate those who demand, gather, analyze, and manage security intelligence, thus establishing a "target-centered" SSI sharing network. This concept is illustrated in Fig. 4.21.

At every stage of SSI production, the key entities aligned with SSM objectives can engage in real-time and in-depth communication, interaction, and intelligence sharing. This enables immediate feedback and progressive evaluation. Armed with the outcomes of such feedback and evaluation, each entity can adjust their collection and analysis strategies in a timely manner. This not only enhances the efficiency of SSI analysis but also leads to the creation of prompt, scientific, and accurate SSI products. It has been analyzed that the failure to effectively share information

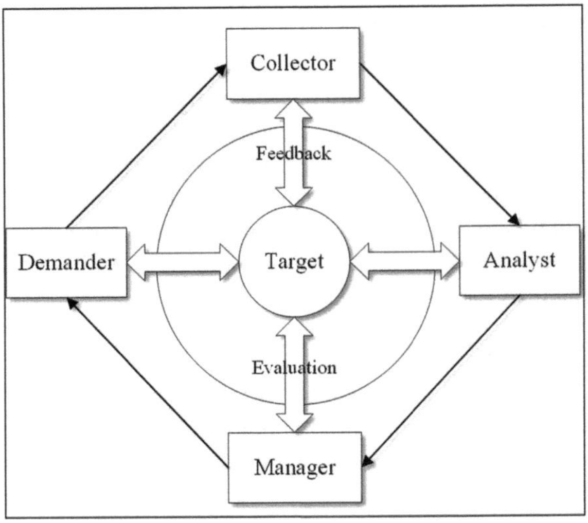

Fig. 4.21 A "Target-centered" SSI sharing network

contributed largely to the "9/11" event. In response, the US has since undergone a drastic intelligence reform, aiming to construct a harmonized and integrated intelligence system. This reform includes the consolidation of intelligence resources and the achievement of seamless information exchange, intelligence sharing, and collaborative efforts among all primary entities, including various intelligence agencies and departments at different levels.

2. "SSI+" and "+SSI" Analytical Methods

Modern SSI analysis often employs a demand (task)-driven method. Managers (users) submit SSM requests to intelligence agencies based on SSM goals or urgent unresolved SS issues. Then intelligence agencies extract relevant intelligence from the intelligence database, conduct intelligence analysis based on the demand (task), and respond to the SSM requests of managers (users). This analytical method allows SSI to play a leading role in SSM. As it provides "passive proactive guidance" for SSM, which is associated with the inverse SSI analysis technique, it is referred to as the "+SSI" analytical method.

It's also important to recognize the presence of "Black Swan" and "Grey Rhino" events in the SS field. The former refers to asymptomatic SS emergencies, while the latter to highly probable yet neglected risky events. For both types of incidents, managers (users) often fail to make precise SSM requests, making the "+SSI" analytical method difficult to achieve the expected effect of SS prediction. In such cases, anomalies and risks can be perceived through intelligence analytical methods that draw clues from an in-depth exploration of SS data. As Gan Yi et al. stated (2018), one of the key tasks of modern intelligence analysis is to seek changes in the unchanging, for outliers are likely to be a signal syndrome, meaning the value of a few processed anomalous data may far exceed that of the rest, which is much larger and of higher

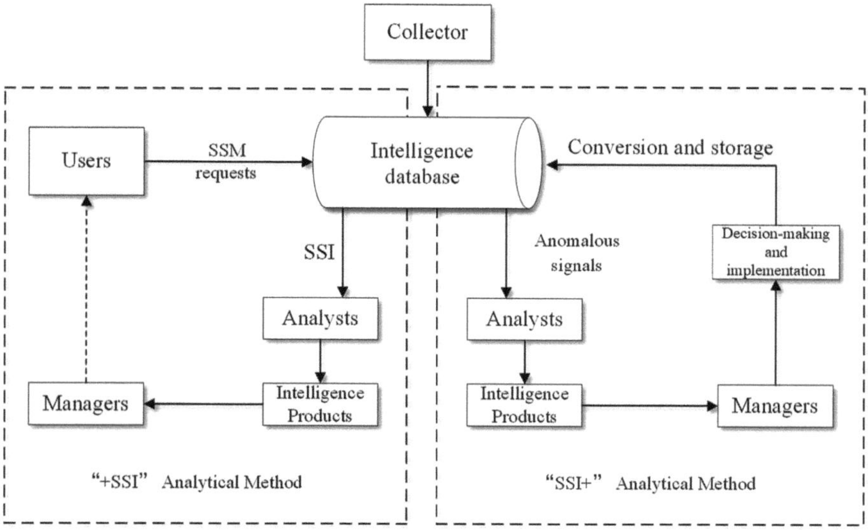

Fig. 4.22 "+SSI" and "SSI+" analytical methods

quality. Therefore, SSI analysis should focus on monitoring and detecting anomalous signals throughout the entire process and within the entire database, carrying out analysis operations on anomalous signal-driven SSI. This type of method, which uses AI technology to track and monitor SS data in real time and uses anomalous signals as the starting point of analysis, belongs to the forward SSI analytical method and is called the "SSI+" analytical method. It can curb events like the "Black Swan" and "Grey Rhino" at their budding stage and enable SSI to warn of SS events early on. The intelligent SSI analysis should harmonize both of the "+SSI" and "SSI+" analytical methods, conducting them in tandem to fully harness the intelligence of SSi in the domain of SSM, prevent and control various SS events in order to conduct intelligent SSM well. The process of the two methods is shown in Fig. 4.22.

4.7.3 Construction of the Overall Model of Intelligent SSI Analysis

Wang Bing and Wu Chao (2019 h) first proposed the concept of SIS and built an overall model of the SIS based on the six connotations of the SIS System. The author constructs an overall model of intelligent SSI analysis by taking the SSI process implementation system in the overall model as a reference and combining it with the application of AI in the field of intelligence. As shown in Fig. 4.23, the overall model of intelligent SSI analysis consists of six main components: the basic support bank, the core support bank, the intelligent dispatching engine, the SS data

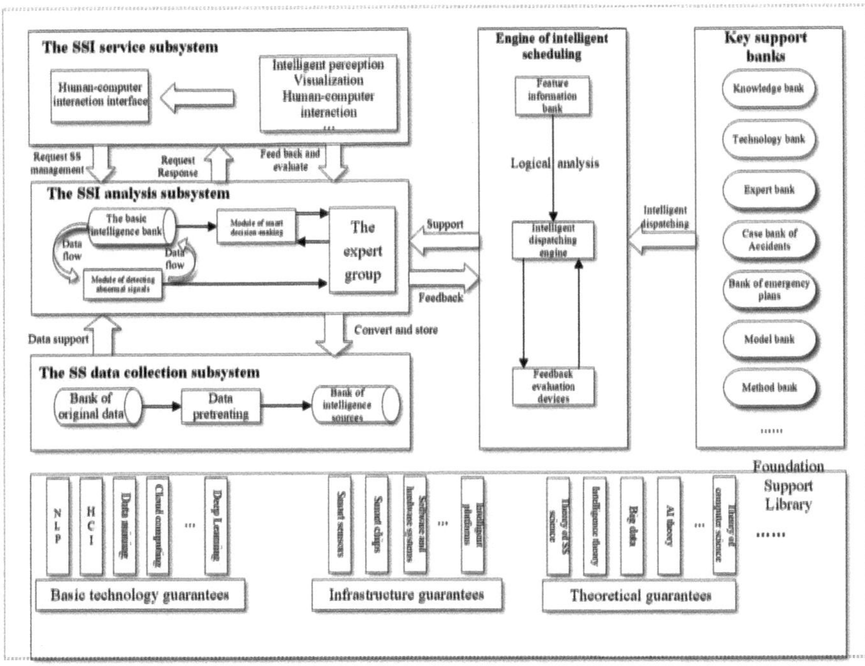

Fig. 4.23 An overall model of intelligent SSI analysis

collection subsystem, the SSI analysis subsystem, and the SSI service subsystem. The sub-models of the three subsystems will be discussed in the next section.

With the storage of abundant resources, the core support bank can support the analysis of the SSI analysis subsystem. However, when facing different management needs of different users, the models, methods, or tools to be used in the analysis become unknown, so is the effect of analysis. Therefore, the author built a bridge—the intelligent scheduling engine, connecting needs with methods and techniques.

The intelligent scheduling engine consists of a feature information bank, an intelligent dispatching engine, and feedback evaluation devices. The feature information bank is used to store user information and situational features intelligently perceived by the human–computer interaction interface. The intelligent scheduling engine establishes a logical connection between models, methods, and tools and user needs, problem situations and so on through logical analysis. By doing so, it precisely dispatches resources in the intelligence analysis support bank to meet the personalized needs of SSI. The feedback evaluation device is used to evaluate the analysis results fed back by the SSI analysis subsystem, and the specificity and precision of the intelligent scheduling engine can be improved step-by-step through feedback.

The basic support bank is to the intelligent SSI analysis what the foundation is to a building: it supports the information exchange and collaborative operation of various parts and provides basic technology, infrastructure, and theoretical guarantees for each part.

4.7.4 Construction of Intelligent SSI Analysis Sub-Model

The SS data collection subsystem, SSI analysis subsystem, and SSI service subsystem are the three major functional subsystems of the intelligent SSI analysis system. To clarify the operational process of the three subsystems, the author has constructed three sub-models for each subsystem.

1. Sub-Model of SS Data Collection

The SS data collection subsystem, seen as the "rear area" of the SSI analysis subsystem, is the prerequisite and foundation for the operation of the SSI analysis subsystem. The sub-model of SS data collection is seen in Fig. 4.24.

Guided by SSM goals, data collectors gather SS data in real time and store them in the original database. Since the SS data in the original database are often complicated

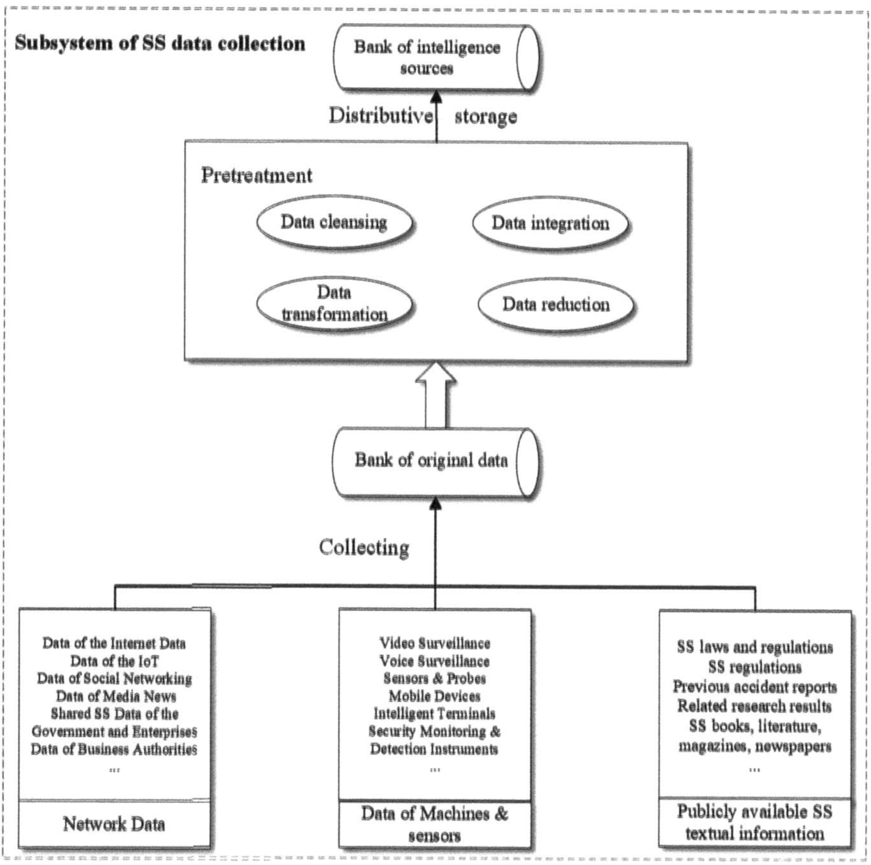

Fig. 4.24 Sub-model of SS data collection

by redundancy, missing data, uncertainty and inconsistency, it is necessary to preprocess the data which flows to the SSI analysis subsystem for the aim of effectively reducing the burden of SS data analysis on the SSI analysis subsystem, avoiding the phenomenon of "evidence illusion" where similar SS data is weighted heavily, and improving the operational efficiency of the SSI analysis subsystem as well as the accuracy and scientific nature of intelligence products.

The stage of pretreatment includes data cleaning, integration, transformation and reduction. Ambiguous and redundant SS data is removed and transformed into machine-readable data, suitable for data mining. In this way, the volume of SS data is streamlined with its quality improved. They will then be transferred into the intelligence source bank for storage.

The technique of distributed storage is employed for treating the SS data of the intelligence source bank by building gradable and distributed storage modules at the national, provincial, city, and county levels, and various industry fields in order to manage massive SS data and improve the efficiency of SS data storage and retrieval.

2. Sub-Model of SSI Analysis

The SSI analysis subsystem is the core of intelligent SSI analysis. The SSI analysis sub-model consists of a basic intelligence bank, a module of detecting anomalous signals, a module of intelligent decision-making, and a group of professional experts (Fig. 4.25).

After being analyzed and processed with various intelligent analytical methods such as statistical analysis, natural language processing, data mining, the SS data of the intelligence source library is then stored in the basic intelligence bank, which

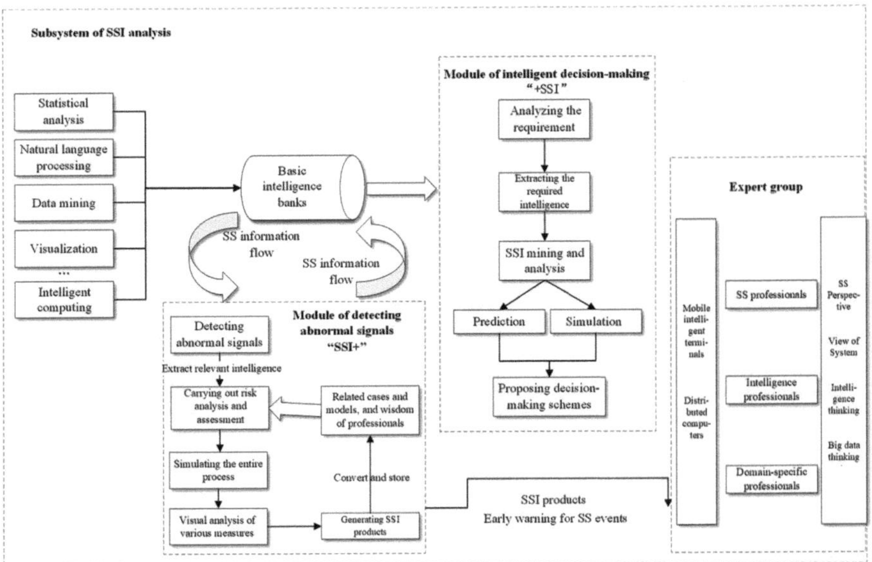

Fig. 4.25 Sub-model of SSI analysis

provides intelligence support for the module of anomalous signal detection and the module of intelligent decision. Table 4.5 lists the main methods of intelligent analysis of intelligence and their representative tools and algorithms.

The module of anomalous signal detection uses a "SSI+" analytical method. This module can achieve real-time and dynamic information exchange with the basic intelligence database through an SS information flow. The anomalous signal sensing device promptly tracks and monitors the dynamic SSi flow passing through. Once an unusual signal in the intelligence database is detected, the computer system immediately extracts relevant intelligence based on this signal and conducts risk analysis and assessment with consideration of related cases and models, catalyzed by the wisdom of professional experts. At the same time, intelligent technology is harnessed to simulate the entire process of an SS event where targeted measures are proposed, and the results of various measures taken can be used to generate reports, simulation animations, visualization models, and other SSI products. In this module, the SSI products are converted into storable ones and saved as data materials; they are also passed on to the group of experts for the role of early warning for SS events to be played, providing support for the decision-making of the expert panel.

The intelligent decision-making module uses a "+SSI" analytical method. After receiving an SSM request from the interface manager of human–computer interaction (user), the first thing this module should do is to analyze the requirements for clarifying SS goals and SSI analysis tasks. After the mission is defined, the module intelligently extracts the required intelligence from the basic intelligence database. The so-called "intelligent extraction" refers to the proactive use of methods of multi-dimensional semantic association and methods of dynamic semantic integration to

Table 4.5 Main methods of intelligent analysis of intelligence

Main methods of intelligence analysis	Main task of the method	Representative tool and algorithm
Statistical analysis	Descriptive statistics, factor analysis, regression analysis, life cycle analysis, time series analysis, etc.	Perl model, SAS, SPSS, Stata, Python, etc.
Natural language processing	Text analysis, semantic analysis, scenario analysis, information retrieval, automatic summarization, etc.	TF-IDF, LDA, Word2vec, CRF model, EM algorithm, etc.
Data mining	Classification analysis, clustering analysis, correlation analysis, etc.	K-means algorithm, FP-growth algorithm, Orange, RapidMiner, etc.
Machine learning	Neural networks, deep learning, system modeling, etc.	/
Visualization analysis	/	Pajek, Ucinet, VOS viewer, Net Draw, etc.

automatically calculate and analyze the value of intelligence within the basic intelligence database according to the problem situation (Feng Qiuyan and Zhu Xuefang 2019). "Intelligent extraction" can serve to reduce the workload of intelligence analysis and improve the efficiency of intelligence product generation. The extracted SSi can predict the whole process of SS events after being analyzed. With the combination of the prediction and the results of simulation analysis, the decision-making system can then intelligently propose decision-making schemes, which together with simulation animations, visualization models, and other SSI products, are passed to the panel, thus playing a leading role in SSM. The panel will then analyze the SSI products and feed the decision back into the intelligent analysis of the intelligence module, thereby optimizing and prioritizing the SS decision-making scheme.

The expert group, composed of SS professionals, intelligence professionals, and domain-specific professionals, guide SSM with their mastery of tacit knowledge of domains, scientific theories, and experiential insights. They are the intellectual guarantee of the SSI analysis subsystem and the key to realizing the "AI + Human in the Loop" approach of human–computer collaborative intelligence analysis. The group interacts with the module of intelligent analysis and the module of anomalous signal detection through mobile intelligent terminals and distributed computers.

3. Sub-Model of SSI Services

The SSI service subsystem is both the application part of the SSI Analysis subsystem and the hub for human–computer interaction. The sub-model of SSI services is shown in Fig. 4.26.

This sub-model consists of a human–computer interaction interface and three major supporting AI technologies: intelligent perception technology, visualization technology, and human–computer interaction technology. The intelligent perception technology can be used to intelligently collect real-time user information, perceive demand scenarios and contextual issues, and subsequently store this information in the feature database of the intelligent scheduling engine. This facilitates the optimal pairing of cutting-edge analysis algorithms and tools by the intelligent scheduling engine for SSI. The visualization technology enables real-time and dynamic delivery of SSI products within the scope of user privileges. And the human–computer interaction technology operates with defined interaction modes, allowing users to access authorized SSI products via specific interfaces and submit SSM requests or feedback.

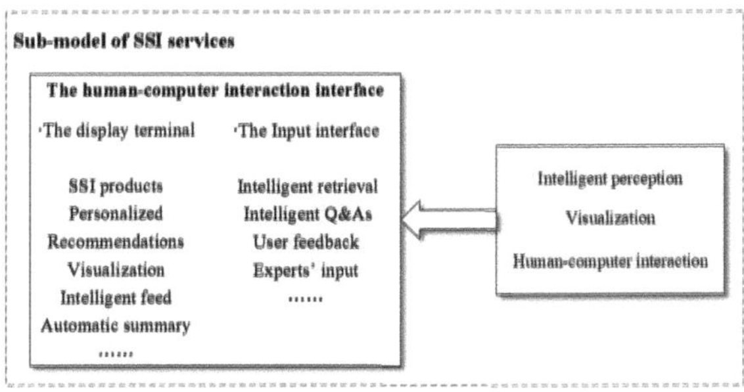

Fig. 4.26 Sub-model of SSI services

The human–computer interaction interface consists of a display terminal and an input interface. The display terminal, based on the user feature database, provides intelligent SSI product services within the range of the user's rights, including accident trend prediction, organizational SS state, accident public opinion analysis, case playback, accident simulation models, etc. To ensure the utility of SSI products, they are conveyed and distributed in various formats, including reports, briefings, videos, illustrative plates, charts, and other media forms. The input interface is directly input by the user, and the user can propose personalized service requests through intelligent retrieval, intelligent Q&As and other methods.

4.7.5 Conclusions

AI technology is key to intelligent SSI analysis. This section takes the basic theory of intelligent analysis of intelligence as a guide, combines the particularity of SSI, emphasizes human–machine collaboration in intelligent SSI analysis, and proposes the basic mode of intelligent SSI analysis. It also constructs an overall model of intelligent SSI analysis and three sub-models: the sub-model of SS data collection, the sub-model of SSI analysis, and the sub-model of SSI services. The overall model of intelligent SSI analysis, centered around the SSI analysis subsystem, employs the intelligent scheduling engine to integrate demands and methodological technologies, ensuring personalized and precise SSI. The two analytical methods of "SSI+" and "+SSI" can fully play the leading role of SSI in SSM and exert its function of providing early warnings for SS events.

4.8 Theory of SSI Service Capability Evaluation

Currently, there is a dearth of in-depth research addressing two major questions: "How should the capability of SSI serving SSM (henceforth referred to as SSI service capability) be assessed?" and "What constitutes the evaluation index system of SSI service capabilities?" The existing gap tends to impede the enhancement of SSI service capabilities. Against this backdrop, this section develops a theory of SSI service capability evaluation. It is grounded in the general framework and methods of intelligence service capability assessment, integrated with the theory and content of SSI and SSM.

4.8.1 Analysis of the Meaning of SSI Service Capability

1. Elements of SSI Service Capability

Research specifically dedicated to SSI service capability is scarce, though studies of counter-terrorism intelligence capability, scientific and technological intelligence capability, and emergency management intelligence capability are comparatively rich. For instance, An Lu et al. (2018), drawing on the intelligence capability theory, identified the supporting elements, business process elements, and target elements of counter-terrorism intelligence capability as components of the counter-terrorism intelligence capability system. Li Hui et al. (2017), based on the structural elements of scientific and technological intelligence service capability, constructed the evaluation index system for scientific and technological intelligence service capability from four aspects: the capability of intelligence resource guarantee, the capability of intelligence interaction, the capability of intelligence analysis judgment, and the capability of intelligence collaborative service. Pan Wenwen and Shen Guchao (2019) suggested that the service capability of the emergency intelligence system can be evaluated from three dimensions: the service construction capability, the service provision capability, and the service iteration capability. Building on these research, this section takes the three stages (preparation, practice, and application) of SSI service activities as a starting point and analyzes the constituent elements of SSI service capability from three dimensions (the assurance of SSI service, the process of SSI service, and the target of SSI service). The system of SSI service capabilities is an organic and complex whole where various resources and capability elements interact with each other, as depicted in Fig. 4.27. Among these dimensions, the assurance of SSI service is the fundamental guarantee for the smooth operation of the SSI service process and the targeting of SSI service; without the interaction and support of elements of SSI service guarantee and process, the SSI service target would be hard to achieve.

2. Definition of the SSI Service Capability

The service capability refers to the degree to which a service system provides its services. The four fundamental elements of the SSI service capability are human resources, information resources, service operations, and technical applications (Chen Xiangzhu and He Yinting 2018). In other words, the service capability is the comprehensive ability of the service system to fully utilize all human, material, and financial resources to meet and satisfy various user needs to the greatest extent possible.

Based on the definition of SSI and service capability, and in conjunction with the basic elements of the capability, the SSI service capability can be defined as follows: on the premise of accurate and effective SSI sources, the SSI personnel use various advanced equipment and technologies to analyze, integrate, organize, and process diverse types of SSI. They provide SSI agencies (users) with SSM services such as advanced and accurate SS forecasting services, scientific and effective SS decision-making services, and timely and accurate SS execution services. This approach aims to maximize user satisfaction and create the value of SSI.

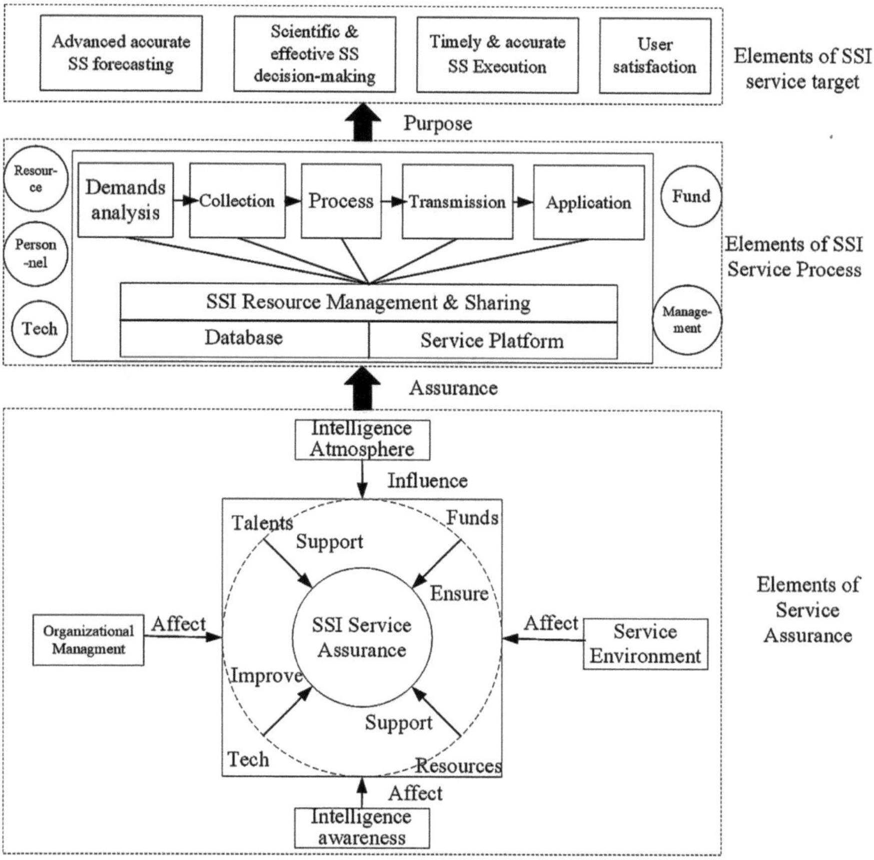

Fig. 4.27 Elements of the SSI service capability

4.8.2 *Construction of the Index System of SSI Service Capability Evaluation*

1. Basis for Construction

(i) Basic Framework of the SSI Service Capability

To clarify the evaluation system, this section begins with the concept of SSI service capability, based on the three elements (the assurance of SSI service, the process of SSI service, and the target of SSI service) of SSI service management. This approach clarifies the basic framework of the SSI service capabilities, as shown in Fig. 4.28. This capability is used to evaluate the assurance of SSI service, and the capabilities of SSI collection, SSI processing, and SSI transmission are used to evaluate the process of SSI service. The SSI application capability is used to evaluate the target of SSI service.

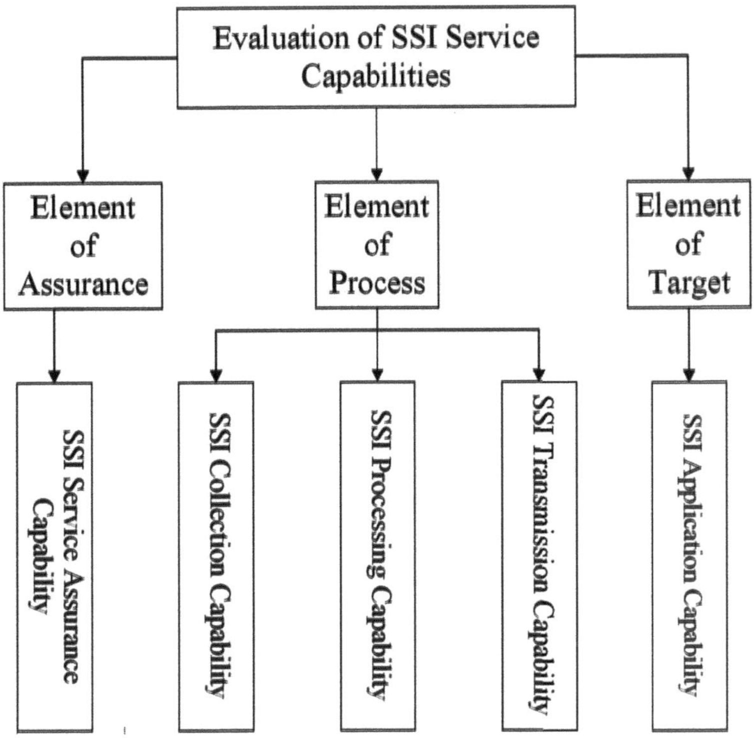

Fig. 4.28 Basic framework of SSI capabilities

(ii) Content of the Construction of the SSI Service Capability

The evaluation of the SSI service capability comprehensively examines and verifies the construction results of the SSI service capability. If the construction results of the SSI service capability are clear, the key points of the evaluation can be discerned. Chen Chengxin and Zeng Qinghua (2018) proposed a new path for developing the intelligence capability of the think tank from the perspective of intelligence research. This path posits that the intelligence service capability of the think tank can only be enhanced by fully upgrading the knowledge service capability of the intelligence service platform, which takes the intelligence assurance capability as the foundation, strengthens the intelligence demand perception capability, and focuses more on the dynamic intelligence collection and analysis capabilities. Li Yang (2019a) recognized that the development of the intelligence capability provides a critical support for urban emergency management. He analyzed the logical relationship between intelligence support and emergency management, and proposed that the intelligence service capability in emergencies can be improved from five aspects: intelligence resources, intelligence personnel, intelligence management, intelligence technology, and intelligence application. Based on the above literature analysis, it is found that the construction of the comprehensive intelligence service capability should be based on high-quality intelligence resources with five elements included: the capability of financial guarantee, the capability of talent support, the capability of technological innovation, and the capability of organizational management. Therefore, the five capabilities can be evaluated from five dimensions: resources, funds, personnel, technology, and management.

(iii) Elements of the SSI Service System

The service capability evaluation assesses the quality and effectiveness of the services and activities provided by the service system, as well as the efficiency of the resources required for the services and activities (Du Liling and Tang Yi 2019; Sun Chengjiang and Li Qi 2015). Therefore, beyond the five aspects of resources, technology, funds, personnel and management, the service capabilities and effects produced by the integration, configuration, and utilization of these five aspects should also be considered. Within the system of SSI process implementation operated by the SSI agency, the overall model of the SSI system (see Sect. 4.6) suggests that the main function of the SSI service subsystem is to collect and analyze SSI, which is transmitted to the SSM agencies and other related departments in a timely manner, or to collect relevant SSI in different forms by various means for the SSI agencies or other related departments to conduct research, which is based on their intelligence needs. Here an SSI service system is constructed in conjunction with the direction of the SSI flow, as shown in Fig. 4.29.

There are four stages in the SSI intelligence service system: collection, processing, transmission, and application, with two service paths. In the first path, the staff of the SSI agency collect, organize, and analyze information obtained from different sources before extracting SSI. Then, the intelligence personnel draw important SSI from the analyzed SSI and screen out SSI that can support SSM by identifying its authenticity, reliability, and usefulness. After that, the SSI is processed into SSI products that can be recognized and learned by SSM agencies and personnel. Among

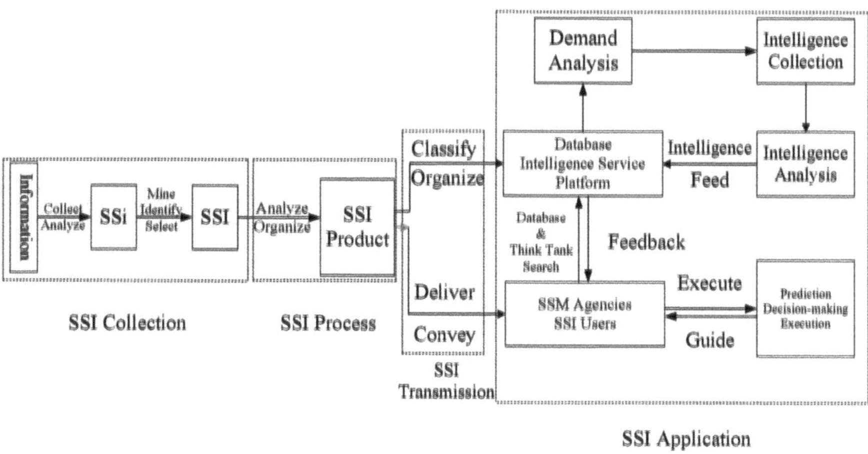

Fig. 4.29 SSI service system

the SSI products, some are distributed and delivered to SSM agencies and SSI users for their SSM activities; some are organized and classified by the staff of the SSI agency and stored in the SSI database for future retrieval and query. The second path is that when the SSM agency and SSI users address SSM operations such as SS prediction, decision-making, and execution, they need SSI as a basis and support. Therefore, the agencies and users will convey the SSI demand to the SSI agency or search for relevant SSI on the intelligence service platform of the SSI agency. In response to the user's demands, the intelligence personnel will search and collect SSI from the established database or service platform. Then, the staff of the SSI agency will analyze and process the collected intelligence and feed it back to the SSM agency and SSI users.

In the SSI service system, it can be observed that the evaluation of service capabilities should not only consider the agency's resources, technology, funds, personnel, and management capabilities but also pay attention to the work capabilities and service effects demonstrated by their joint efforts at each stage of the service, as shown in Table 4.6. The collection of SSI requires the SSI agency to complete the work of collection, organization, analysis, identification, and screening. The collected SSI can also be transformed into an SSI product after being analyzed and organized. The created SSI product needs to be classified, distributed, and delivered to realize its value, and the effects of each stage of the SSI service are evaluated by work efficiency and product quality.

2. Thoughts of Construction

Guided by the SSI service capability and adhering to the principles of selecting comprehensive, objective, feasible indicators that guide evaluation, and starting from the three elements of the SSI service capability, capabilities of SSI service assurance, SSI collection, SSI processing, SSI transmission, and SSI application are selected as the first-level evaluation indicators of the SSI service capability. According to the content of the construction of the SSI service capability, the secondary and

Table 4.6 Work capabilities and service effects that need to be evaluated at each stage of the SSI service

Stage of SSI service	Evaluation of work capabilities	Evaluation of service effects
SSI service assurance	Capabilities in resources, funding, personnel, technology and management assurance	Service level of security intelligence agencies
SSI collection	Abilities in gathering, organizing, analyzing, identifying, and filtering	Quality and efficiency of collection
SSI processing	Capabilities in interpreting and organizing intelligence	Quality of SSI products
SSI transmission	Abilities in categorizing, organizing, and distributing intelligence	Quality and efficiency of the intelligence service platform
SSI application	Application capabilities	Effectiveness of SSM and user experience

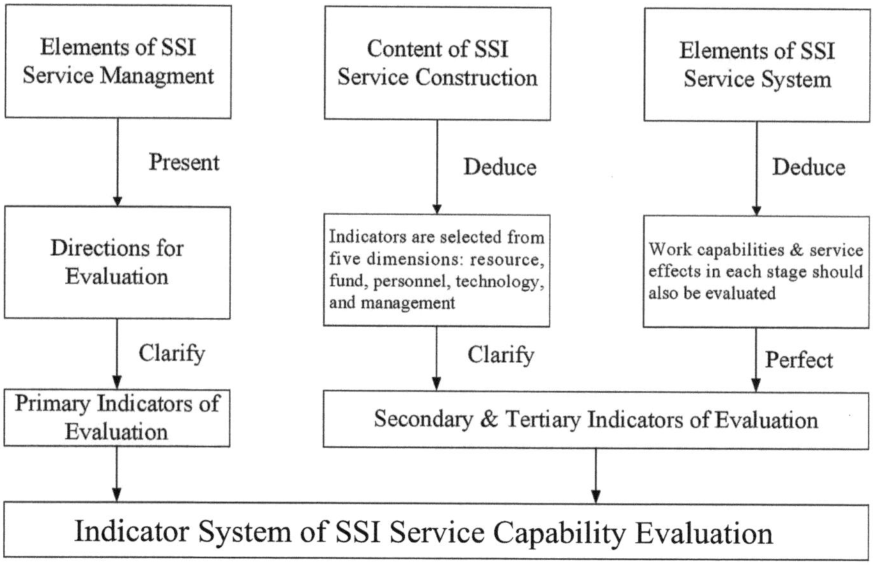

Fig. 4.30 Construction approach of the indicator system of SSI service capability evaluation

tertiary indicators should be further divided in the five dimensions of resources, funds, personnel, technology, and management, with the consider of the ability and effect of completing various tasks in the process of SSI services. The construction idea of the indicator system of SSI service capability evaluation is shown in Fig. 4.30.

3. Construction of the system

Based on the foundational framework, construction content, and constituent elements, an index framework of SSI service capability evaluation is built and an index system of SSI service capability evaluation is proposed, which includes 5 primary indicators, 17 secondary indicators, and 65 tertiary indicators (see Table 4.7).

4.8.3 Analysis of the Index System of SSI Service Capability Evaluation

1. Capability of SSI Service Assurance

The capability of SSI service assurance, as the foundation of high-quality SSI services, can ensure and maintain the smooth, orderly, and efficient operation of the SSI services (Li Hui et al. 2017). After analyzing the elements of SSI service assurance in Fig. 4.27, it is found that when carrying out SSI service activities, SSI agencies must invest sufficient funds, collect enough SSI resources, and introduce highly professional SSI talents. Only then can they provide the highest quality SSI

Table 4.7 Index framework of SSI service capability evaluation

Primary indicator	Secondary indicator	Tertiary indicator
SSI service assurance capability	SSI resource assurance capability	Diversity of SSI sources
		Degree of SSI network construction
		Basic reserve capacity of SSI resources
		In-depth development capacity of SSI resources
		Collaborative sharing capacity of SSI resources
		Sustainable development capacity of SSI resources
	SSI personnel support capability	Number of staff in SSI agencies
		Number and proportion of SS professionals
		Working conditions of the SSI service team
		Educational background structure of SSI staff
	SSI technology assurance capability	Completeness of infrastructure
		Usage of advanced equipment
		Big data processing and analysis technology
		Establishment of proprietary databases
		Working capacity of think tanks
		Degree of construction of SSI service platforms
	SSI funding assurance capability	Daily budget of SSI agencies
		Average salary of SSI personnel
		Costs for purchasing advanced equipment and technology
		Amount and proportion of special funds invested
SSI collection capability	SSI demand perception capability	SSI awareness
		Sensitivity to SSI collection
	Collection agencies	Diversity of SSI collection channels
		Comprehensiveness of SSI collection scope
		Intelligence of SSI collection methods

(continued)

Table 4.7 (continued)

Primary indicator	Secondary indicator	Tertiary indicator
	Professional quality of SSI collection personnel	SSI demand analysis capability
		SSI mining capability
		SSI identification capability
		SSI screening capability
	Collection capabilities of SSI collection Tools	Frequency of SSI updates
		Number of updated SSI items
		Quality of collected SSI
SSI processing capability	SSI demand response capabilities	Resource allocation and integration capabilities of the SSI service system
		Organizational and collaborative capabilities of the SSI service system
		Time to output required SSI
	Analysis Capabilities of SSI Processing Tools	Efficiency of the analysis of SSI processing tools
		Thoroughness of the analysis of SSI processing tools
		Completeness of the analysis of SSI processing tools
		Classification of SSI
		Organization and storage of SSI
	Processing capabilities of SSI personnel	Sensitivity of SSI personnel
		Observational and analytical capabilities of SSI personnel
		Creative analytical capabilities of SSI personnel
	SSI product quality	Richness of SSI service products
		Advancement of SSI service products
		Simplicity of SSI service products
		Reliability of SSI service products
		Value of SSI service products
		Targeted nature of SSI service products
		Innovativeness of SSI service products
SSI transmission capabilities	Service capabilities of SSI service platforms	Openness of SSI service platforms
		Convenience of SSI user queries

(continued)

Table 4.7 (continued)

Primary indicator	Secondary indicator	Tertiary indicator
		Efficiency of SSI user queries
		Number of SSI user queries
		Quality of SSI feedback from service platforms
	Dispatch capabilities of SSI agencies	Number of published SSI items
		Completeness and smoothness of the dispatch chain
		Efficiency of SSI product dispatch
	Communication capabilities of SSI agencies	Exchange with similar intelligence agencies
		Exchange with SSI users
		Exchange with collaborative institutions
SSI application capabilities	SSM outcomes led by SSI	Whether the SS prediction is forward-looking and accurate
		Whether the SS decision is scientific and effective
		Whether the SS execution is timely and in place
	SSI users	Satisfaction of SSI users

services with the support from advanced technology, that is, the SSI service capability requires resources, talents, technology and funds. Therefore, this capability can be deconstructed from four aspects: the capability of SSI resource assurance, the capability of SSI talent support assurance, the capability of SSI technology assurance, and the capability of SSI fund assurance.

(i) Capability of SSI Resource Assurance

The SSI resource assurance capability refers to generating, storing, developing, and sharing SSI from different sources, in order to integrate heterogeneous resources and complete the construction of the SSI network, aiming at providing diversified channels for the SSI collection and massive data for SSI analysis. Based on this, the diversity of SSI sources, the degree of SSI network construction, the basic reserve capability of SSI resources, the development capability of SSI resources, the sharing capability of SSI resources, and the development capability of SSI resources are used as the evaluation basis for the SSI assurance capability. The reasons are as follows:

From the perspective of information theory, the SSI source is equivalent to the information source. A stronger information field is formed with richer information sources, which provides enough and selectable SSI to the information destination (SSI users). That is, when there are more comprehensive SSI sources, more SSI can be collected, processed, transmitted, and applied, contributing to the construction of a higher degree of SSI network, leading to more effective support provided by SSI for SSM. Therefore, the diversity of SSI sources and the degree of SSI network construction are used to examine the SSI resource assurance capability. Li Yang (2019b) constructed the assurance capability of intelligence resources in emergency situations, indicating that the basic reserve capability of intelligence resources is the "source capability" of the system of intelligence assurance capabilities; the SSI obtained from the SSI source, the selected SSI, and the processed SSI products all need to be organized and stored for SSM personnel to use at any time. The greater the reserve capability, the more comprehensive the SSI. Therefore, the basic reserve capability of SSI resources is used to examine the capabilities of SSI storage and dispatching assurance. But the ultimate purpose of SSI is not to possess SSI but to use SSI for SSM. The overall value of SSI stored in the resource pool can be evaluated by the ability to exploit SSI resources. As the collaborative sharing of SSI resources is inevitable in the development of SSI services in the era of information, only through resource sharing can the collaborative cooperation of various SSI sources in the SSI network be realized, and the efficient integration and rapid flow of SSI resources be ensured. Besides, SSI resources are not static. On the one hand, SSI is more than a single text form now, as it has already been transformed into images, sounds, physical objects, actions, or their composite forms (such as video images). On the other hand, laws and regulations, various incidents, and solutions pertaining to SS will keep updating. Without incorporating them into the SSI resource pool, the provided SSI cannot be easily obtained and adapted for contemporary SSM undertakings. Therefore, it is necessary to take the capability of SSI resource development as one of the evaluation indicators of the capability of SSI resources assurance.

(ii) Capability of SSI Talent Support

The SSI staff are the main body to carry out SSI service activities and realize its value. They are the key to ensuring and providing high-quality SSI services (Fraser-Arnott 2017). Jiang Dan (2013) pointed out that the unstable structure of the SSI team and the lack of professional training for intelligence staff are the reasons why intelligence talents cannot display their abilities for intelligence to play its role. Those intelligence personnel who know SSM or SSI, or who have rich working experience in SSI, tend to have a higher level of SSI literacy. In the process of intelligence collection, screening, and processing, they can be keener in capturing SSI which is beneficial to SSM. Therefore, the structure of the SSI service team, the knowledge, specialties, and skills possessed by the SSI staff at SSI agencies directly affect the level of SSI service capabilities. Providing professional knowledge and skills training for the SSI personnel within the organization, together with arranging full-time professional SSI personnel at necessary positions, can directly improve the quality and efficiency of services. Therefore, in this book, the number of staff in the SSI agency, the number and proportion of professional SSI personnel, the knowledge level of SSI personnel, and the educational structure of SSI staff are used to assess the support capability of SSI talents in the SSI service team. In detail, the number of staff in the SSI agency is used to measure the overall scale of the agency; the number and proportion of professional SSI personnel, as well as the knowledge level of SSI personnel, is used to measure the professional talent reserve capability; the educational structure of SSI staff is used to determine the overall quality of SSI personnel.

(iii) SSI Technical Support Capability

Reasonable application of technologies often contributes to the achievement of twice the result with half the effort, and SSI service activities are no exception. Technology is an important tool and means for the development of SSI capabilities. Du Liling and Tang Yi (2019) found that the number and completeness of basic equipment greatly affect the work efficiency of the SSI service team. With the rise of high-tech technologies such as big data and AI, advanced equipment and technology have become key factors in enhancing SSI service capabilities. Big data processing and analysis technology, proprietary databases, think tanks, and service platforms ensure high-quality and high-level SSI services. Therefore, the completeness of infrastructure, the use of advanced equipment, big data processing and analysis technology, the establishment of proprietary databases, the working ability of think tanks, and the construction level of SSI service platforms are used to reflect the technical support capability of SSI services. The completeness of infrastructure can be used to effectively measure the strength of the SSI agency in their daily operations. The more complete the basic equipment (like computers), the more orderly the SSI service can be advanced; the other indicators are used to judge the use of contemporary advanced technology by the SSI agency.

(iv) Capability of Financial Support

As the economic base determines the superstructure, the maintenance of the SSI agency and the orderly development of SSI service can only be ensured by the orderly investment of funds. The fulfillment of capabilities of SSI resources, talents, and technology assurance requires sufficient financial support. Special funds need to be invested in key areas of the service to ensure the efficient completion of SS service activities. Therefore, the financial support capability is characterized by the daily budget of the intelligence agency, the average salary of intelligence personnel, the cost of purchasing advanced equipment and technology, and the amount and proportion of special fund investment.

2. Capability of SSI Collection

(i) Capability of SSI Demand Perception

In the information age, SSI personnel need to perceive users' needs and capture valuable SSI which can be developed into SSI with certain intelligence awareness, in order to eliminate redundant interference information from complex SSI sources and select information that meets users' needs. The collection efficiency can be greatly improved with superb perceptual abilities. Therefore, both the SSI awareness and the sensitivity of SSI collection are used to assess the capability of SSI demand perception.

(ii) SSI Collection Capability of SSI Collection Agencies

Diverse intelligence sources, types of intelligence, and comprehensive intelligence materials determine the hardware conditions for intelligence collection (Ding Lulu et al. 2019). With the development of technology, SSI sources, instead of being singular and rigid, become more diversified and complex; the scope of collection, beyond the form of text, encompasses pictures, images, recordings, physical objects, and other materials. The methods of SSI collection are no longer obtained with the SSI personnel's mental labor and experience but with the intelligence of SSI collection. Therefore, the diversity of SSI collection channels, the comprehensiveness of the SSI collection scope, and the intelligence of the SSI collection methods are used as the basis for examining the SSI collection capability of the SSI collection agency.

(iii) Professional Qualifications of SSI Collection Personnel

Intelligence collection is a process that can be completed with both technical means and talents' wisdom. That is to say, the assessment of the SSI collection capability cannot be separated from the examination of the professional qualifications of SSI collection personnel. Seen from the two paths of the SSI service system, there are both normal and anomalous situations in the stage of SSI collection. The normalized SSI collection is to excavate, identify, and select valuable SSI from the SSI source, organize it into the database, and publish it on the SSI service platform for SSI users; the anomalous SSI collection is a demand-driven approach taken by SSI personnel who determine specific collection tasks based on the needs of SSI users, and consciously collect required SSI from different databases, data libraries, big data platforms, and think tanks based on their knowledge, experience, and possessed intelligence literacy. Based on the above process, the ability to analyze SSI needs, excavate SSI, identify SSI, and select SSI are used as indicators to examine the professional qualifications of SSI collection personnel.

(iv) Collection Capability of Tools in SSI Collection

Tang Mingwei et al. (2018) found that using the technology of "big data + artificial intelligence" to complete different stages of the intelligence collection process (such as data mining, multi-source collection, data integration, data preprocessing, intelligent classification, three-dimensional presentation, and report compilation) is a key factor in improving the intelligence collection capability. In recent years, platforms of intelligence collection supported by high technology have been widely used as effective tools for intelligence collection because a large amount of intelligence can be pooled within a short time. Regarding the wide scope and difficulty which are two major issues in the SSI collection process for SSI service activities in the information age, SSI collection tools can be used to bulk collect, identify, and classify output information from SSI sources, reducing the workload of SSI collectors and improving collection efficiency. Therefore, the efficiency of available collection can be used to characterize the collection ability of tools, and the efficiency can be indicated by the frequency of SSI updates and the number of updated SSI. It is worth noting that in the era of "big data + Internet," not all data collected is authentic, effective, and applicable, because there is a wide scope of different data sources, and the total amount obtained is large, making it difficult to distinguish. Therefore, the quality of SSI collected with tools is also included in the evaluation index system.

3. Capability of SSI Processing

In the process of SSI intelligence processing, SSI personnel, based on the SSI users' needs, and the collaboration with various departments of the SSI processing system, employ SSI analysis tools to analyze and organize the collected raw intelligence, which is transformed into SSI analysis products for SSM. In this process, the SSI processing capability is a capability system that includes at least four elements, namely, the demand response capability of the SSI processing system, the analytic capability of the SSI processing tools, the intelligence processing capability of the SSI personnel, and the quality of the SSI products.

(i) Capability of SSI Demand Response

To respond to SSI demands, SSI processing personnel, guided by tasks and problems, should conclude the most capable SSI for problem-suffering SSI users through searching the SSI sources or extracting and analyzing the reserved SSI. This process requires full allocation of various resources and multi-party coordination, so the resource allocation and integration capability of the SSI service system and the organizational coordination capability of the SSI service system are used to characterize the response level of the SSI system; the time to output the required SSI is used to measure the efficiency of the response.

(ii) Analytic Capability of SSI Processing Tools

With the popularization and development of computers, intelligent computer software can help SSI personnel complete SSI analysis online, which not only reduces the workload but also improves the efficiency of analysis. However, due to its limitations, the analysis results need to be examined for their completeness, comprehensiveness and depth. Thus the efficiency, thoroughness, and completeness of the analysis, together with the classification of SSI and the situation of SSI sorting into the database are used to evaluate the analytic capability of SSI processing tools. The analytic efficiency of SSI processing tools measure the efficiency of analysis, the thoroughness and completeness of analytic operations, classification of the SSI, and the quality of the analysis judged by the sorting of SSI into the database.

(iii) Processing Capability of SSI Personnel

The sensitivity, observational analytic ability, and creative analytic ability of SSI personnel are used to assess their processing capability. This is because excellent SSI personnel can always capture important SSI, discern useful SSI, and discover valuable SSI from the details in the quick manner, thanks to their ultra-high sensitivity and observational analytic ability. The ability to connect and organize scattered intelligence into clear and complete SSI or to discover new intelligence based on the similarity and relevance of SSI is assessed by creative analytic ability. For example, in slope monitoring, although a large amount of data obtained from precise instruments such as GPS, liquid level, and corner sensors can be used by the monitoring personnel for safety analysis, it is still difficult to judge the state of the slope. Therefore, the obtained data still requires analysts to combine on-site soil structure, slope front sliding space, and other information for comprehensive analysis, and thus it can make a prediction of the state of the slope. This requires intelligence analysts to have excellent sensitivity, observation skills, and other professional capabilities.

(iv) Quality of SSI Products

SSI is used to support and serve SSM. The intelligence-led SSM model divides the content of SSM into two major parts: "SSI operation" and "implementation of SSM behaviors based on SSI." This means that SSI operation is an integral part of SSM, always running through the entire process of the SSM behavior. The completion of SSI operation is marked by the output of SSI products, which is the basis for SSM personnel to execute SSM. The richness, advancement, simplicity, reliability, value, specificity, and innovation of SSI products directly or indirectly affect the level of SSM. Therefore, they are used to assess the general quality of SSI service products. Richness refers to the diversity of SSI product forms; advancement measures whether SSI products keep up with the times; simplicity refers to whether SSI products are simple and easy to obtain; reliability refers to the credibility of SSI products; value is proposed for the utility of SSI products for SSM; specificity refers to whether the output SSI products are effective for special SS problems; innovation refers to the innovative ability of SSI products.

4. Capability of SSI Transmission

The delivery of SSI is the process of presenting SSI products already treated to SSM agencies (personnel). In this process, SSI personnel directly communicate with SSI users to provide SSI services, which can directly improve the capability of SSI services. This section assesses the delivery capability of SSI from three dimensions: the service capability of the SSI service platform, the intelligence-sending ability of the SSI agency, and the communication ability of the SSI agency. Firstly, starting from the SSI service system, SSI products, on the one hand, are published on the SSI service platform and delivered to SSI users through users' queries. On the other hand, these products are classified and delivered to SSI users by SSI agencies. Therefore, the service capability of the SSI service platform and the intelligence-sending ability of the SSI agency are included in the assessment. Based on the usefulness and service quality, the service capability of the SSI service platform is measured by five indicators: the openness, the convenience of SSI user queries, the efficiency of SSI user queries, the number of queries, and the quality of the SSI feedback from the SSI service platform. The delivery capability is generally measured by the quantity and efficiency of the delivery, so the quantity and efficiency of the SSI regularly delivered to the SSM agency are included in the evaluation. The delivery of SSI products is subject to external environmental (signal noise) interference, thus SSI loss and omission are inevitable. Since a complete and smooth delivery chain can help reduce the occurrence of such phenomena, it is necessary to include the completeness and smoothness of the delivery chain as one of the evaluation indicators of the delivery ability of SSI agencies. Besides, SSI agencies must maintain close communication with similar intelligence agencies, SSI users, and collaborating agencies to ensure the continuity of the delivery process, so the communication ability of the SSI agency is also used as an evaluation indicator of the ability in SSI delivery.

5. Capability of SSI Application

The capability of SSI application is used to measure the effect and quality of SSI services. SSI supports and serves SSM behaviors, only when SSM demonstrates its utility is the value of SSI shown. From the perspective of the constituent elements of SSI service management, the ultimate purpose of SSI is to achieve accurate SS prediction, reasonable SS decision-making and timely SS execution to the satisfaction of SSI users. Therefore, the effect of SSM led by SSI and the experience of users are used to evaluate the service effect of SSI. For example, in the remote wireless monitoring system of a boiler, the system can monitor the operational state of the boiler online and can measure real-time steam temperature, pressure, flow, and other information. With the help of AI software, the system can warn of the dangerous state of the boiler (SS prediction), discover the dangerous factors, and take necessary measures to remove the danger (SS decision and execution) in time. This section has explored the effect of SSM from three dimensions: SS prediction, SS decision, and SS execution, and examined SSI users from the degree of satisfaction.

4.8.4 Conclusions

The study of the SSI service capability is an important content of SSI study oriented toward SSM. This section analyzes the connotation of SSI service capability, explains its basic framework, the content of its development, and the constituent elements of the SSI service system. On this basis, an index system of SSI service capability evaluation is constructed. It includes five primary indicators: the capability of SSI assurance, the capability of SSI collection, the capability of SSI processing, the capability of SSI delivery, and the capability of SSI application. Each primary indicator can be further divided into 17 secondary indicators and 65 tertiary indicators. It should be pointed out that this section only constructs an index system of initial evaluation of the SSI service capability. Subsequent research should focus on the quantitative research of the initial indicators to determine their relative weights.

4.9 A System Model of Intelligent SSI Services

In the current landscape, research on SSI services is predominantly concentrated on the field (or intelligent emergency intelligence services) of anomalous SSM (or emergency management), with little attention given to normal SSM processes. Notably, there is a gap in research regarding SSI services that encompass the entire spectrum of SSM (including normalized and anomalous SSM). Therefore, there is an imperative need to develop an intelligent SSI service system that covers the entire SSM process, thereby creating a comprehensive framework for intelligent SSI services. Against this backdrop, this section, grounded in theoretical foundations and oriented toward the entire SSM process, probes into the hierarchical framework of the intelligent SSI service system and constructs and analyzes the model of the system.

4.9.1 Analysis of the Hierarchical Framework of the Intelligent SSI Service System

Literature analysis elucidates the hierarchical structures of the intelligent SSI service system and their interrelationships. The search with "advanced search" in the CNKI database with the theme "intelligence + intelligence + service system" results in a total of 262 Chinese-language papers, from which 39 representative works were selected. This study primarily focuses on these 39 papers, analyzing their content, synthesizing and refining the relevant insights and findings on the intelligent SSI service system, and delineating the levels of the intelligent SSI service system (Zhang Xu 2019; Guo Jiguang and Huang Sheng 2017; Hong Liang et al. 2018; Zhang Xingwang et al. 2014; Dong Lei 2019), as depicted in Table 4.8.

Based on the aforementioned analysis, the hierarchical framework of the intelligent SSI service system is depicted in Fig. 4.31. The intelligent SSI service system covers four layers: the support layer, the SSI resource layer, the SSI service layer, and the user layer. The support layer involves the technical methods and hardware facilities of the intelligent SSI service system, underpinning its construction and service functions—it is the "magic weapon" for realizing intelligent services; the SSI resource layer, built upon the support layer, supports and influences the service

Table 4.8 Levels of the intelligent SSI service system

Level	Specific explanation
Support layer	Keywords: big data, cloud computing, etc. This layer provides the technological foundation for the intelligent SSI service system, employing new technologies to realize intelligent services within the system
SSI resource layer	Keywords: intelligence, data resources, spatio-temporal geographic information, etc. This is the foundation for conducting intelligent SSI services, encompassing various SSI resources necessary for supporting SSM. It requires the construction of SSI resources within the intelligent SSI service system
SSI service layer	Keywords: intelligent information services, smart knowledge services, etc. This layer embodies the intelligent services of the intelligent SSI service system, using support for SSM as the entry point to achieve the functions of intelligent services, thereby forming the intelligent SSI service system
User layer	Keywords: government, public, etc. This primarily refers to the recipients and users of SSI. It aims to meet the SSI needs or the SS requirements of the user layer, promoting the construction of the intelligent SSI service system

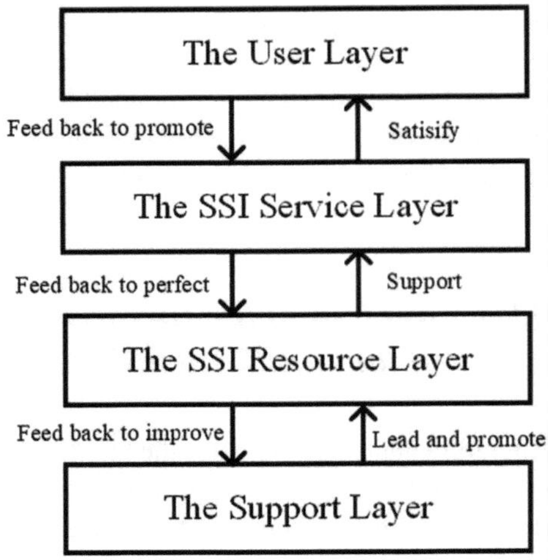

Fig. 4.31 Hierarchical framework of the intelligent SSI service system

functions of the intelligent SSI service system; the service functions of the SSI service layer, which reflect the value of SSI, are dependent on the SSI provided by the SSI resource layer and can be enhanced through the feedback mechanism within the service process; the realization of the SSI service layer's functions meets the needs of the user layer, which in turn influences and enhances the service capabilities of the SSI service layer, indirectly promoting the development of the SSI resource layer. Being interconnected and mutually influential, these four levels form an integrated whole, supporting the smooth operation of the intelligent SSI service system.

4.9.2 Research on the Construction of the Intelligent SSI Service System

1. Construction of the Intelligent SSI Service System

The Intelligent SSI Service System is constructed in line with the framework of the intelligent SSI service system (Fig. 4.32).

(i) Support Layer

The support layer serves as the foundation of the intelligent SSI service system and the key to implementing intelligent services. It is divided into two components: the module of technical support and the module of hardware support. The former includes big data technology and cloud computing, while the latter mainly involves network infrastructure (wired broadband networks, wireless networks, 4G and 5G, etc.), high-performance computing devices, and massive data storage devices, which are the basis for the application of high-tech. The combination of software and hardware in the support layer enables the intelligent SSI service system to fulfill the following five functions:

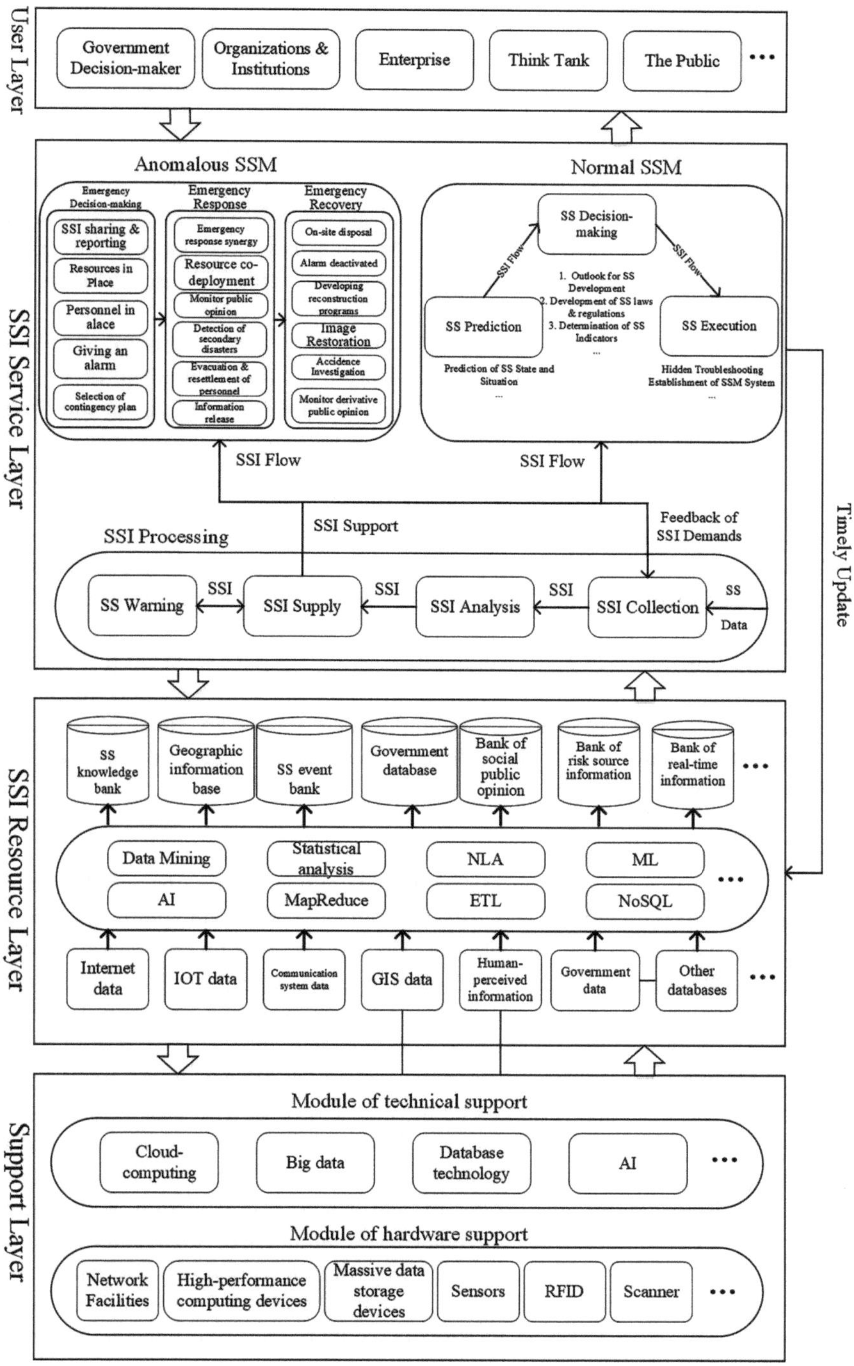

Fig. 4.32 Intelligent SSI service system

Comprehensive and real-time SSI collection: Sensors, scanners, and other devices are used to obtain real-time field information, the technology of big data collection is employed to gather data generated online by the Internet, communication systems and so on. The information collected through various means contains comprehensive SSI required for SSM, which can prevent the omission of SSI in SSM.

Storage of massive SSI data: The support layer provides large-capacity, high-performance storage devices for the distributed storage of SSI collected real-time and historical SSI, such as information of a hazard source, and coordinates the integration of data from different types of heterogeneous systems in the SSI resource layer (Bai Mei, etc. 2017).

Intelligent SSI analysis: The collected SSI is initially parsed by employing intelligent analytical methods such as machine learning, natural language processing, image analysis, message analysis, time series analysis, etc. (Guo Jiguang and Huang Sheng 2017) And the intelligent SSI service system can obtain optimal SSI with various intelligence analytical methods, relying on high-performance computing devices.

Efficient transmission of SSI data: The intelligent SSI service system requires real-time SSI, which is transmitted via high-bandwidth networks, and the data transmission at each link of the intelligent SSI service system also needs to be conducted through the network.

Sufficient utilization of SSI: The application of big data retrieval technologies such as correlation methods and big data retrieval technologies such as the parallel retrieval technology, the MapReduce technology, etc. can be leveraged to collect the required SSI from all sides, which effectively avoids the "heavy storage and light use" of SSI resources.

(ii) SSI Resource Layer

The SSI resource layer is the informational foundation for the intelligent SSI service system to conduct SSM and is the outcome of a processing platform of cross-departmental data sharing and integration. The SSI resource layer covers both static and dynamic SSI resources. The former includes SS event reports, standard SS legal and regulatory documents, various emergency plans, materials of past public opinions, SSI and SS knowledge accumulated, and materials about SS resource allocation and so on. The latter includes real-time SSI generated, perceived, and uploaded by IoT, the Internet, communication systems, and sensor devices, as well as potential danger information, SS event information (information of personnel casualty, spatio-temporal information, environmental information, etc.), and information of government affairs (information of department business, information of city operation, etc.). These resources, such as graphics, text, audio, and video, and other types of data, assemble SSI obtained in various ways into a comprehensive data service bank of big data, establish standardized, serialized SSI resource pools such as an information bank of SS risks, an SS knowledge bank, etc., and manage and maintain them according to the standards and requirements of big data management by means of different ways including knowledge management, content management, etc. (Huang Lanqiu 2012).

(iii) SSI Service Layer

The SSI service layer, or the application of SSI to SSM, is a manifestation of the "intelligence" in the intelligent SSI service system. The SSI service layer consists of three parts: the module of SSI processing, the module of normal SSM, and the module of SSM. The module of SSI processing includes indication of early warnings, and SSI collection, analysis and supply, which are achieved through the integration of big data and human wisdom. The module of SSM is divided into emergency decision-making, response and recovery. The normal SSM primarily includes SS prediction, decision-making and execution. The intelligent SSI service system can provide comprehensive and precise SSI services (such as decision-making schemes, action guidelines, etc.) (Li Yang 2019b), or proactively collect relevant SSI based on SSI needs to solve SS problems and complete SSI service activities. With technical support, the collection, analysis and use of SSI for SSM is equivalent to completing the process from information-to-intelligence and then to wisdom. At the same time, observable quantities (such as SSM effectiveness, features and so) generated in the SSM process becomes SSI resources through a series of transformation processes, further enriching the SSI resource layer and promoting the perfection of the intelligent SSI service system.

(iv) User Layer

The user layer is positioned at the top of the intelligent SSI service system and embraces government decision-makers, organizations, think tanks, the public, and enterprises and so on. This layer plays two roles in the system. First, it delivers the SSI products and SSI generated in the SSM process to the user layer after collection, analysis and processing, enabling the user layer to obtain the latest SSI and ensuring the openness, transparency, and digitization of SSI results (Wang Xinyan and Wang Xiaohui 2019). It is not only the receiver and user of SSI but also its producer. It can feed back the received SSI, so the intelligent SSI service system can seek valuable SSI information from the feedback. Second, since safety and security is related to all aspects of society and requires full participation in necessary situations, by delivering SSI to the society as a whole, the society can cooperate and coordinate with SSM, thereby improving the SS performance of the intelligent SSI service system.

2. Analysis of the Operation Process of the Intelligent SSI Service System

(i) SSI Collection

The acquisition of SSI in the intelligent SSI service system follows this procedure: collection and pretreatment → storage and management → computation, analysis and mining. Various organizational databases store a large amount of structured data, and the Internet, IoT, etc., and generate a large amount of unstructured data online. The first step in big data collection is to collect SS data from massive structured and unstructured data using "ETL" for offline collection, "Flume" for real-time collection, "Crawler" for Internet collection and other techniques. After pretreating the data through extracting, transforming, loading in the data collection process, the data will be integrated to improve its quality (usability and reliability). The collected data is stored in distributed file systems and databases (such as HDFS, NoSQL database systems) for further computing, analyzing and mining with neural network algorithms, genetic algorithms, and so on. In this way, valuable SSI contained in the data can be properly explored and stored, and the SSI can be intuitively displayed with the technology of visualization. Here, the SSI collection aims at transforming the big data of SS into SSi.

(ii) SSI Analysis

The SSI analysis refers to the further processing of SSi, enabling it to transform into SSI. In the process of SSI analysis, the collected SSi is sequenced, systematized, and hierarchized through the integrated analytical means, namely, manual analysis and machine analysis, aided by the knowledge of information discrimination, relevance judgment, metric analysis, and professional knowledge from the fields of natural sciences and social sciences. The process is often guided by SS science theories including the iceberg theory, the causation theory of SS events and the system theory of similarity in SS. With the analysis of the collected data through proper means, the characteristics, causes, essence, and consequences of SS issues are also anatomized and evaluated, and the results of the evaluated issues are formed. The SSI about the SSM capability analysis, analysis of the strengths and weaknesses of SSM, and the degree of occurrence and development of SS events are also acquired.

(iii) Analysis of SSI Warning

The SSI is further interpreted for conducting intelligent analysis of SS warning in this stage. Two methods are employed for the functioning of SS warning of the smart SSI service system. One is to judge whether the SS warning threshold is reached on the basis of experts' analysis (SSI is presented to experts in various forms for them to make professional judgments in accordance with their knowledge and experience: should an SS warning alarm be triggered? Is it time to enter into the state of emergency preparedness?). The second is to give SS warnings with the aid of formation processing technologies like big data.

As the method of giving SS warnings based on information processing technologies like big data is comparatively complex, illustrations need to be made. This is a method that combines the technology of big data with SSI, proposes an online processing method based on prior knowledge, models, algorithms, and real-time SSI, and produces two schemes. The first one is the SS warning based on the associative relation, which models or writes algorithms for the intelligence of original SS events, in order to obtain models or algorithms for different events. After the real-time SS data is processed into SSI, it is analyzed with its association with historical SSI, then relevant models/algorithms are selected and input, and issues like the current state, future development of SS events, and whether the threshold is reached are judged, and abnormal situations are identified through real-time SSI. The second one is the SS warning based on artificial intelligence, which outputs the feature parameters of real-time SSI, establishes data functions or models based on machine learning (ML), AI, pattern recognition, and performs calculation analysis, which can also determine whether the threshold is reached and identify abnormal situations, as shown in

Fig. 4.33. In practice, the models available for different SS events include inference mechanisms of chain events, and the algorithms include comparative methods represented by the California algorithms and algorithm TSC7, event sequence algorithms, immune algorithms, etc. SS warning analysis is part of the SSI processing, and the results can be used for SS decision-making. The model/algorithm of SS warning analysis is constantly adjusted and updated in the operation of the intelligent SSI service system.

(iv) Anomalous SSM (Emergency Management)

The module of SSI processing is capable of delivering comprehensive risk identification, analysis, and evaluation for SS events. This includes insights into the direction, severity, and likelihood of secondary events, which are crucial for emergency management. The module offers standardized and systematic prevention and control recommendations for both SS events and SSI. Drawing on the aforementioned SSI data, the intelligent SSI service system facilitates emergency management services through the following approach: it utilizes characteristic parameters of real-time SSI to perform associated searches, aligning with historical emergency plans and selecting appropriate responses for past SS events, as well as managing resource allocation and other emergency response strategies that include a range of decision-making and execution schemes. Furthermore, the system integrates the geographical coordinates of SS events, SS resource allocation, and personnel deployment—such as the arrival of rescue teams, on-site rescue operations, evacuation, and placement—into map applications for route planning, incorporates SS knowledge and expert insights, and performs scenario matching to devise immediate emergency response plans (Fig. 4.34). When executing emergency management services, the intelligent SSI service system utilizes network technologies including big data to

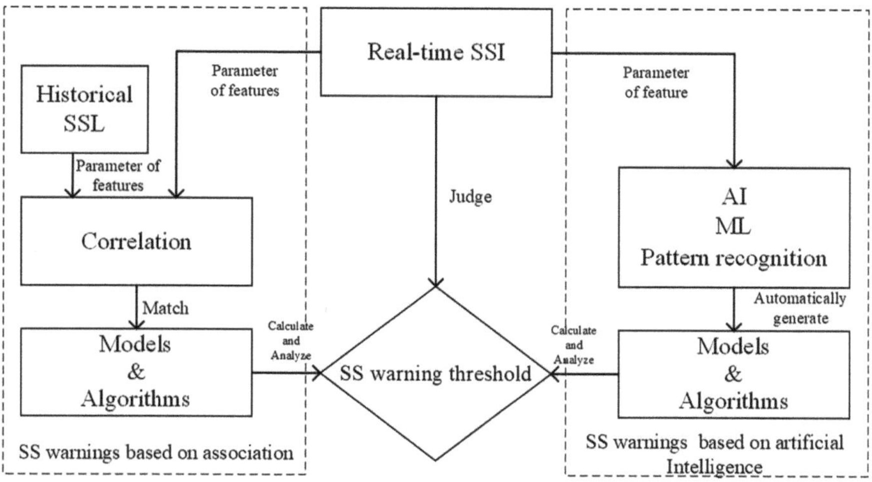

Fig. 4.33 SS warning method based on prior knowledge, models, algorithms, and real-time SSI

rapidly generate an initial emergency response plan guided by real-time SSI. This plan covers the life cycle of emergency management, allowing for real-time reporting, optimal resource allocation, and swift response.

(v) Normal SSM

The normal SSM constitutes the core service of the intelligent SSI service system.

(1) SS prediction: Following SSI analysis, the serialized, systematic, and standardized SSI and its portrayal of SS issues are used for SS prediction. The theories of SS science is employed in dissecting and applying SSI in the analysis and prediction of the state of hazardous sources or medium to long-term SS trends. Intelligent modeling or simulation can also be achieved by utilizing SSI, forecasting the future state of the system, analyzing the reasons behind changes of the SS state of the system, and understanding how the SS state might evolve if factors impacting the system safety change. In essence, SS prediction aims to make advanced, accurate, scientific, and precise predictions about safety and security on the basis of comprehensive SSI and thorough analysis about the SSI. Serving as the foundation of SS decision-making, it does not only reflect the current state of the system but also its future development.
(2) SS decision-making: SS decision-making entails the formation of both strategic and tactical SS decisions following a thorough examination and assessment of the intelligence about SS predictions. Primarily based on medium to long-term SS trend forecasts, these decisions are crucial for formulating SS vision planning and the top-level design of the SS system. Tactical SS decisions, on the other hand, address specific, localized, or individual SS concerns, representing

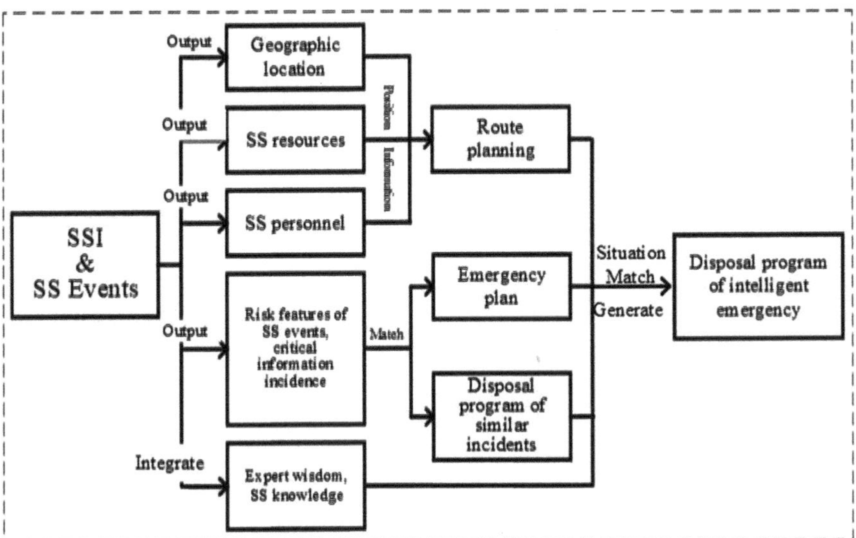

Fig. 4.34 Generation of the scheme for intelligent emergency management

concrete steps to achieve SS strategic goals (for example, outcomes from SS modeling or simulation in SS prediction can guide the determination of SS indicators or the establishment of SS warning thresholds).

(3) SS execution: Guided by the SSI derived from SS decisions, SS execution is conducted strictly in line with clear objectives and precise requirements. This includes hazard identification and management, SS inspections, and the establishment of SS protocols. SS execution demands a sense of responsibility in maintaining safe and secure, which is to be realized through in-depth analysis of SSI, comprehension of SS decisions, exhaustive investigation of SS issues, implementation of SS measures.

Normal SSM emphasizes the prevention of SS events. It considers factors impacting the system's SS state as informed by SSI, and intervenes early according to various influencing factors in order to preempt SS events at their onset. Moreover, it aims to cultivate an environment conducive to averting SS events. Normal SSM not only mitigates existing SS risks through the aforementioned three behaviors of system SS but also forestalls the emergence of new SS risks.

3. Service mode of the Intelligent SSI Service System

The service mode of the intelligent SSI service system include both intelligence-proactive and user-proactive modes. The intelligence-proactive mode concerns about the SSI obtained from real-time perceived information after the processing of a series of analytical work, and the historical SSI that is identified as related with real-time SSI after being mined and analyzed from the SSI resource layer. It detects the signs of SS events, identifies their characteristics and trends, enters a preparatory state in advance, swiftly allocates personnel and resources for SS events, executes immediate emergency plans, prevents the situation from escalating, minimizes the consequences of SS events, and fulfills early warning functions. The intelligence-proactive service mode can also be applied in conducting routine SSM, investigating hidden dangers and risk factors, eliminating conditions for the emergence of SS events, reducing economic and manpower investment, maximizing the utilization of limited resources, and achieving preventive functions. In the user-proactive mode, the SSM personnel utilize big data retrieval technology to analyze the SSI collected from the SSI resource layer according to SSM needs, thereby fulfilling its early warning and preventive functions.

4.9.3 Analysis of Key Issues About the Intelligent SSI Service System

1. Issue of SS Collaboration

The planning, construction, and operation of the intelligent SSI service system should be government-led, with information being shared across departments in collaboration. Although cross-department SSM is operated under the unified leadership of the government, departments within the current system are still fragmented, with each one working according to its own organizational function. The vertical and horizontal linkages across departments tend to be weak, making it challenging to integrate a large amount of SSI resources, which is likely to cause "information silos" and impedes cross-departmental SSM. Besides, cross-organizational collaboration is to be sought in the coordination of multiple interests. As the powers, resources and needs of departments vary, it becomes critical to explore the construction of a mechanism of well balanced interests. At present, the informatization of SSM in China is not fully developed, and there is considerable scope for enhancing cross-departmental collaboration. Although the deployment of big data technology can enhance collaboration between organizations, research and practical application are to be conducted for establishing effective mechanisms for inter-departmental SSi sharing and cooperation, building oversight and incentive systems, and legal protections, and elucidating the dynamics between SS collaboration levels, departmental involvement, and SS performance.

2. Issue of SSI Management

(i) Issue of SSI Collection

The collection of SSI for the intelligent SSI service system is heavily dependent on the technology of big data, which in fact carries certain ambiguities. As there is a wide array of entities involved in information dissemination, and these entities exhibit varying levels of cognitive abilities, the understanding of information may be subject to biases. Moreover, the expansion of communication channels has amplified the ability to spread information, which increases the likelihood of false information emerging during transmission. Therefore, a comprehensive collection of SSI becomes essential for the effective performance of the intelligent SSI service system. However, as the SSI consists of personal data and proprietary business information, etc., it is of paramount importance to avoid harming others' rights and interests in the operation of the intelligent SSI service system. And it is imperative to establish and maintain privacy protection mechanisms (like a data desensitization process), data access protocols, and SSI correction mechanisms.

(ii) Issue of SSI Processing

The majority of the big data of SS is unstructured and semi-structured data, with unstructured data posing a particular challenge in terms of standardization and comprehension. The interplay within SS big data is complex, mirroring the intricacies of SS issues, which complicates the representation of SS problems through SSI. The processing of SSI demands precision, because even a minor error can lead to substantial consequences. Given that current AI models and algorithms are not fail-safe, there is an ongoing imperative to investigate how best to leverage information technology to process SS data accurately and convert it into SSi and further into SSI.

(iii) Issue of SSI Storage

A defining feature of big data is its sheer volume, often starting at a computing unit of petabytes (PB) or more, and covering a variety of types and sources, which contribute to a swift rate of data generation. Data storage becomes not just extensive but also intricate in structure and is high-dimensional, which amplifies the challenge of managing SSI data storage. Thus the efficient storage of vast SS data is far from a trivial concern.

3. Issue of Personnel Development

Talented people can energize the intelligent SSI service system with fresh thoughts and new understandings, improving the performance of the entire system. In the collection, analysis, processing, and application of SSI within the system, the work team is required to interpret SS data, SSi and SSI at all stages. As complex as the iceberg theory, SS issues demand the awareness of discovery, exploration, and innovation from the personnel. As the SSI must be precise, the personnel who treat SSI should be proficient and knowledgeable. And as the intelligent SSI service system is an intersection of information science, intelligence studies, and SS science, its personnel ought to have cross-disciplinary knowledge in big data, intelligence studies, and SS science, though there are relatively few who possess the aforementioned interdisciplinary knowledge, awareness, and capabilities in obtaining and utilizing SSI. It is imperative to enhance the development of personnel.

4.9.4 Conclusions

In the era of intelligence, the concept of "smart + " is gaining traction, with applications like smart healthcare, smart tourism, and smart campuses among others continuously emerging. Intelligence is utilized for shaping the future of SSM and SSI initiatives. Thus, the study of intelligent SSI service systems is of paramount importance. This section probes into the theoretical underpinnings and examines the entire life cycle of SSM, outlining an intelligent SSI service system that embraces a support layer, an SSI resource layer, an SSI service layer, and a user layer. Supported by technologies of big data, cloud computing, and artificial intelligence, the intelligent SSI service system facilitates comprehensive collection, analysis, and application of SSI. The service model of the intelligent SS system makes early warning and prevention possible with intelligence-proactive and user-proactive methods. The three critical issues concerning the development and operation of the intelligent SSI service system are the issue of SS collaboration, the issue of SSI management, and the issue of personnel development.

4.10 Causal Model of SIF

Research and practice of SSI which is oriented toward SSM have identified the deficiency of SSI as one of the critical causes of SSM failure. Even with SSI support, SS managers still need to deal with SS events, where SIF (safety and security intelligence failure) should be responsible. It is the top significant cause of SSI deficiencies. Thus, examining SIF is critical for addressing SSI shortages in SSM.

There is a lengthy history in the study of intelligence failure. Traditional research has primarily focused on national SS and military operations (research shows that SIF originated from the "Pearl Harbor Event" in World War II). The American scholar Roberta (1962) is the first to propose the "signal–noise" theory of intelligence failure, introducing concepts of "signal" and "noise" from the field of communications to the field of intelligence studies, where "signal" refers to an indicator of an impending event, and "noise" is misleading information affecting intelligence analysis and judgment (Zhang Yang and Jin Pinxuan 2014). Building on this, the American scholar Handel (1989) proposed the "triple noise" theory of the intelligence failure, explaining the three sources of noise that lead to intelligence failure. Since then, research on intelligence failure has been ongoing (Turner 2005; Gao Jinhua 1998; Zhang Xiaojun 2007), establishing it as a significant area of study within intelligence research. According to the intelligence failure theory, the causes of SSI errors are multifaceted and can occur at any stage of an SSI operation. Therefore, investigating the causes of SSI errors is both important and necessary. Regrettably, the academic community has not yet conducted specific research on SSI errors, leading to a dearth of theoretical foundation and methodological guidance for preventing SSI errors in SSM.

In light of this, this section integrates intelligence failure theories, investigates the connotations and causes of SSI errors, and constructs a causal model of SSI errors as oriented toward SSM. It will also propose related strategies of avoiding SSI errors. The aim of the section is to provide a theoretical basis and reference for preventing SSI errors in SSM.

4.10.1 Connotation of SIF

Seen from its composition, the SIF is an academic concept arising from the integration of SSI and intelligence failure (intelligence failure refers to the inability to timely and accurately collect, analyze, make decisions, and utilize intelligence (Betts 1978), an importance concept born in the intersection of SS science and intelligence studies. Based on the fundamental definitions of SSI and intelligence failure, the definition of SIF is as follows:

The SIF refers to the phenomenon within the entire SSM process where comprehensive and accurate collection, rigorous and scientific analysis, stable and precise communication, and timely and decisive use of SSI are not achieved, leading to SSM failures (such as the occurrence of SS events). As a concept combining SSI and intelligence failure, the SIF represents a phenomenon where an error in a certain process leads to SSM failures. Its connotations include:

Essence of SIF: The SIF is an unfavored phenomenon caused by one or more reasons and leads to SSI failures in the SSM process. Reasons for the phenomenon and nodes in its occurrence are various and complex.

Scope of SIF: The concept is termed "safety and security intelligence failure" from a holistic SS perspective, considering that the objects of SIF include failures concerning both "safety" and "security."

Nodes where SIF is likely to occur throughout the process: SIFs may occur throughout the life cycle of SSI products, namely, in the process of collection, analysis, communication, and use of SSI, thus a full-cycle prevention and control should be carried out for SIF;

Consequences of SIF: SIFs can lead to SSM failures, thus SSI plays a critical role in preventing and controlling SS events and provides indispensable support for SSM. Efforts should be made to avoid SIF problems.

Additionally, from the perspective of SSI studies, a comparative analysis of general (traditional) intelligence failure and SIF is conducted (see Table 4.9).

Table 4.9 Comparison between general (traditional) intelligence failure theory and SIF

Content of comparison	General (traditional) intelligence failures	SSI failures
Research field	Limited to national SS and military warfare, with a relatively narrow and limited scope of study	Applicable to SSM in all fields, with a broad scope of study, oriented toward comprehensive SS
Feature of content	Featuring a distinct political hue and strong confrontational nature, focusing mainly on military intelligence activities between adversaries and allies, with intelligence failure factors often leaning toward enemy factors, international environment, and one's own factors	SSI failure research from an overall and broad security perspective, with universality and general applicability, failure factors mainly considered from three aspects: intelligence itself, intelligence environment, and intelligence personnel
Significance of signals	Indicating a large amount of intelligence information showing the enemy's real intention to attack, with extremely important significance of military SS and unavoidable limitations	Indicating SSi with intelligence value (i.e., SSI), this "signal" can serve SSM in any field, providing "omens and clues" for early warning of SS events
Significance of noise	Indicating false intelligence that cancels out and weakens the real signal, directing the enemy's intentions toward the wrong direction, the enemy can deliberately release related "noise" for military deception	Indicating various internal and external factors affecting the quality of SSI, which can be understood as the main reasons leading to SSI failures

4.10.2 Analysis of the Causes and Mechanisms of SIF

(i) Causes of Failure

The SIF in essence is a phenomenon. Understanding its underlying causes is crucial for devising scientific, rational, and potent strategies mitigating SIFs.

The systematic methodology of "WSR (Wuli-Shili-Renli for 物理-事理-人理, or the Way of things - the Way of Matters - the Way of People)," which targets management (not management of personnel or the technical system) as its research object and scope of application, emphasizes the adaptive use of diverse system management theories, methods, models, techniques, and tools to tackle a variety of management problems across different contexts. It is neither a method or a technique. Based on this methodology, the primary causes of SIF can be drawn from the three aspects of the thing, the matter and the people with a view to the essence of the SIF concept (Table 4.10). The causes are the factor of SSI (the level of the way of things), the factor of SSI environment (the level of the way of matters), and the factor of the SSI personnel (the level of the way of people).

(ii) Mechanism of Generation

In intelligence-led SSM, the content of SSM can be divided into two parts: SSI operation and the performance of SSM behaviors based on SSI. The SSI operation is carried out by the SSI agencies, supporting the performance of SSM behaviors

Table 4.10 Primary causes of SIF based on the "WSR methodology"

No.	WSR methodology	Reason for SSI failures	Analysis of connotations
1	Way of things	Intrinsic Factors of SSI	Primarily refers to factors at the level of SSI itself, or rather, the intrinsic problems of SSI, such as the attributes of timeliness, ambiguity, and cultural differences of intelligence
2	Way of matters	Environmental factors of SSI	Mainly refers to the impact factors at the level of the broader social environment in which SSI agencies operate, including issues such as the lack of mechanisms, imperfect systems, and an unfriendly environment related to SSI
3	Way of people	Personnel factors in SSI	Mainly refers to factors at the level of intelligence personnel in SSI activities, such as technical capability issues during SSI operation, and the subjective consciousness and cognitive biases when analyzing SSi and making decisions based on SSI

by supplying SSM organizations with SSI. According to the SSI process, these two stages are in reality the production stage and application stage of SSI. Besides, SSI errors may occur as SSI products are transmitted for use upon completion (or the transmission stage of SSI). To sum up, there are three stages in the SSM process where SSI errors may occur. A model of the mechanism of SIF occurrence is hereby constructed, as shown in Fig. 4.35, considering the three primary causes of SSI errors, and the impact of SSI errors on the three stages of SSM as the structural unit. According to Fig. 4.35, the construction idea of the model of the mechanism of SIF occurrence is as follows: vertically, the three stages of SSI (also known as the SSM process) are unfolded from top to bottom; horizontally, the three causes of SIF (namely, the factor of SSI itself, the factor of SSI environment, and the factor of SSI personnel) are presented.

In this model, SI-P represents the production phase of SSI, SI-T the transmission phase of SSI, and SI-A the application phase of SSI. The corresponding subscripts 1 ~ 3 represent influences of three causes. The interpretation of each structural unit

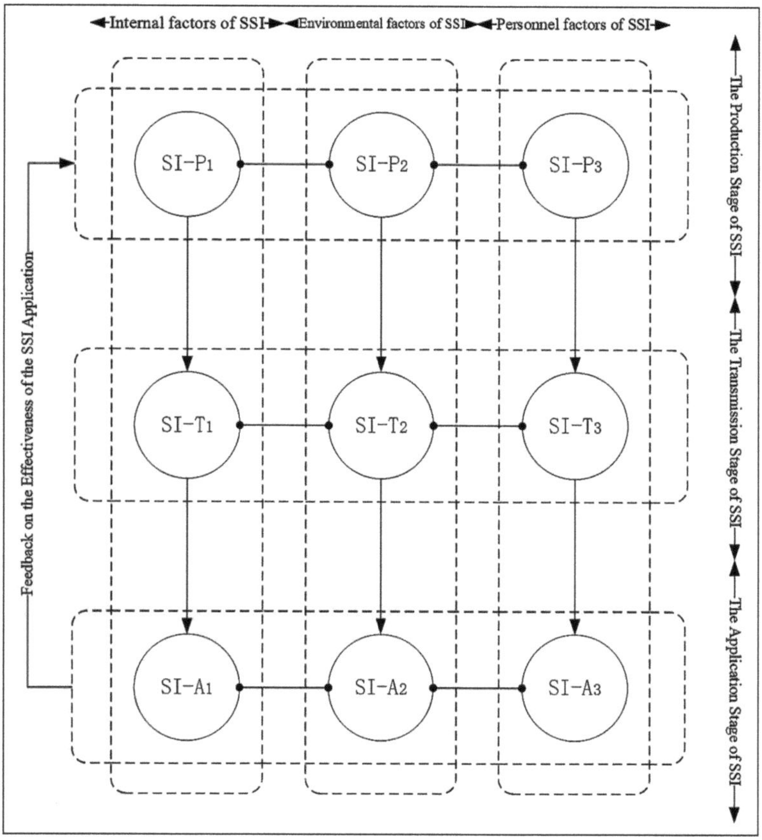

Fig. 4.35 Model of the mechanism of SIF occurrence

is shown in Table 4.11. From the "signal–noise" perspective, the factor of SSI itself, the factor of SSI environment, and the factor of SSI personnel are factors of "noise." Each factor interferes with the three stages of SSM to varying degrees, affecting the generation, transmission, and application of "signals (or SSI)," which causes SIFs. It should be noted that the three causes of SIF are interrelated, influencing each other and forming a holistic cause of SIF (see "Special Note" in Table 4.11).

Table 4.11 Interpretation of the meaning of the structural units in the model of the mechanism of SIF occurrence

No.	Structural unit	Explanation	Special note
1	SI-P_1	It indicates SIFs in the production stage due to intrinsic factors of the SSI itself	When SI-P_1, SI-P_2, and SI-P_3 occur in pairs or together simultaneously, it indicates that multiple factors have concurrently led to SIFs during the production stage of SSI
2	SI-P_2	It indicates SIFs in the production stage due to intrinsic factors of the SSI itself	
3	SI-P_3	It indicates SIFs in the production stage due to intrinsic factors of the SSI itself	
4	SI-T_1	It indicates SIFs in the transmission stage due to intrinsic factors of the SSI itself	When SI-T_1, SI-T_2, and SI-T_3 occur in pairs or together simultaneously, it indicates that multiple factors have concurrently led to SIFs during the transmission stage of SSI
5	SI-T_2	It indicates SIFs in the transmission stage due to intrinsic factors of the SSI itself	
6	SI-T_3	It indicates SIFs in the transmission stage due to intrinsic factors of the SSI itself	
7	SI-A_1	It indicates SIFs in the application stage due to intrinsic factors of the SSI itself	When SI-A_1, SI-A_2, and SI-A_3 occur in pairs or together simultaneously, it indicates that multiple factors have concurrently led to SIFs during the application phase of SSI
8	SI-A_2	It indicates SIFs in the application stage due to intrinsic factors of the SSI itself	
9	SI-A_3	It indicates SIFs in the application stage due to intrinsic factors of the SSI itself	

4.10.3 Proposal of the Causal Model of SIF

According to the event tree principle (Tan Qinwen et al. 2015), there are two possibilities (success and failure) for every link that any object goes through from the initial cause to the final result. Based on the entire process of intelligence-led SSM operations (including the stage of SSI production, the stage of SSI transmission, and the stage of SSI application) and considering the causes of SIF, the results of success and failure at every link are analyzed, the key nodes of SIF occurrence are specified, and the causal mechanism of SIF is obtained. The causal model of SIF is built as in Fig. 4.36.

In Fig. 4.36, Y denotes success, N denotes failure, and the numeric subscript signifies the various processes in the cycle of SSI operations. Therefore, in the stage of SSI production, Y11 represents successful SSI planning, and N11 represents a failure in SSI planning; Y12 shows the success of SSI collection, and N12 shows the failure; Y13 indicates the success of SSI analysis, and N13 indicates a failure in the analysis (SSI planning, collection, and analysis are basic components of the process of SSI production process). In the stage of SSI transmission, Y21 is successful SSI transmission, while N21 is a failure in SSI transmission. In the stage of SSI application, Y31 denotes successful SS prediction, while N32 denotes a failure in SS prediction; Y32 indicates successful SS decision-making, while N32 indicates a failure in it; Y33 is successful SS execution, while N33 is failure in the execution (SS prediction, decision-making, and execution constitute the content of implementation supported by SSI). It is clear that this model takes SSI planning as the initial task in a series of SSM operations. When all steps have been successfully taken, there will be no failure in SSI, so the SSM operation led by intelligence runs smoothly, the noise affecting SSI is eliminated, thus the system remains stable; the failure of any step causes miscues in SSI, which in turn affects the effectiveness of SSM. Therefore, the three stages in intelligence-led SSM are prioritized in the prevention and control of SIFs. It can be seen that the causal model of SIF is built upon the basis of two

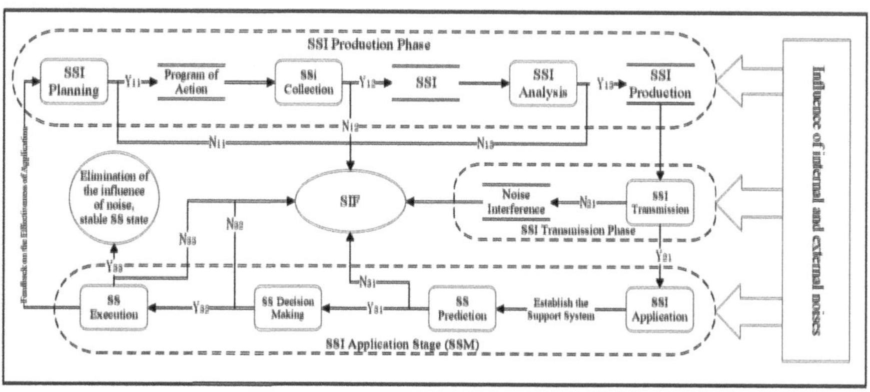

Fig. 4.36 Causal model of SIF

dimensions: the time dimension (i.e., the three stages of SSI) and the cause dimension (i.e., the reasons for the three levels of SIF), which can explain and analyze the entire cycle of SIF at all levels.

In order to understand the connotation of the causal model of SIF, it is necessary to thoroughly analyze the causal model of SIF (Fig. 4.36) according to the three levels of causes. Therefore, based on the model of the generation mechanism of SIF (Fig. 4.35), a full-process unfolded causal model of SIF (Fig. 4.37) is constructed, and each level in this model has its corresponding causal chain of SIF (the elements in the model have been explained in detail in the previous content), as shown in Fig. 4.37.

(1) Causal chain of SIF at the level of SSI itself: The inherent delays and ambiguities within the SSI process pose significant challenges in preventing SSI failures. For instance, during the production phase of SSI, personnel are only able to initiate collection and analysis activities once the situation has evolved to a certain extent and pertinent SSI has been produced. Thus the SSI products generated are inherently behind the curve of the development of the situation, which skews decision-makers' judgment of the SS state and causes the SS events to deteriorate. It is evident that SIF caused by the factor of SSI itself represents a flaw in SSM, so the personnel of SSM must mitigate the negative impact of this factor by enhancing the quality and timeliness of SSI.
(2) Causal chain of SIF at the level of SSI environment: SSI operations are carried out by SSI agencies, which support SSM actions by delivering SSI to SS management agencies. In this sense, SSI can be understood as a specific knowledge and achievement acquired by an organization. As an organization, an SSI agency is not unregulated by regulations, systems, or institutions (Li Peiling 2019). Causes of SIF at the level of SSI environment include an unsound mechanism governing SSI agencies, prevalence of bureaucracy, unhealthy competition, and absence of platforms for sharing. For example, in the stage of SSI application, there is a lack of an SSI sharing mechanism among various departments (departments of security, emergency, public security, firefighting, health, etc.), which weakens the synchronicity and relevance of related intelligence, hinders the work of responding to SS events, and results in the adverse consequences of missing the best prevention and control opportunities. Therefore, the larger environment in which an SSI agency exists exerts an important impact on SSI, as it provides a sound and objective operating environment and social environment that are needed by SSI.
(3) Causal chain of SIF at the level of SSI personnel: "Intelligence = information + analysis" signifies that intelligence is the outcome of knowledge that emerges from the systematic analysis of information through human thinking and it addresses specific requirements. In terms of SSM, the operational competencies and cognitive frameworks of SSI personnel—encompassing the technical challenges they face in intelligence operations, as well as the subjective biases and cognitive lapses that arise during the analysis of SSI and the formulation

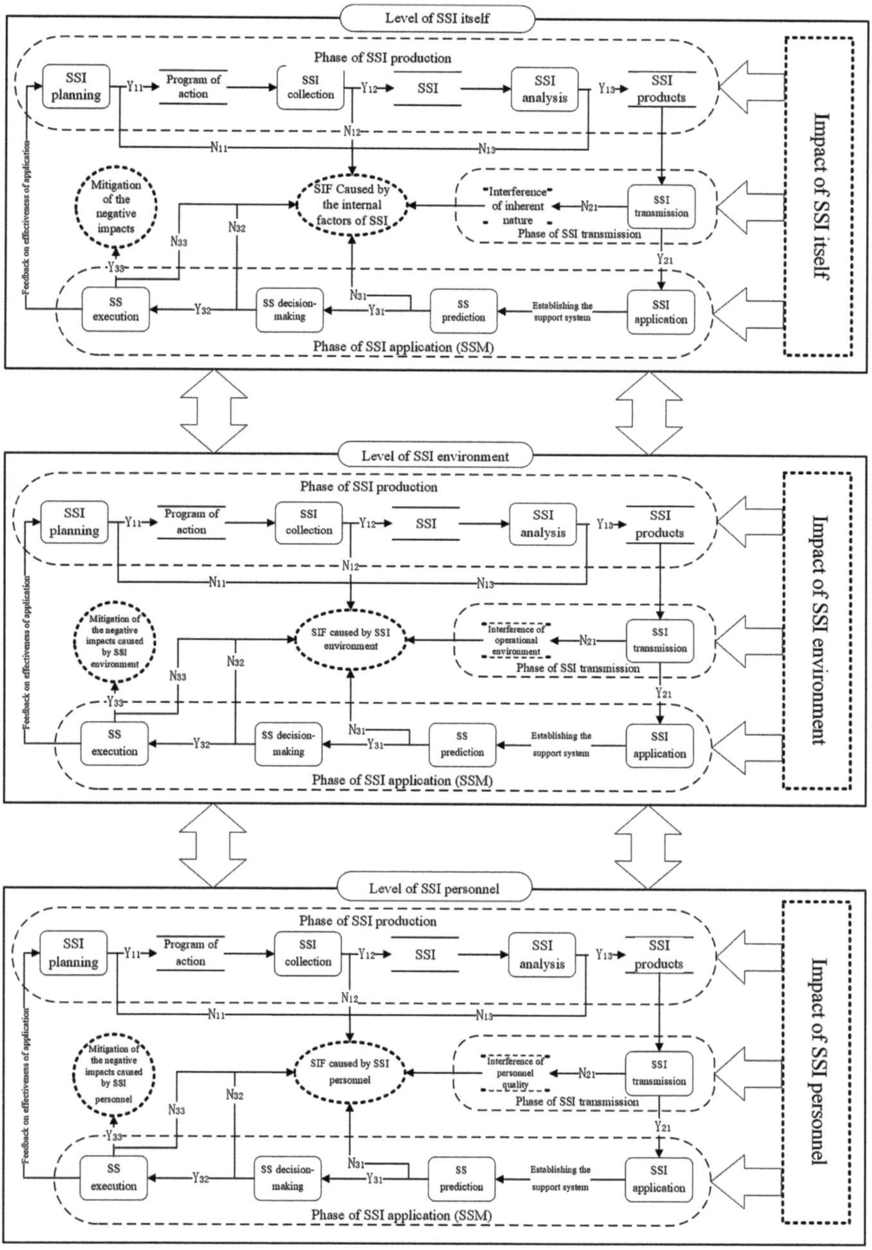

Fig. 4.37 Unfolded full-process causal model of SIF

of decisions—play a critical role in shaping SSI outputs, which in turn influence targeted SSM actions. For instance, during the phase of SSI dissemination, people may be influenced by "group think" as they may suppress their genuine thought under the pressure of a group consensus, and passively comply with the group. They convey the team's unified stance (which may be flawed) to decision-makers instead of voicing their own opinions and standpoints, which can result in the distortion of SSI products as perceived by decision-makers and cause SIFs. It is evident that the role of SSI personnel is a significant subjective element of SIF, therefore the SSI departments need to enhance the professional acumen of personnel and enhance their capacities of critical thinking.

4.10.4 Analysis of SIF Avoidance Strategies

The SIF causal model primarily elucidates and dissects the cause of SIF from three dimensions and across three operational phases, offering a comprehensive and multi-tiered analysis. Utilizing this framework, a succinct examination of the strategies designed to mitigate SIFs is conducted and a strategic system aimed at SIF avoidance is formulated (refer to Fig. 4.38). As depicted in Fig. 4.38, the trio of SIF causal factors can induce "noises" within the three operational stages of SSI, with the potential for their impacts to be cumulative (that is, various elements may exert influence concurrently). This scenario is highly detrimental to the orderly execution of SSI operations and significantly increases the risk of SIF occurrence. Addressing the intervention of these three SIF causal factors is a critical issue in the circumvention of SIFs.

As depicted in Fig. 4.38, three key elements of SIF can be drawn from the causal dimensions of SIF, which include the intrinsic characteristics of SSI, the operational

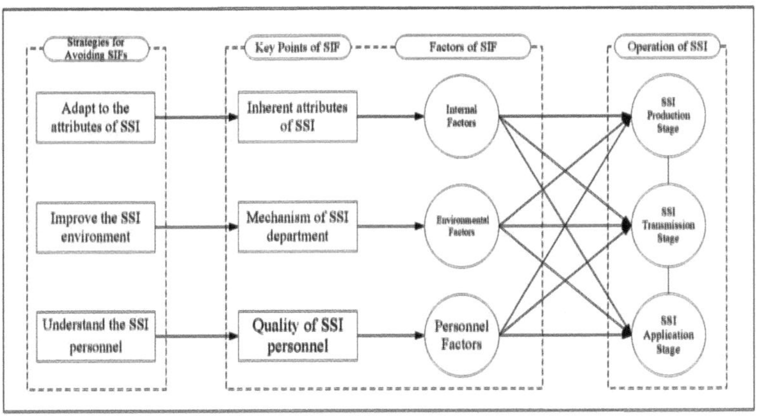

Fig. 4.38 System of strategies for circumventing SIFs based on the causal model of SIF

mechanisms of SSI departments, and the caliber of SSI personnel. In response to these three key points, strategies of circumvention are proposed as follows:

(1) Adaptation to SSI Attributes: A key element of failure related with SSI itself lies in the inherent attributes of SSI. They are objective and reflect the operational pattern of SSI operations, so they cannot be altered by other means. Therefore it is needed to adapt to the attributes of SSI. The SSI departments, while accommodating these attributes, can address the inherent deficiencies by seeking possible changes from three aspects including enhancing the speed of SSI collection to solve the problem of delaying in the process of SSI collection, enhancing the precision of SSI to compensate for its inherent ambiguities; addressing cultural disparities by enriching cultural elements of SSI and providing associative knowledge.

(2) Enhancement of the SSI Environment: A key element of failure related with the environment of SSI lies in the mechanism governing SSI departments. The mechanism established to govern SSI departments must address the issue of how to effectively schedule SSI personnel to accomplish tasks in a timely, scientific, and efficient manner. Enhancing the SSI environment involves improving the overall operational milieu of SSI, which includes refining various internal and external mechanisms, curbing unhealthy departmental climates and so on to foster a conducive environment for SSI operations. Specific initiatives include establishing a collaborative mechanism of "transfer-share-supervise" within SSI agencies to foster a healthy operating environment and reduce SIFs, and advancing the institutional reform of SSI agencies, curbing unhealthy climates, streamlining SSI agency procedures, eliminating bureaucratic red tape so as to avert SIFs.

(3) Understanding SSI Personnel: A key element of failure related with SSI personnel lies in their caliber, including qualities like professional expertise, cognitive patterns, and cultural literacy. As SSI personnel are inherently subjective beings, their subjective thinking naturally influences the operation of their professional tasks. Understanding them entails recognizing their inherent weaknesses, including cognitive biases and differences of ability for every intelligence officer. With this understanding, management personnel should assist intelligence personnel in enhancing their competencies across various dimensions to prevent SIF stemming from human factors. The measures include: firstly, improving their professional capabilities and personal qualities and refining their technical skills and cultural competence; secondly, fostering critical thinking and innovative mindsets among intelligence officers, aiding them in overcoming cognitive pitfalls, and effectively managing complex and dynamic intelligence scenarios.

4.10.5 Conclusions

In recent years, intelligence-led SSM has garnered increasing academic interest about the importance of research on SIF, as it is a significant contributor to the failure of SSM. This section offers a definition of SIF based on established SSI theories, explains the causes and mechanisms behind SIF, constructs a causal model of SIF, and provides a brief analysis of strategies for mitigating SIFs. The section aims to establish a theoretical foundation for SIF research and to expand the scope of intelligence-led SSM studies. What's more, intelligence-led SSM holds substantial potential for growth and promising prospects. SS workers ought to "remove barriers" for the operation of the SSM model, address pertinent issues related to SIF, and enhance the effectiveness and proficiency of SSM practices.

4.11 Chapter Summary

This chapter probes into the foundational theories of SSI pertinent to SSM for the aim of solidifying the theoretical underpinnings of SSI research and application. Nine insights should be noted about the chapter:

(1) The R-M approach to SSI acquisition and analysis underscores the necessity of grounding SSI efforts within the execution of SSM. This method integrates qualitative and quantitative aspects, acquiring and analyzing SSI from the vantage points of "risk module" and "management module";
(2) The SSI perception and analysis model centers on the process itself, elucidating the primary components of SSI operations at every stage. It constructs a task-oriented, intelligence-assisted framework for SSI perception and analysis, advocating for the development of an SSI ecosystem to foster the positive evolution of SSI operations;
(3) The fusion of SSI serves as an analytical tool that amplifies the value density of SS data, extracting value from multiple dimensions. Under the auspices of big data, the SSI fusion system can be categorized into five types of "fusion": the integration of SSI needs with SSI systems, the convergence of SSI analysis with the technology of big data, the fusion of SSI analysis, the synthesis of SSI applications, and the union of human–machine interaction;
(4) The foundational elements of SSI literacy encompass SSI theories, awareness, acquisition skills, analytical capabilities, proficiency of utilization, and SSI ethics. It is the synergistic effect of these six elements that constitutes a robust SSI literacy framework;
(5) The SSI system represents an original SSM and intelligence system that operates in concert with the broader SSM framework. Its basic structure can be encapsulated as "one organization, one intelligence source, two institutions, two major systems, and three networks";

(6) The comprehensive model of intelligent SSI analysis, with the SSI analysis subsystem at its core, employs an intelligent scheduling engine to bridge the gap between needs and methodologies, and ensure the customization and precision of SSI. The two types of analysis methods can fully leverage the decision-making guidance and SS warning of SSI and safeguard organizational integrity;
(7) The assessment of SSI service capability can be approached from five dimensions: the capability of SSI assurance, the capability of SSI collection, the capability of SSI processing, the capability of SSI transmission, and the capability of SSI application;
(8) The intelligent SSI service system encompasses four layers: the support layer, the SSI resource layer, the SSI service layer, and the user layer. There are two types of service mode: intelligence proactive and user proactive. Key issues within this system involve SS collaboration, SSI management, and the development of personnel;
(9) SIF primarily stems from the intelligence itself, the intelligence environment, and intelligence personnel. The junctures at which it occurs include the generation, transmission, and application of SSI. Strategies to circumvent SIFs should focus on adapting to the attributes of SSI, enhancing the SSI environment, and comprehending the dynamics of SSI personnel.

Chapter 5
Intelligence-Led SSM and Typical Applications

5.1 Introduction

SSM is one of the critical issues that are highly concerned in the contemporary society. As SSI is designed to support and serve SSM, playing a leading role in conducting SSM, the research and application of the methods of intelligence-led SSM are significant in both theory and practice. This chapter primarily introduces intelligence-led SSM and exemplary applications. It conducts a qualitative and quantitative analysis of the mechanisms and value of SSI in SSM, then proposes intelligence-led SSM methods, and introduces a series of fundamental issues related with the methods. At the end of the chapter, it discusses empirical applications of intelligence-led SSM methods in fields of urban SSM (including smart city SSM), emergency prevention and control, and SSM of mineral resources.

5.2 Mechanism and Value of SSI in SSM

SSI is aimed at serving and supporting SSM, as it is the foundation of SSM. It also serves as the focal point in SSM, undertaking roles that are indispensable. The absence of SSI in SSM is like water without a source. Hence, the study of SSM from the perspective of intelligence should be central to SSI research, which is of theoretical and practical significance and warrants attention. It is acknowledged that current SSI research remains in its nascent stage without being implemented or integrated in SSM. From an academic perspective, the primary reason is the lack of clarity regarding the connotations and mechanisms of SSI within SSM (which mainly includes the essence of SSM from an intelligence perspective, as well as the mechanism and value of SSI in SSM).

In response, this section will approach SSI and SSM from a holistic perspective, encompassing the entire domain of SSM. It will also integrate the practical aspects

B. Wang and C. Wu, *An Introduction to Safety and Security Intelligence Science*,
https://doi.org/10.1007/978-981-96-7261-5_5

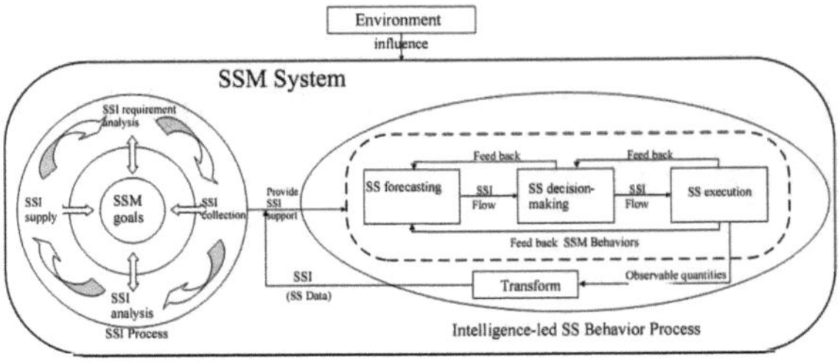

Fig. 5.1 Model of "SSI-SSM behaviors"

and characteristics of intelligence and SSM operation to analyze the mechanism of SSI in SSM. On this basis, the exploration of how to analyze and implement the role of SSI in SSM is undertaken, aiming to provide a universally applicable theoretical foundation for SSM research and practice from the perspective of intelligence studies.

5.2.1 *Analysis of the Mechanism of SSI in SSM*

The "SSI—SSM behavior" model is constructed and analyzed to elucidate the mechanism of SSI in SSM.

5.2.1.1 Model Construction

It can be inferred from the above analysis that the ultimate goal of generating SSI is serving and supporting SSM instead of possessing SSI, and the essence of SSM lies in executing SSM tasks with SSI. Seen from the perspective of intelligence, SSM ought to cover two major aspects: the execution of the SSI procedure and the procedure of intelligence-led SSM behaviors. The model of "SSI-SSM Behaviors" is constructed (Fig. 5.1) with reference to general intelligence procedures (Yao Leye and Fan Wei 2014; Qu Zhikai et al. 2017) and Yovits' generalized intelligence system model (Ye Ying and Wu Yishan 2012; Yovits and Kleyle 1993).

5.2.1.2 Interpretation of the Model

From the Model of "SSI-SSM Behaviors," it is understood that from an intelligence perspective, SSI operation is an integral part of SSM, and it runs through the entire process of SSM behaviors. SSI and SSM behaviors mutually reinforce (i.e., they coordinate and influence each other). The entire "SSM System" consists of two core components: the "SSI procedure" and the "procedure of intelligence-led SSM behaviors. The core of the SSI procedure is intelligence operation, which is the prerequisite and foundation for implementing the intelligence-led SSM process; the core of the procedure of intelligence-led SSM procedure is the SSM behaviors based

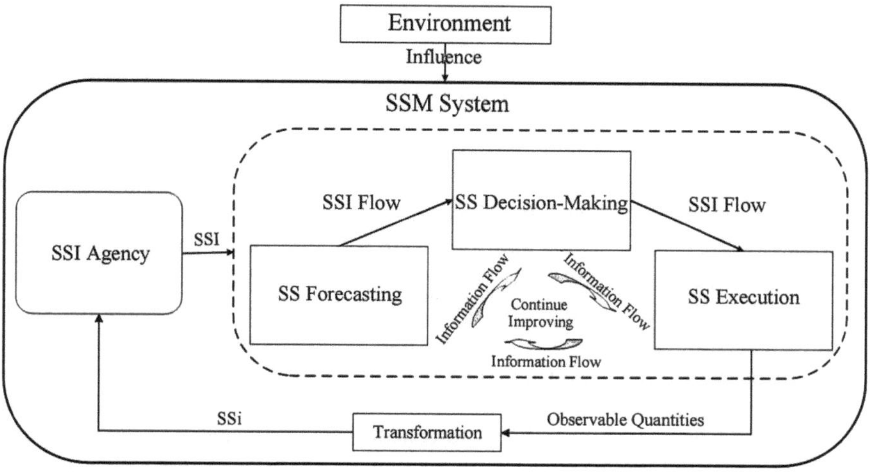

Fig. 5.2 SSM model from the SSI perspective

on SSI (including behaviors of SS forecasting, decision-making, and executing). Additionally, the entire SSM system is influenced by external environmental factors. Here the SSI procedure and the procedure of intelligence-led SSM behaviors will be explained.

1. SSI Procedure

As depicted in Fig. 5.2, from the SSM perspective, the SSI procedure possesses two main features:

(1) The basic steps of the SSI procedure are akin to the four fundamental steps of a general SSI procedure (Qu Zhikai et al. 2017). The first step is "analysis of SSI needs," which involves determining SSI needs based on the SSM issues faced or raised by SSI users (referred to as "SS managers in the narrow sense") and issuing tasks for SSI collection and analysis. The second step is "SSI collection," where SSI collectors carry out the collection based on the received tasks. The third step is "SSI analysis," where analysts discern, summarize, and synthesize relevant SSi, combining historical SS data and their own experience to analyze and assess related SS factors. And the fourth step is "SSI provision," where analysts provide the finished SSI products to SSI users.
(2) Participants in intelligence-led SSM activities (referred to as "SS managers in the broad sense," including SSI collectors, analysts, and users) establish an SSI network centered around specific SSM objectives, enabling real-time sharing and mutual exchange of SSI. These participants can join in real-time and in-depth exchange and interaction of SSI around central SSM goals, ensuring that the collection, analysis, and utilization of SSI information are more targeted and purposeful. This effectively promotes the fluidity of SSI information, breaks

down barriers between different SSi silos, and makes the collection, analysis, and utilization of SSI more effective, timely, and efficient.

2. Procedure of Intelligence-Led SSM Behaviors

The realization of the value of SSI relies on advocating and practicing the concept of "intelligence-led SSM," which involves establishing and executing the procedure of intelligence-led SSM behaviors. SSI exerts its utility and produces observable outcomes by serving the behavioral procedure of SSM (Ma Feicheng 1984). The procedure of intelligence-led SSM behaviors centers around the "SSM behaviors" following the receipt of SSI, leveraging the effects of SSI, and observing its impact on SSM behaviors. On this basis, by determining various parameters and variables of the SSI process, one can form certain perspectives and methods for studying the utility and value of SSI. It is particularly important to note that the support and service of SSI to SSM behaviors, namely SS forecasting, decision-making, and executing, are not necessarily influenced sequentially, they three are more likely characterized by a relationship of concurrent random selection.

The SSI operation, which is the implementation of the SSI process, should be conducted by SSI organizations. SSI is provided by these organizations to SSI users (SS managers in the narrow sense). The SS managers utilize the acquired SSI to select the optimal path for SSM behaviors and transform the SSI into observable SS data during the execution of SS behaviors. SSI then re-enters the organization from the external environment and through the internal feedback. SS managers can use the feedback of SS data information to adjust their SS forecasting, decision-making, and executing behaviors repeatedly and cyclically.

5.2.2 *Value Analysis of SSI in SSM*

As inferred from the above analysis, the SSI-SSM model primarily analyzes the mechanism of SSI in SSM from a qualitative standpoint (i.e., the interaction and connection between SSI and SSM behaviors), yet it does not quantitatively assess the value of SSI in SSM (i.e., the support and service of SSI for SSM behaviors). In light of this, a quantitative model for analyzing the value of SSI in SSM is established according to the model. During the building process, utility functions, ex-ante post-hoc analysis methods (Zhang Lingying 2002), relevant mathematical knowledge (such as Bayes' theorem and the total probability formula, etc.), and intelligence value analysis models from other domains (Zhang Lingying 2002; Jiao Jian 2008) are consulted.

5.2.2.1 Analysis of Expected Utility of Using SSI

Assume that the basic elements of an issue of SSM are as follows: (1) all possible SS states of the system, or the system SS state space is $S = \{ s_j \} \quad (j = 1, 2, 3, \cdots, n)$;

(2) the probability distribution of the system SS state space S is $P = \{p_j\}$ $(j = 1, 2, 3, \cdots, n)$. P can represent both the subjective probability of the SS manager and the objective probability; (3) the set of various SSM schemes (paths), i.e., the space for SSM behaviors is $B = \{b_i\}$ $(i = 1, 2, 3, \cdots, m)$; (4) the set of execution results of different schemes of SSM behavior, i.e., the resultative space for SSM behaviors is $R = (r_{ij})^{m\times n}$ (in other words, R is the benefit value that SSM leads to, or referred to as SSM performance). The above assumptions are shown in Table 5.1. Obviously, Table 5.1 can present a model of SSM behavior analysis.

In addition, if we suppose the utility function of SS manager is $E = (*)$, the goal of SSM is to maximize the expected utility of the SS manager. For the SSM behavior scheme b_i, its expected utility value $E = (b_i)$ is expressed as

$$E = (b_i) = \sum_{j=1}^{n} p_j \bullet E(r_{ij})(i = 1, 2, \cdots, m; j = 1, 2, \cdots, n) \tag{5.1}$$

Clearly, much SSi is involved (such as the probability values of the system in various SS states, etc.) in the above issue of SSM. For the convenience of description in the following text, the SSi originally possessed by the SS manager, such as SS experience, SS knowledge, and SS records, can be referred to as previous SSi (Zhang Lingying 2002; Jiao Jian 2008). When the SS manager has no intelligence available for the future development and changes of the system's SS state, the goal of the management is to select the plan of SSM behaviors with the maximum expected utility value as the plan, that is, to select scheme $\overline{b}$ of SSM behaviors. Its expression is

$$E\left(\overline{b}\right) = \max_{1\leq i\leq m} E(b_i)\ (i = 1, 2, \cdots, m) \tag{5.2}$$

In fact, it is more difficult to make accurate predictions about the future SS state of the system because of the uncertainties of the future SS state, especially when the probability values of each system's SS state (p_j) are very close. Therefore, in the operation of SSM by SS managers, the key of directly affecting the results of SS behaviors lies in the estimated accuracy and precision of the probability distribution of the system's SS state. In addition, when the future SS state of the system differs

Table 5.1 Model of SSM behavior analysis

S B	s_1 p_1	s_2 p_2	s_3 p_3	…	s_n p_n
b_1	r_{11}	r_{12}	r_{13}	…	r_{1n}
b_2	r_{21}	r_{22}	r_{23}	…	r_{2n}
b_3	r_{31}	r_{32}	r_{33}	…	r_{3n}
…	…	…	…	…	…
b_m	r_{m1}	r_{m2}	r_{m3}	…	r_{mn}

from the system's SS state under the highest expected utility value, the results brought about by the operated SSM according to formula (5.2) may be failure rather than improvement in SSM performance. Therefore, before implementing SSM, the SSI that is related to changes in the system's SS state should be collected as much as possible to control the dynamics of the state, ensuring the practice of the best SSM adapted to changes of the SS state. On the other hand, as the collection and acquisition of SSI consumes SS costs, if the costs are being too high, the practical value of SSM can be notably weaken or even lost.

In summary, if SS managers can obtain SSI reflecting the probabilities of system SS states in certain way before executing SSM behaviors, the obtained intelligence can be used for updating previously gained SSi, then SS managers can acquired new probabilities of the occurrence of system SS states [which can be referred to as "posterior SSi" (Zhang Lingying 2002; Jiao Jian 2008)]. Theoretically, posterior SSi is more accurate and reliable than previously gained SSi (Zhang Lingying 2002; Jiao Jian 2008). What's more, from the intelligence perspective, the most critical question throughout the SSM process is: how much SSI do SS managers need to acquire in order to achieve the optimal balance between the cost of acquiring SSI and the risk of SSM behaviors, thereby maximizing the utility value of the resulting SSM behaviors? To address this question, a method of preliminary post analysis is employed here (i.e., using posterior SSi to conduct a preliminary post analysis of the expected utility of SSM behaviors) (Zhang Lingying 2002) to determine the expected utility value of SSI in SSM behaviors. Based on this, it is assessed whether SS managers need to acquire more SSI to ensure that the final SSM behavior actions are optimal.

If C_h represents the hth ($h = 1,2,\cdots, k$) SS state in terms of the future SS state of the system shown by the SSI used by the SS managers, and P ($C_h \mid s_j$) represents the probability that the future system will appear in the jth ($j = 1,2,\cdots,$ n) SS state and the SSI used by the SS manager C_h is shown as the hth SS state. Then

$$P(C_h|s_j) = \begin{cases} T_1 & (h = j) \\ 1 - T_1 & (h \neq j) \end{cases} \tag{5.3}$$

In the formula, T_1 represents the reliability of the SSI, while $1 - T_1$ indicates the unreliability of SSI. According to the Bayesian formula and the total probability formula, it can be concluded that when the SS manager uses the SSI C_h, the system's future SS state will appear the hth SS state, but the actual SS state of the system is the jth SS state, the conditional probability P ($s_j \mid C_h$) is

$$P(s_j|C_h) = \frac{p_j \bullet P(C_h|s_j)}{\sum_{j=1}^{n} p_j \bullet P(C_h|s_j)} \tag{5.4}$$

According to formula (5.1), with the prerequisite that the SS manager uses the SSI C_h, the expected utility value of each scheme of SSM behaviors is

$$E\left(b_i^h\right) = \sum_{j=1}^{n} P\left(s_j \middle| C_h\right) \bullet E\left(r_{ij}\right) \tag{5.5}$$

According to formula (5.2), provided that the SS manager uses the SSI C_h, the scheme of SSM behaviors with the maximum expected utility value b^h must satisfy:

$$E\left(b^h\right) = \max_{1 \le i \le m} E\left(b_i^h\right) \tag{5.6}$$

In addition, within the SSI used by the SS manager, the probability of the occurrence of the SSI C_h is

$$P(C_h) = \sum_{j=1}^{n} p_j \bullet P\left(C_h | s_j\right) \tag{5.7}$$

It can be seen that if an SSM behavior is executed with SSI (i.e., executing SSM behaviors on the basis of posterior SSi), the maximum expected utility value of this SSM behavior is

$$E\left(\hat{b}\right) = \sum_{h=1}^{k} P(C_h) \bullet E\left(b^h\right)(h = 1, 2, \cdots, k) \tag{5.8}$$

5.2.2.2 Value Analysis of SSI

If G represents the SSI used by the SS manager, then V(G), the value of SSI G in SSM, is the difference between the maximum expected utility value obtained by the SS manager in executing SSM behaviors on the basis of posterior SSi and the one obtained by the SS manager in executing SSM behaviors on the basis of prior SSi, that is,

$$V(G) = E\left(\hat{b}\right) - E\left(\overline{b}\right) \tag{5.9}$$

To ensure the rational execution of SSM, it is necessary to analyze the value of SSI in the context of SSM behaviors. According to Eq. (5.9), the value of V(G) will clearly result in one of two outcomes: $V(G) \le 0$或$V(G)>0$. (1) If $V(G) \le 0$, it indicates that the SSI does not have practical value for SSM, and will not increase the expected utility value of SSM actions. (2) If $V(G)>0$, it suggests that obtaining SSI can enhance the expected utility value of SSM behaviors (meaning that the SSI has utility value for SSM), thus a preliminary value analysis can be conducted to determine whether to acquire additional SSI.

Assuming that the SS cost of obtaining SSI G is D(G), then the value analysis formula of SSI G for SSM is

$$\Delta(G) = V(G) - D(G) \quad (5.10)$$

Similarly, according to formula (5.10), it is clear that the value of V(G) will also lead to two results: $\Delta(G) \leq 0$ or $\Delta(G)>0$, the specific analysis is as follows:

1. If $\Delta(G) \leq 0$, the SS cost of the obtained SSI by the SS manager is higher than the additional expected utility value generated by the SSI for SSM. In this case, the overall benefit decreases instead of increases, and the cost of obtaining SSI is too high considering that there is also time cost for collecting and obtaining SSI. For an organization, when $\Delta(G) \leq 0$, obtaining SSI has no practical significance for SSM. There is no need for the SS manager to attain SSI, who needs to work out a scheme of SSM behaviors according to the prior SSi, that is, according to formula (5.2).
2. If $\Delta(G)>0$, the cost of the SSI obtained by the SS manager is lower than the additional expected utility value generated by the SSI for SSM. This indicates that the SSI obtained by the SS manager can generate additional expected utility for SSM. For an organization, when $\Delta(G)>0$, there is practical significance in obtaining SSI for SSM. In this case, it is necessary to collect and obtain relevant SSI timely to assist SSM in order to improve SSM performance. The additional expected utility value is $\Delta(G)$, and the size of the $\Delta(G)$ value is directly proportional to the value (utility) of SSI for SSM. However, it should be noted that after SS managers use the SSI (posterior SSi), if the best scheme of SSM behaviors is still the best scheme of SSM behaviors made on the basis of the prior SSi, then even if $\Delta(G)>0$, obtaining SSI has no practical significance for SSM. Therefore, there is no need to attain SSI.

In summary, the value of SSI in SSM can be clearly known from a quantitative perspective. Furthermore, by calculating and analyzing the value of SSI in SSM, SS managers can easily, rationally and economically obtain and utilize SSI for the best SSM behaviors to achieve the best performance.

5.2.3 Conclusions

This section establishes and analyzes the model of "SSI-SSM Behavior" for elucidating the mechanism of SSI in SSM. With utility functions and relevant mathematical knowledge, a value analysis model of SSI in SSM is built to clarify the role and mechanism of SSI in SSM and to propose basic methods and approaches for analyzing and realizing the value of SSI in SSM, providing a theoretical basis and reference for the research and practice of SSM from an intelligence perspective. The research on the value and the theory of functions of SSI in SSM in this section is oriented toward the domain and process of SSM and takes a theoretical perspective on SSM science and intelligence studies, it can provide universal theoretical support and reference for the research and practice of SSM from an intelligence viewpoint (i.e., intelligence-led SSM). Following the macro-level discussion in this section, subsequent research needs to probe into the specific content of each module and their interrelationships as proposed in this section.

5.3 Intelligence-Led SSM Methods

As previously mentioned, SSI, which is the foundation of SSM, is aimed at serving and supporting SSM. It also serves as the focal point in SSM, undertaking roles that are indispensable, after all, the absence of SSI in SSM is like water without a source. Meanwhile, intelligence-led practical methods [such as "intelligence-led policing" (Peterson 2005), "intelligence-led competition" (Zhao Bingfeng 2014), etc.] are prevailing methods in recent years. For these reasons, SSI researchers believe that the realization of the value of SSI depends on advocating and practicing "intelligence-led SSM" (Wang Bing and Wu Chao 2018d; Yao Leye and Fan Wei 2014; Dong Yin and Liu Qianli 2016; Liska 2015). The intelligence-led SSM has the potential to become the most influential driver of SSM revolution in the twenty-first century, and is able to represent the future mode of SSM operation in the information age, particularly the intelligence era.

However, it is regrettable that the academic community has not yet clarified the universality and fundamentally theoretical issues of intelligence-led SSM, leading to a lack of basic theoretical foundations and methodological guidance for its research and practice. In light of this, this section employs literature analysis and speculative methods to conduct an in-depth analysis of the ground for establishment, definitions, and models of intelligence-led SSM, aiming to lay a theoretical foundation for its research and practice.

5.3.1 Philosophical Basis of Intelligence-Led SSM

From a basic conceptual perspective, "intelligence-led SSM" is a combination of the academic concepts of "SSI" and "SSM," which originally belongs to two independent academic fields (i.e., intelligence studies and SS science). Yet do they possibly have the basic conditions for integration? Can that be effective? These questions can be taken as the first philosophical proposition to establish intelligence-led SSM. From a philosophical standpoint, the philosophical foundation for the integration of two or more entities is that there should be some similarities between them, that is, points of convergence. Therefore, from a philosophical perspective, identifying and determining the points of convergence between the concepts of "SSI" and "SSM" is the philosophical premise and foundation for proposing intelligence-led SSM. Drawing from the research of Wang Bing and Wu Chao (2019e), Yao Leye and Fan Wei (2014), Dong Yin and Liu Qianli (2016), and Liska (2015), the points of convergence between the two academic concepts of "SSI" and "SSM" or their commonalities, are primarily reflected in the following eight aspects.

5.3.1.1 Centered on SS

Both concepts adhere to the concept of "centering around SS." As a series of management activities carried out to enhance system SS, SSM operations always stay committed to the concept (or goal) of "centering around SS." In the same vein, SSI adheres to the concept of "centering around SS" (in other words, "centering around SS" is the substantial difference and characteristic of SSI, compared with other types of intelligence), and its concept is proposed to provide new elements, perspectives, and paradigms for promoting SS.

5.3.1.2 Focus and Emphasis on Solving SS Problems

Both concepts focus on and emphasize the resolution of SS problems. The concept of intelligence is centered on problem-solving, and likewise, the concept of SS intelligence also highlights the importance of addressing SS problems, which means that SSI efforts should always be directed toward dealing with SS concerns. In terms of SSM, tackling SS issues is both the starting point and the ultimate goal, and the identification, analysis, assessment, and response to SS problems constitute the core content and essential components of SSM.

5.3.1.3 SSI as the Basic Resource

Both concepts emphasize the fundamental role of SSi. SSI is SSi analyzed and processed, thus a complete and accurate collection of information characterizing the system's SS state forms the basis for obtaining and producing high-quality and effective SSI. Since SSi is the rudimentary element and resource of SSM, Wang Bing and Wu Chao (2017) made an important assertion: "The road to safety and security must pass through safety and security information."

5.3.1.4 Emphasis on "Appropriate SS Scenarios" and "Moderate SS Costs"

First, both concepts emphasize "appropriate SS scenarios." Both are in reference to the scenario of system SSM within certain time and space (such as a well-structured SSM system, SS risk factors of a system, and the SS culture environment inside and outside a system), which is the boundary for both concepts. Second, both concepts emphasize "moderate SS costs": the cost of obtaining SSI and achieving SSM goals must be moderate. If it is unacceptable, the activities carried out around SSI and SSM would lose their practical significance and value.

5.3.1.5 Concern for the SS Environment

Both concepts support the view that "a safe and secure environment is a system open up" and concern about the impact of the internal and external environment on system SS (such as a system's response to changes in the external SS environment, as well as SS threats, challenges, and SSM opportunities to a system which are brought about by changes in the SS environment), in order to better understand and intervene in the factors causing and driving the course of SS events.

5.3.1.6 Consideration of the Entire System SS

Both concepts are grounded at the systemic level, focusing on the SS operation and development of the system and its related subsystems. SSI aims to comprehensively collect and provide SSI that can enhance the SSM of the entire system under the guidance of the system's SSM objectives. Similarly, SSM emphasizes systemic (comprehensive) approach, which means managing SS factors that could possibly impact the entire system.

5.3.1.7 "SS Managers in the Broad Sense" as Subjects of Implementation

The content of both concepts refers to a series of activities, and the implementation subjects of these activities can be summarized as "SS managers in the broad sense." The "SS managers in the broad sense" mainly include intelligence professionals (also referred to as "information professionals" in SSM) and SS professionals. That is to say, whether it is operations of SSI or SSM, it requires the cooperation of intelligence professionals and SS professionals. Neither can effectively complete the tasks of SSI and SSM alone, but if some SS professionals possess necessary intelligence skills, the tasks can be fulfilled by SS professionals alone.

5.3.1.8 Focus on "Future SS"

SSI aims to analyze and produce intelligence based on past and present SSi, utilize the produced intelligence to predict the future SS state of a system, and guide and improve the future SSM of the system. As one of the important characteristics of SSM, "futurity" emphasizes the idea of "Prevention is the first concern" and "Preventing before problems occur," aiming to ensure the "future SS" of the system.

The series of common ground possessed by the above "SSI" and "SSM" concepts are the key to integrating the two, as they constitute the philosophical basis for proposing the concept of "intelligence-led SSM." In addition, since SSI serves as SSi that influences management, there is a clear cause-and-effect relationship between the concept of the two (the former is the cause, and the latter the effect). Furthermore,

SSM and SSI support each other in interaction (SSI supports and serves SSM, which is the basic "carrier" of SSI as SSI originates from SSM). The above characteristics form an essential philosophical basis for the integration of the two.

5.3.2 *Theoretical Basis of Intelligence-Led SSM*

In the process of SSM, SSi is always in flux, while SSI is the uncertain SSi that directly confronts SSM issues and are valuable and meaningful for SSM (Wang Bing and Wu Chao 2018d; Yao Leye and Fan Wei 2014). It is evident that SSI permeates all aspects of the entire SSM process, with its purpose of serving and sustaining SSM, for it provides support, focal point, and essential elements for SSM. It is crucial to SSM as the fountainhead is to a river.

The ultimate goal of SSI is not to possess it, but to utilize it, which emphasizes service and support for SSM behaviors to achieve "proactive prevention," "keen awareness," and "targeted execution." From an intelligence perspective, the essence of SSM is the utilization of SSI by SS managers to perform SSM behaviors. On this basis, an SSM model within the purview of SSI is constructed (as shown in Fig. 5.2) according to Yovits' model of general intelligence system (Ye Ying and Wu Yi Shan 2012; Yovits and Kleyle 1993), and Wang Bing and Wu Chao's (2017b) SSM behavior model.

Shown by Fig. 5.2, from the perspective of intelligence, SSI operation is an integral part of SSM, running through the entire process of SSM behaviors. SSI and SSM are reciprocal, that is, they are mutually influential (it should be pointed out that the support and service of SSI for SSM behaviors, i.e., the operations of SS prediction, decision-making and execution, do not necessarily influence each other in a linear sequence, but more like a parallel and random selection that occurs among the three). Therefore, the content of SSM can be divided into two sections, namely "SSI operation" and "performance of SSM behaviors based on SSI (intelligence-led SSM behaviors)." Among them, SSI operations are conducted by an SSI organization, which provides SSI for the SSM division to support the execution of SSM. It should be noted that the SSI organization and the SSM division can be arranged as two different units or work as one (i.e., the generalized SSM organization). SSI plays a role by serving SSM and engendering observables (Yovits and Kleyle 1993), which are transformed into SSi (data), which is then transmitted to the SSI organizations as information resources for the production of SSI, thus a cycle of SSM mode is formed. In addition, the entire SSM system is influenced by the environment.

5.3.3 *Practical Basis of Intelligence-Led SSM*

In terms of practice, the direct goal of intelligence-led SSM is to achieve the organic integration of intelligence, while in terms of theory, the interwork between intelligence operation and SSM endeavors is the basis and premise for integrating the two. First, they share a common goal of supporting SSM, especially SSM decisions, through the analysis of relevant information (including data), or in other words, both depend on an information analysis process. Additionally, intelligence professionals can enrich, promote, and extend the connotation, category, and boundary of intelligence operation; they can also develop new ways of understanding SSM practices from the perspective of intelligence. In the meantime, the thought of intelligence professionals on SSM practices under the perspective of intelligence can assist them to explore new ways of SSM practices and improve them. Secondly, the performance of intelligence operation and the practice of SSM are similar. The similarities of the two are compared according to the basic model of intelligence operation (Peng Zhihui 2016) and the basic model of SSM (Luo Yun et al. 2016; Wang et al. 2017) (as shown in Fig. 5.3), on the basis of the research of Wu Chao and Wang Bing (2018b); Luo Yun et al., (2016); Dong Yin and Liu Qianli (2016); Liska (2015); Peng Zhihui (2016); Wang et al. (2017). The analysis is as follows:

1. Both emphasize the application of a "task (goal)-centered" operation mechanism and process (Wang et al. 2017; Wu Subin et al. 2013). In intelligence operation and SSM practices, all participants working around the same task (goal) in respective fields can achieve real-time, in-depth information exchanges, ensure that the operation is more goal-oriented and targeted, break down information barriers among members, and boost cooperation, thus making the operation more effective, real-time, and efficient.
2. The central task of the first step for both is "determining the scope, content, and needs of the work." In the step of "intelligence planning" of intelligence operation, it is important to clarify the intelligence needs of the service objects

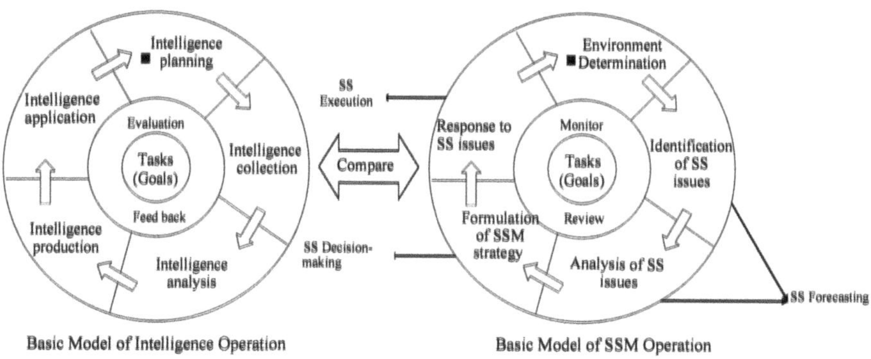

Fig. 5.3 Comparison of SSI operation and SSM practice

(intelligence users) of intelligence operation, which include the specific field, content, and needs of intelligence operation. In the step of "environment determination" of SSM, it is important to determine the requirements, scope, tasks, and key points of SSM, which is based on the clarification of system environment (background) (such as the goals, boundaries, criteria, requirements, constraints, key tasks of SSM, etc.).

3. The focus of the second step for both is "information collection." In this step of intelligence operation, it is important to identify and gather information from various sources. In the step of "identification of SS issues" of SSM practices (part of "SS prediction"), the key task is the identification and collection of SSi. In addition, they both emphasize the reliability and credibility of the information collected during the collection process.
4. The focus of the third step for both is "analysis." In this step of intelligence operation, information is analyzed and transformed into intelligence. In the step of "analysis of SS issues (including assessment)" SSM practices (part of "SS prediction"), it is important to analyze and evaluate the features, causes, nature, and consequences of SS issues based on a systematic analysis of the collected SSi, and then produce the results of analysis and evaluation.
5. The objective of the third step for both is to yield "products" that meet users' needs. In the step of "intelligence production" of intelligence operation, the priority of intelligence is firstly confirmed, then the intelligence is "packaged" into an authentic "intelligence product" which is to be supplied to intelligence users for application. In the step of "SSM strategy formulation" (equivalent to "SS decision-making") of SSM practices, the goal is to formulate feasible SSM strategies (essentially a kind of "intelligence product" from an intelligence perspective) based on the SS issue analysis and evaluation, and then to confirm the priority of SSM strategies.
6. The task of the fifth step for both is "application." In the step of "intelligence application" of intelligence operation, intelligence users utilize intelligence products to roll out decision-making and other activities. In the step of "responding to SS issues" (equivalent to "SS execution") of SSM practices, SS managers employ SSM strategies to respond to and tackle with SS issues.
7. The mechanism of "feedback regulation" runs through the entire process of intelligence operation and SSM practices. Both share extremely similar functions and goals in terms of the section of "evaluation and feedback" of intelligence operation and the section of "monitoring and reviewing" of SSM practices, that is to timely adjust and improve the content and methods of operation through the mechanism of "feedback regulation."
8. Both of them run as a cycle. They are always repeated and continuous; neither is a "one-off." Both emphasize "ongoing improvement and perfection," which demands constant repetition of every section and procedure for achieving continuous improvement of performance.

5.3.4 Meaning of Intelligence-Led SSM

Based on the above analysis, the crux of SSI research and practice is that the realization of the value of SSI in SSM depends on actively advocating and practicing the concept of "intelligence-led SSM," that is, conducting "intelligence-led SSM." It should be noted that although SSi belongs to SSI, the two should not be mixed up (SSi is to SSI what information is to intelligence) (please refer to Hua Bailin and Zheng Yanning 2012, for more details). "SSi" cannot lead SSM as it is often messy and disorganized; only "SSI" can take the lead, hence the title "intelligence-led SSM" (rather than "information-led SSM").

Here, a specific definition of intelligence-led SSM is given. The "intelligence-led SSM" refers to an SSM method, model, and philosophy about "widely collecting SSi, and comprehensively and thoroughly analyzing and processing it to produce SSI products for SS managers to guide SSM activities from different angles including the overall SS strategy, specific SSM strategies and measures." In other words, intelligence-led SSM means collecting and analyzing SSi, generating "SSI products" through processing, and using "SSI products" to lead SSM behaviors, thereby improving the overall effectiveness of SSM. In short, intelligence-led SSM means "harnessing SSI to guide the entire process of SSM."

As for the term "led" in this concept, the denotation is: to lead, guide, penetrate, and promote overall development; to play an important guiding role (Lu Xuemei 2008). Therefore, the basic feature of intelligence-led SSM is that SSI is central to SSM (in detail, SSI serves as an objective SSM tool in intelligence-led SSM, and the analysis and interpretation of SSI as the basis of SSM). In addition, intelligence-led SSM emphasizes SSi sharing and cooperation. It promotes the maximization of the leading role of SSI in SSM through coordinating various SSI functions and effectively integrating SSI into SSM, enabling SSI to serve SSM better, and achieving the goal of influencing SSM at all levels with the use of SSI.

From the practical perspective, the intelligence-led SSM aims to highlight the leading role of SSI in SSM, that is, "accurate prediction with foreknowledge, reasonable decision-making with scrutiny, swift execution (response) with prevention." Achieving the goal of "prevention first" is where the essence and advantage of intelligence-led SSM lies. To achieve the "-led" effect in the intelligence-led SSM, it is necessary to strengthen SSI operations to ensure a comprehensive and extensive interpretation and analysis of the obtained SSi, thereby enhancing the support and service capabilities of SSI in SSM. In detail, one ought to have excellent capabilities of intelligence operation and make accurate predictions, correct decisions, and speedy execution on the basis of SSI.

Compared with traditional SSM methods and models (such as accident-based SSM, problem-based SSM, and statistics-based SSM), the intelligence-led SSM has at least seven major advantages (Wang Bing and Wu Chao 2018d; Wang et al. 2017): First, the ground of SSM is the fact (as can be known from the SSi chain, the initial prototype of SSI is facts), which helps to enhance the accuracy and effectiveness of SSM; second, it can deal with SSI missing in the SSM process as much as

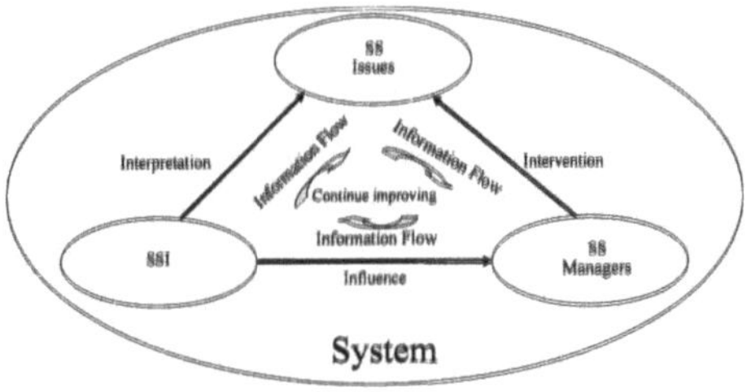

Fig. 5.4 Basic elements of intelligence-led SSM

possible; third, the error correction mechanism is introduced into SSM (the intelligence operation flow involves feedback), which helps to improve SSM; fourth, it helps to share SSi; fifth, traditional types of SSM (such as accident-based SSM, problem-based SSM, and statistics-based SSM, etc.) can be effectively integrated with the intelligence-led SSM; sixth, SSi is the most important SSM resource in the information age, especially in the era of big data, thus the intelligence-led SSM can make full use of SSi resources; seventh, as SSI operation is directly aimed at solving SS problems, the intelligence-led SSM helps to enhance the pertinence and practicality of SSM. In summary, the intelligence-led SSM is more valid, effective, targeted and practical than the traditional methods and modes, and has a huge potential to become the greatest innovation in SSM in the era of big data.

In addition, according to the definition of intelligence-led SSM, three basic elements of intelligence-led SSM can be refined (as shown in Fig. 5.4), namely SSI, SS issues, and SS managers. The relationship among the three is as shown in Fig. 5.4. The function of SSI in dealing with SS issues is interpreting (including identify) SS issues, that is, identifying and analyzing SS issues according to SSI; the function of SSI for SS managers is influencing, that is, managers use SSI to support their SSM activities; the role of SS managers play in tackling SS issues is solving SS issues. These three communicate and cooperate through the information flow to achieve continuous improvement of SSM performance.

5.3.5 *Intelligence-Led SSM Model*

In order to provide overall theoretical and methodological guidance for the research and practice of intelligence-led SSM, the conceptual model (emphasizing the theoretical level) and execution model (emphasizing the practical level) of intelligence-led SSM are constructed according to the basic theories and meaning of intelligence-led SSM.

5.3.5.1 Conceptual Model

The conceptual model of intelligence-led SSM is shown in Fig. 5.5. The model shows that intelligence-led SSM focuses on how to achieve system SS through effective SSM (i.e., predicting, deciding, and executing accurately. SSM cannot be fulfilled without support from leadership, organization, coordination, control, and evaluation in general management), for it depends on punctual, timely, and comprehensive SSI. And the generation of punctual, timely, and comprehensive SSI cannot be fulfilled without the support of SSI operation (its workflow is shown in Fig. 5.5; it also needs auxiliary activities like leadership, organization, coordination, control, and evaluation of general management). In addition, the execution of intelligence-led SSM needs theoretical guidance from theories and methodologies of system SS science, which mainly includes:

1. Theory of "Four Flows in One" in System SS

This theory holds that SSi plays an integrative function, as the three flows of materials, energy, and SS behaviors within a system can be integrated into an SSi flow. Therefore, SSi can be used to represent the overall SS state of the system (Wang Bing and Wu Chao 2018c). As SSI originates from the processing and analysis of SSi, a complete and effective supply of SSI operation is the prerequisite for obtaining comprehensive and accurate SSI. Therefore, this theory provides one of the core theoretical foundations for executing intelligence-led SSM.

2. Methodology of "Way of Things—Way of Matters—Way of People" of SS in System SS

This methodology is proposed on the basis of the system methodology of "Way of things—Ways of matters—Way of people" in the field of system science (Gu Jifa et al. 2007). Its theoretical significance for executing intelligence-led SSM is to guide the building of a system SSI framework. According to Yao Leyuan and Fan Wei, (2014), from the perspective of building a system SSI framework, the "Way of

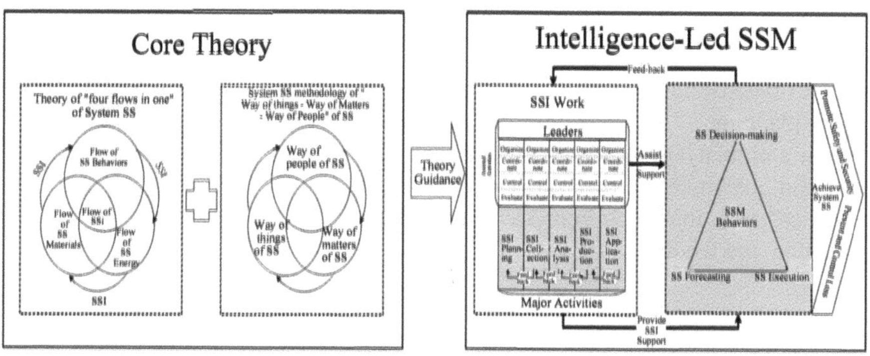

Fig. 5.5 Conceptual model of intelligence-led SSM

things" of SS includes SSI resources and methods, and a system of SSi technologies, among which SSI resources constitute the body and SSI methods, and the system SSi technologies provide auxiliary support. The purpose of studying the "Way of things" of SS lies in improving the hardware capacities of SSI collection and analysis. The "Way of matters" of SS covers relevant system SSM organizations, cooperation mechanisms established among various organizations, and so on, with its objective of improving the efficiency of SSI operations. The "Way of people" of SS refers to the staff involved in SSM, including personnel of SS prediction, SS decision-making and SS execution, SS professionals, personnel of intelligence and operation, etc., and its goal is to appropriately distribute SSI to various SSM roles of personnel, so as to fulfill accurate prediction, effective decision-making, in-place execution, personnel linkage, etc.

5.3.5.2 Implementation Model

According to the practical foundation of intelligence-led SSM (or "the fit between intelligence operation and SSM practice"), despite the difference of specific objectives (tasks), contents, and specific tasks and objectives of each link in intelligence operation and SSM practice, their overarching concepts and general approaches are consistent with each other. That is to say, from a practical perspective, intelligence operation and SSM practice are quite similar in terms of objectives (tasks), central links, and the overall process. Being a highly comprehensive interdisciplinary field, SSM remains fairly open to other disciplines, constantly absorbing theories and practical experiences from other disciplines. Thus it is feasible to allow the intervention of SSI activities in SSM, and the two can be integrated to form a model of intelligence-led SSM practice. The execution model of intelligence-led SSM (as shown in Fig. 5.6) is constructed on the basis of Fig. 5.5, taking the similarities between intelligence operation and SSM practice as the starting point and integrating the core elements of both, thereby providing methodological guidance for the intelligence-led SSM practice.

As can be seen from Fig. 5.6, intelligence-led SSM practice is the organic combination of intelligence operation and traditional SSM practice. Centering on certain SSM objectives (tasks), it brings in a basic workflow of reciprocating cycles (the specific objectives and tasks of each link have been illustrated above), which consists of "SS planning," "SSi collection," and "SSI analysis and production" and "identification and analysis of SS issues" (the last two influence and interact with each other and constitute the core links of intelligence-led SSM), "formulation of SSM strategies" and "intervention in SS issues," and perfects SSM practices with timely evaluation and feedback.

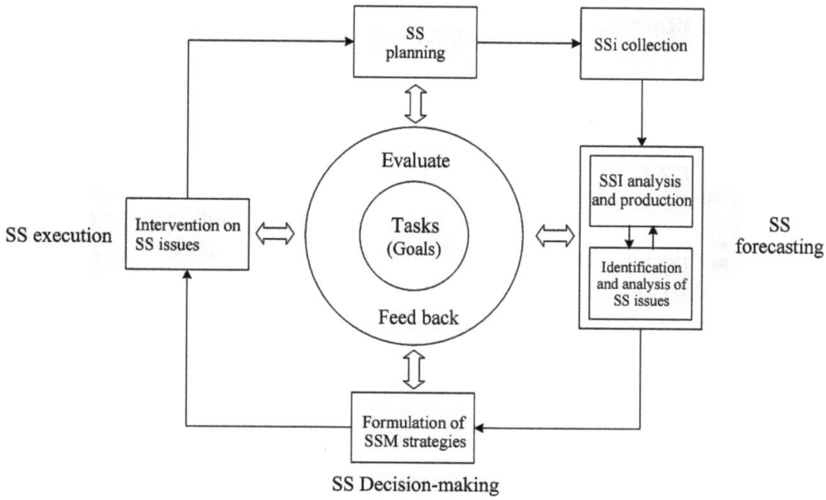

Fig. 5.6 Execution model of intelligence-led SSM

5.3.6 Conclusions

This section discusses the foundation for establishing intelligence-led SSM with methods of theoretical deduction and speculative research. Three levels are included in the discussion: philosophical, theoretical, and practical levels. The meaning of intelligence-led SSM is analyzed and two models of intelligence-led SSM (i.e., the conceptual model and the execution model) are constructed and interpreted. The section shows that there is a solid basis for methods of intelligence-led SSM to form. Intelligence-led SSM refers to commanding and leading the entire process of SSM with SSI, and its conceptual model and execution model can provide general theoretical and methodological guidance for its research and practice. Intelligence-led SSM is a program with a broad field of research and practice. Moreover, intelligence-led SSM focuses on SSI as well as SSM, with its essence being the transformation of the mode of SSM—the transformation and enrichment of intelligence operation as well as the evolution of SSM mechanism and methods. Thus intelligence-led SSM is a domain with enormous space of research and practice for future exploration, inviting more scholars to engage in research to bring about fruitful outcomes, and this section has already made a start.

5.4 Intelligence-Led Urban SSM

Cities serve as the fundamental units of societal governance, and urban SS has remained as a major concern within the social fabric. In the contemporary era which is characterized by rapid urbanization, burgeoning population and the expansion of urban functions and scale, numerous new hazard sources and SS risks keep showing up, complicating the urban operational system. The issue of urban SSM becomes increasingly important, and the global concern about urban SS is gaining prominence (Frevel 2013). On January 7, 2018, the General Office of the CPC Central Committee and the General Office of the State Council of China issued *Opinions on Promoting Urban Security Development*, which aims to guide and foster the development of urban security. Since safety and security is a cornerstone of modern cities and a non-negotiable baseline for urban development, this is an opportune moment for conducting research on urban SS.

The field of urban SS is inherently interdisciplinary, which necessitates collaborative efforts across various academic domains. In recent years, the intelligence community has turned its gaze toward intelligence research for urban SSM. A review of the literature reveals that the current research centers on urban emergency management. Notable studies include Li Yang's (2019a) examination of urban emergency intelligence capabilities, the investigation of Guo Hua et al. (2016) into an intelligence system of urban emergency management, and Li Gang and Li Yang's (2016a; 2016b) exploration of the application of smart city intelligence engineering for emergency decision-making, as well as the development of an intelligence system of smart city emergency decision-making. There are two critical stages in urban SSM: "normal urban SSM" (emphasizing preemptive control) and "anomalous urban SSM" (i.e., urban emergency management, focusing on post-incident rescue and mitigation). The emphasis of urban SSM should be placed on "normal urban SSM," which explicitly reflects the fundamental requirement of the policy of "prevention first" in SS operation. Although the support from intelligence is needed in both "normal urban SSM" and "anomalous urban SSM," current research on intelligence in urban SSM has overlooked the segment of "normal urban SSM," resulting in the absence of SSI research oriented toward the entire process of urban SSM and affecting both the scale and depth of the current research. Therefore, there is a pressing need to transit from focusing solely on urban emergency management to the inclusion of the entire spectrum of urban SSM.

This section investigates the overall process of urban SSM and introduces the concept and methodology of "intelligence-led SSM" (Wang Bing and Wu Chao 2019g) [this approach is regarded as an innovative and effective method of SSM in the information age, particularly within the context of big data, and it is closely aligned with the concept of smart SSM (Wang Bing and Wu Chao 2019g). It has been successfully applied in the SSM of mineral resources (Wang Bing et al. 2019)] into urban SSM, aiming at establishing a theoretical framework for intelligence-led urban SSM (encompassing analysis of the intelligence connotations of urban SSM, examination of intelligence operation in urban SSM, and modeling of intelligence-led urban SSM), thereby laying a theoretical foundation and offers methodological guidance for subsequent research and practical application in intelligence-led urban SSM, as well as for intelligence research and practice in the whole process of urban SSM.

5.4.1 An Overview of Urban SS

5.4.1.1 Analysis of the Meaning of Urban SS

From a systemic perspective, SS is defined as the state in which a system is protected from unacceptable adverse internal and external influences (Piètre-Cambacédès and Chaudet 2010; Wang Bing and Wu Chao 2017a), and urban SS can be defined as the state in which the urban system is shielded from unacceptable adverse internal and external influences. The "adverse influences" are manifested by disruption in urban operation, degradation, or paralysis of urban functions, occurrence of SS events or losses of a city, while the "acceptable level of adverse influences" can be gauged by the "acceptable urban SS risks."

The connotation of urban SS can be analyzed as follows with a view to its definition:

1. Essence of urban SS: A typical issue of system SS (Guo Ru and Tang Hong 2013). From the perspective of system SS, a system like a city can be regarded as an object in the study and management of urban SS, thus urban SS is a typical issue of system SS. To put it differently, from the perspective of system SS, the concept of "urban SS" is essentially a shorthand for "urban system SS," thus the management and research of urban SS should address the object "city" in the form of "system."
2. Content of urban SS: A typical issue of SS integration. As urban SS risks cover both safety and security hazards, the content of urban SS includes both levels of safety and security, which are closely connected and mutually influencing each other.
3. Core theory and primary method of urban SS promotion (management): The system SS theory and method. With regard to the essence of urban SS, the core theory and primary method of urban SS promotion (management) should be those of system SS.

5.4.1.2 Analysis of the Complexity of Modern Urban SS

As previously discussed, urban SS is essentially a typical issue of system SS. Modern cities have prominent features of complex giant systems, for they are essentially typical complex giant systems (Wang Bing and Wu Chao 2018f; Guo Ru and Tang Hong 2013). It is the combination of these two aspects that establishes "modern urban SS as a typical issue of the SS of complex giant systems" (Wang Bing and Wu Chao 2018f; Zhou Ganzhi 2002). It is necessary to analyze the SS traits of those systems concerning modern urban SS for understanding the "theory of the SS of complex giant systems for modern urban SS." The features are manifested in three dimensions (the dimension of traits, the dimension of content, and the dimension of elements (Wang Bing and Wu Chao 2018f). Here is the analysis:

1. From the dimension of traits, the traits of complex giant systems for modern cities (i.e., multi-element, multi-subject, multi-level, high-dimensional, multi-scale, non-linear, open, dynamic, etc.) Wang Bing and Wu Chao (2018f) determine that the research object of modern urban SS is a complex giant system. Research on its safety and security should be classified as an SS issue of complex giant systems. This forms the foundation for developing the "theory of the SS of complex giant systems for modern urban SS."
2. From the dimension of content, modern urban SS covers the content about the interaction, interconnection, and transformation between safety and security. Both aspects can be further divided into content of urban SS concerning a series of specific fields or industries, which also influence each other and can be mutually transformed.
3. From the dimension of elements, in modern cities, the flows of people (behavior), economic transactions, materials, energy, and information are highly interconnected and constantly influence the smooth operation of the urban system. They are also the fundamental resource elements necessary for ensuring urban SS. Each of these flows is a complex giant system with frequent interactions and close connections.

5.4.2 Analysis of the Intelligence Implications of Urban SSM

From the perspective of management science, the core function and task of intelligence is to provide effective support for management, particularly in decision-making. Therefore, intelligence can be understood as information that influences management (Chen Chao 2017). Generally speaking, intelligence refers to information that is useful for management. Thus SSI covers all SSi that affects SSM. In modern urban SSM in the era of big data, a huge amount of dynamic and static SSi (including SS data) is being generated, which includes basic static SSi related to urban SS facilities, equipment, and hazard sources, as well as dynamic SSi resulting from changes in urban hazard sources, operation of urban systems, and urban SSM activities—all of these form a bit-space of urban SS and. Urban SSM is always accompanied by the flow of relevant SSi. In terms of urban SSM, SSI is a type of SSi that directly serves and supports urban SSM, which is extracted and analyzed from a large pool of urban SSI and holds significant value and importance for urban management. It runs through the entire process of urban SSM, with its core task and goal being providing effective service and scientific support for urban SSM. Clearly, it is the "the ears and nose, or the pioneer and adviser" of urban SSM, which would be like "a tree without roots" without it.

Building on the aforementioned analysis, the connotation of urban management from an intelligence perspective can be delineated. Urban SSM refers to the process of managing the urban system, with its focus and objective of being maintaining urban SS. It is based on basic SSi (including SS data) and involves the collection, analysis, production, and application of urban SSI to serve and support a series of behaviors and activities aimed at ensuring urban SS (Wang Bing and Wu Chao 2018f). It is important to note that, according to the theory of "four flows in one" of system SS (Wang Bing and Wu Chao 2018c), SSi has an integrative function for the material flow, behavioral flow, energy flow, and economic flow within the system can be integrated and represented by the SSi flow. Thus the overall SS state of the system can be characterized by SSi. Additionally, since SSI is the SSi that is processed and activated, the basic SSi of a city is the fundamental raw material for producing SSI that serves urban SSM.

The meaning of urban management from an intelligence perspective shows that the essence of urban SSM is utilizing SSI to execute urban SSM behaviors and activities; SSI operation is at the core of urban SSM endeavors. According to the meaning and main content of urban SSM, three core dimensions of urban SSM can be formed with reference to Hall's three-dimensional structure (Hall 1969). They are the temporal dimension (reflecting the phases of urban SSM), the logical dimension (reflecting the steps of urban SSM), and the intelligence dimension (reflecting the support required from SSI operation in urban SSM) (Wang Bing and Wu Chao 2018f). Thus, a three dimensional analysis model of "temporal dimension—logical dimension—intelligence dimension" for urban SSM is established (as shown in Fig. 5.7). The specific meanings of each dimension are detailed in Table 5.2. From Fig. 5.7, it is evident that these are the three fundamental dimensions of urban SSM. Among them, the urban SSM from the intelligence dimension serves as the basis and support for the urban SSM from the other two dimensions; various types of SSi (including SS data) generated in the urban SSM from the temporal and logical dimensions become the raw materials for SSI production from the intelligence dimension through feedback, influencing the urban SSM from the intelligence dimension.

In the modern era of urban SSM, it is highly important to comprehend the intelligence-driven nature of urban SSM, particularly in the context of the big data era and the development of smart cities. The construction of smart cities places significant emphasis on the support and application of big data. Utilizing big data for urban SSM is becoming a reality. However, the scientific understanding, the thinking of intelligence studies, theories, and methods are essential for the technology of big data; one cannot rely solely on big data thinking and technology in urban SSM. Without the support of the thinking of intelligence studies, theories, and methods, urban SSM practitioners may not know which SS data need to be collected, what collected information is intended for, or what urban SS issues and risks should be prioritized. The big data would end up being inefficient in urban SSM and urban SS research, or even lead to unforeseen adverse effects (such as the prevalence of irrelevant urban SS data and the absence of useful SSI for urban SSM).

Fig. 5.7 A three-dimensional analysis model of "time dimension—logic dimension—intelligence dimension" of urban SSM

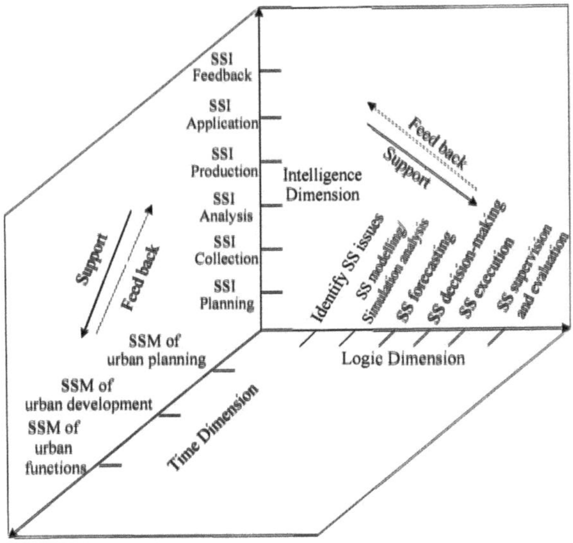

Table 5.2 Connotation of "time dimension-logic dimension-intelligence dimension" in urban SSM

No	Dimension	Connotation
1	Time	Urban SSM is primarily divided into three phases: early-stage urban planning SSM, mid-stage urban construction SSM, and late-stage urban operation SSM. Among these, the key focus of urban planning and construction SSM is to strengthen the root cause governance of urban SS issues. The key focus of urban SSM is the prevention and control of SS risks during the operation process. The relationship among these three phases is not linear, as there is a mechanism of mutual influence and interaction among them; they are in constant dynamic changes
2	Logic	According to the system SSM process, the urban SSM process includes several key stages: identifying SS issues, conducting system SS modeling and SS simulation analysis, making SS predictions, formulating SS decisions, executing SS measures, and carrying out SS supervision and evaluation
3	Intelligence	The intelligence dimension refers to the necessary support of intelligence operation for conducting urban SSM activities (i.e., urban SSM activities in the time and logic dimensions). The process of intelligence operation includes intelligence planning, intelligence collection, intelligence analysis, intelligence production, intelligence application, and intelligence feedback (Wang Bing and Wu Chao 2019g; Peng Zhihui 2016)

Comprehending the intelligence-driven nature of urban SSM is beneficial for urban SSM and urban SS research; it should become the fundamental value in urban SS research and management, lest people's pride in current urban SSM turns into a disaster or confusion in the future.

5.4.3 *Analysis of Intelligence Operation in Urban SSM*

5.4.3.1 Trends and Basic Requirements of Intelligence Operation in Urban SSM

Cities were comparatively small and simple in traditional times, even though their scale has grown as they develop, they remain as simple giant systems. Thus the issue of traditional urban SS is the issue of simple giant systems. In terms of intelligence operation in traditional urban SSM, the reductionist approach of traditional theories of system science (which emphasizes decomposition and simplification) has been relatively effective in guiding urban SSI operations. With this approach, urban SS managers partition the urban system and its SSM system into several subsystems, gathering and analyzing the SSI of each subsystem to support urban SSM efforts.

However, the traits of modern urban SS and their SSM systems as complex giant systems are becoming increasingly prominent, and it is difficult to decompose and simplify numerous sub-issues of urban SS problems and subsystems of urban SSM for they are generally interwoven and overlapping. What's more, even when they are decomposed, they are not the same as traditional sub-issues and subsystems. It is challenging to integrate the analysis and research of those segments, and the optimization of innumerable subsystems may not necessarily lead to the overall SS optimization of the urban system. Because of these, the intelligence operation of urban SSM faces a series of problems like fragmentation, multiple command, linkage failures, and a lack of coordination and unity. Therefore, traditional approaches of decomposition and superposition are no longer suitable for guiding intelligence operation in modern urban SSM, as they will ultimately result in failures in intelligence acquisition, analysis, and application.

The objective of addressing modern urban SS and its management as typical issues of SS and SSM of complex giant systems is to ensure the SS of each subsystem and the synergy and cooperation among the SS of each subsystem, the overall SS and the goals of management of the urban system. It is encouraging that the present information technology, especially the big data, has facilitated information exchange and sharing among urban SSM departments, contributing to the formation of an interactive model of intelligence operation in urban SSM within the "large department" framework and to the application of a mode of urban SSM featuring coordination among various organizations and personnel. In the intelligence operation of modern urban SSM, it is crucial to grasp the orientation and precise focus of the efforts, that is, theories and methods of complex giant systems, information technology, big data, and so on should be harnessed to strengthen the integration and collaboration of intelligence operation in urban SSM, which is a necessary requirement for addressing the problems of modern urban SS and catering to the features of complex giant systems of modern urban SSM systems. Additionally, the "SS theory of complex giant systems of modern urban SS" calls for a shift in the thinking modes of intelligence operation in modern urban SSM, from linear to non-linear, reductive to holistic, entity SS to relational (associated) SS (one of the core thinking mode of SS big data), and static to dynamic. In summary, compared to the intelligence operation in traditional urban SSM, the intelligence operation in modern urban SSM should be a notable leap from "simple to complex, small to large, local to overall, linear to non-linear, entity-based to associated, static to dynamic, narrow to broad, temporary to sustainable, and digital (data) to intelligent," as the traditional recognition of "expansion of scale and scope" is no longer suitable.

5.4.3.2 Method of Intelligence Integration in Urban SSM

Given the trend and fundamental requirements of intelligence operation in urban SSM, it is necessary to employ appropriate theories and technologies as support for integrating and coordinating intelligence operations in urban SSM. In 1990, Qian Xuesen proposed the theory of complex giant systems based on system engineering practices, incorporating interdisciplinary knowledge from systems science, complexity science, and control theory. This theory emphasizes analyzing complex giant systems with holistic and reductionist approaches. It highlights that the only effective way to address complex giant systems is through an integrated qualitative-quantitative method, known as "metasynthetic engineering" (Yu Jingyuan and Liu Yi 2002). This method involves three steps: (1) proposing empirical hypotheses through qualitative integration; (2) employing the means of human–machine integration to perform qualitative-quantitative integration for the aim of achieving integration from qualitative to quantitative; (3) deriving scientific conclusions through qualitative-quantitative integration (Qian Xuesen et al. 1990). In recent years, this method has been widely applied in many researches and processes of complex giant systems, yielding significant results. As an essential methodology for intelligence operation in urban SSM, it can provide solid theoretical and methodological support and guidance for this issue.

The fundamental approach to applying the integrated method to intelligence operation in urban SSM involves integrating thinking, wisdom, experience of safety and security from urban SS managers, experts, and the public, and other SS materials and diverse SS data. Under the guidance of intelligence experts, scientific and information-based methods such as data mining (especially the "trendy" big data mining), text mining, model mining, and mining of experts' opinions are used to guide urban SSM. Quantitative understanding is formed on the basis of multi-faceted qualitative understanding, then the qualitative-quantitative analysis results (from the perspective of intelligence studies, the essence of conclusions and suggestions obtained through qualitative-quantitative analysis is SSI) are integrated for directing urban SSM. As IT-driven SSM informatization becomes an innovative theme of SSM in the new era, it is applied to the field of urban SSM too. Urban SS managers should seize the opportunity of developing urban SSM informatization to integrate expert systems, SSM information systems, and SSI systems for collecting, analyzing, and producing high-quality SSI to serve urban SSM. The structure of the intelligence integration system in urban SSM is depicted in Fig. 5.8, which presents a modern urban SSM support system, where SSI serves as the fundamental medium

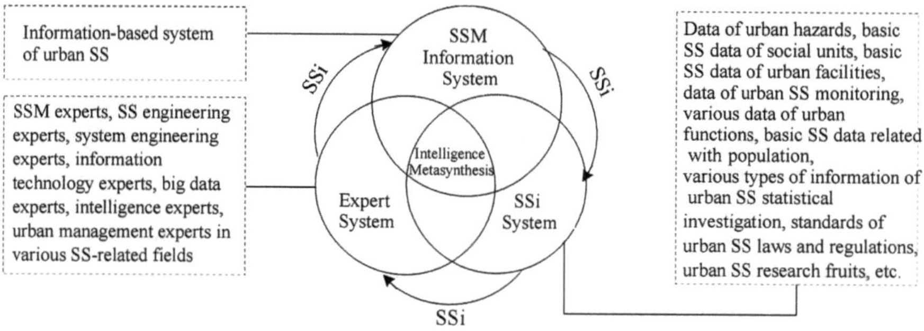

Fig. 5.8 Structure of the intelligence integration system in urban SSM

for achieving interoperability between the expert system, SSM information system, and SSI system. According to the methodology of metasynthesis explicated by Qian Xuesen (1990) and Yu Jingyuan and Liu Yi (2002), the procedure for urban SSI operation on the basis of metasynthesis can be derived, as shown in Fig. 5.9.

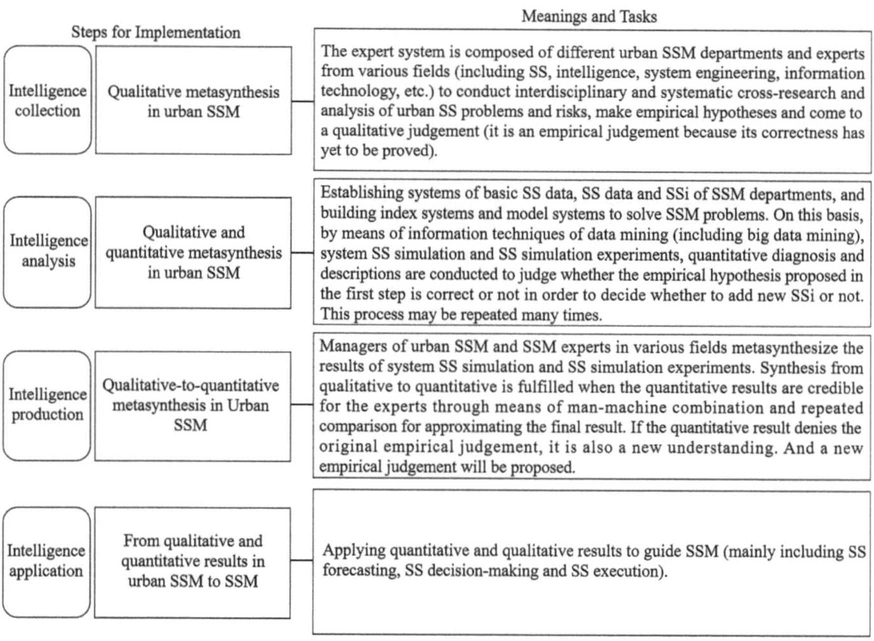

Fig. 5.9 Procedure of the method of intelligence metasynthesizing in urban SSM

5.4.4 Construction of an Intelligence-Led Urban SSM Model

The embodiment of the value of SSI hinges on promoting and implementing the concept and methodology of "intelligence-led SSM" (Wang Bing and Wu Chao 2019g), and the manifestation of the value of intelligence and its operation in urban SSM lies in advocating and executing intelligence-led urban SSM. Considering the definition of the intelligence-led SSM approach (Wang Bing and Wu Chao 2019g), intelligence-led urban SSM involves guiding the entire process of urban SSM with SSI as the foundation. Drawing from the framework of the procedure of the metasynthetic method proposed by Qian Xuesen (1990), a model of intelligence-led urban SSM (Fig. 5.10) is developed according to the procedure of the method of intelligence metasynthesizing in urban SSM (Fig. 5.9).

As illustrated in Fig. 5.10, intelligence-led urban SSM represents the fusion of intelligence operation in urban SSM and traditional urban SSM operation, or the fusion of "metasynthesis of intelligence" (the upper part of Fig. 5.10) with "application of intelligence" (the lower part of Fig. 5.10), aiming to establish a mechanism for urban SSM that "employs the metasynthetic method to acquire and analyze intelligence for guiding urban SSM." The following sections provide a detailed analysis of the connotations of the model of intelligence-led urban SSM.

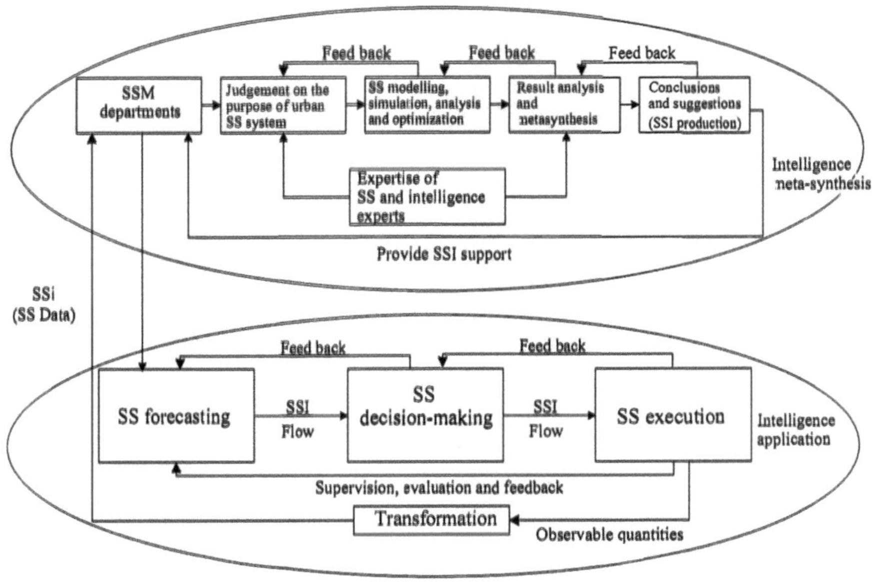

Fig. 5.10 Model of intelligence-led urban SSM

5.4.4.1 Connotation of the Portion of "Metasynthesizing Intelligence"

The primary objective of this portion is to acquire and analyze intelligence that supports urban SSM. The fundamental steps include determining the purpose of the mechanism of urban system SS (equivalent to intelligence planning)—SS modeling, simulation, analysis, optimization, result analysis, and metasynthesis (equivalent to intelligence collection and analysis)—drawing conclusions and making suggestions (equivalent to intelligence production). The details of each step are depicted in Fig. 5.9. It can be known that the essence of this portion is the operation of urban SSI.

5.4.4.2 Connotation of the Portion of "Intelligence Application"

The purpose of this portion is to utilize intelligence to support and guide the entire process of urban SSM (i.e., all the aspects and activities of urban SSM) in order to achieve the desired outcomes of urban SSM. Generally speaking, the main aspects and activities of urban SSM include urban SS prediction (identifying urban SS risks or problems), urban SS decision-making (developing urban SSM plans and specific strategies), and urban SS execution (implementing decisions, i.e., managing urban SS risks or addressing SS problems with the developed plans and strategies). From this perspective, the critical issue of this portion is the performance of intelligence-led urban SSM behaviors and activities based on intelligence.

5.4.4.3 Other Connotations

First, urban SSM departments, including intelligence agencies of urban SSM, are joint entities responsible for conducting "intelligence metasynthesizing" and "intelligence application." They also serve as intermediary organizations that facilitate the integration of these two tasks. Second, "intelligence metasynthesizing" forms the foundation for executing intelligence-led urban SSM for it produces necessary SSI. Third, "intelligence application" utilizes the SSI generated with the metasynthesis method to guide the entire urban SSM process, ensuring that the synthetic intelligence embodies its value in urban SSM practices. Fourth, "intelligence application" can generate new intelligence by providing SS data and information resources for intelligence metasynthesizing in urban SSM.

5.4.5 *Conclusions*

Urban SS is currently a hot issue, and intelligence is essential for urban SSM. Research on urban SS requires the support of intelligence studies, thus Intelligence-led urban SSM is not only a topic of theoretical and practical significance but also a timely concern. On the theoretical level, this section focuses on the entire urban SSM process with the analysis of urban SS connotations and the complexity of modern urban SS as a basis. It examines the intelligence connotations of urban SSM, the operation of intelligence in urban SSM, and the construction of a model of intelligence-led urban SSM. It is found that the essence of urban SSM is to execute urban SSM behaviors and activities on the basis of SSI; the methodology for intelligence operation in urban SSM is the metasynthesis method; and intelligence-led urban SSM is an organic combination of “intelligence metasynthesizing” and “intelligence application.”

The aim of this section is to lay the groundwork for research and practice in intelligence-led urban SSM, thereby advancing intelligence research and practice across the entire urban SSM process, which calls for future studies of urban SSI systems, urban SSI capabilities, and the construction of urban SSI platforms for the entire urban SSM process. Additionally, insights drawn from the research in this section include: intelligence research and practice in urban SSM must be based on the theory of complex giant systems; it is urgent for intelligence-led urban SSM to be elevated to a strategic level; intelligence operations in urban SSM must be fully coordinated and collaborated; it is necessary to establish a mechanism of prediction, decision-making, and execution for urban SSM on the basis of intelligence; and it is important to conduct information engineering (including big data engineering) for urban SSM and intelligence operation in urban SSM.

5.5 Intelligence-Led Smart City SSM

In 2009, the International Business Machines Corporation (IBM) firstly introduced the concept of “Smarter Planet” (Li Deren et al. 2010), marking a significant transition of the digitization of the earth from “Digital Earth” to “Smarter Planet.” A smarter planet hinges on the development of “Smart Cities” (Wu Xibo and Yang Zaigao 2010). However, a smart city must first be a safe and secure one. In other words, urban SS is a prerequisite for building a smart city, primarily because SS issues are closely tied to personal safety and security, property, and well-being of residents in the earth; they are the sources of a city’s vitality and the foundation of orderly functioning, and urban SS (Ba Zhongtan 2008) serves as a microcosm of national SS.

In the context of the rise of the information age and the era of big data, urban management is becoming more integrated, intelligent, and collaborative, and the concept of urban SS is evolving in a more intelligent manner. The methods of urban SSM are to be innovated, and smart cities, encountering unprecedented opportunities and challenges, are going to fulfill historical responsibilities. However, there are still many issues to be addressed in developing and researching the system of smart city SSM. Urban SSM (especially that of smart cities) requires the assistance of SSI.

It is both urgent and practically significant to establish an intelligence-led smart city SSM system. In this section, led by the concept of smart cities, an intelligence-led urban SSM model is developed and an intelligence-driven urban SSM system is established. The objective is to build an urban SSM system featuring scientific prevention and control, accurate early warning, and efficient response. This will promote the transformation of urban SSM from experience-based to intelligence-based, and from passive response to proactive prevention.

5.5.1 A Cycle Model of Intelligence-Led Smart City SSM Operation

What urban SSM needs in the context of smart cities is not a one-size-fits-all SSI, but valuable SSI, or SSI derived from "processed" and "activated" data. Based on the general paradigm for solving SS problems, namely the SSM behavior chain of "SS prediction—SS decision-making—SS execution" and the general process of intelligence operation, a model of urban SSM is constructed, as shown in Fig. 5.11.

The model consists of two loops: the inner loop and the outer loop. The inner loop is the SSM cycle, which embraces four stages: SSI perception, SS analysis and prediction, SS decision-making, and SS execution, represented by P, A, D, and E respectively, thus the cycle of intelligence-led smart city SSM operation can be abbreviated as the "PADE" cycle. The outer loop is the SSI cycle, which also includes four stages: SSI collection, SSI analysis, SSI production, and SSI application. The main connotations of this model are twofold:

Firstly, the SSI loop supports the SSM loop. That is, from the perspective of SSI, urban SSM is conducted with the collection and application of SSI at its core. Without the overall guidance of SSI, there would be no functioning methods in urban SSM

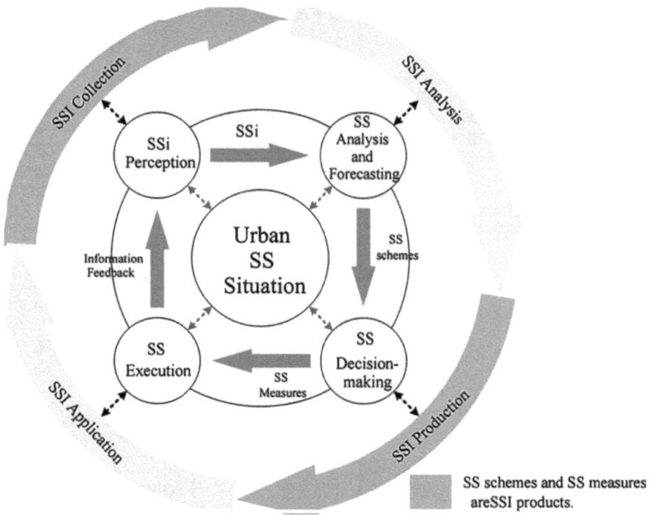

Fig. 5.11 A cycle model of intelligence-led smart city SSM operation

and the efficiency of operation would be reduced. Therefore, intelligence-led city SSM needs to serve as the fundamental method as well as the general principle of smart city SSM.

Secondly, each time a PADE cycle is completed, the urban SS state changes, with unresolved issues carried over to the next PADE cycle. If any link in the PADE cycle fails, the urban SS state cannot be improved in a timely manner, and urban SS cannot be guaranteed. Thus, urban SSM is highly dependent on SSI in the context of emerging new IT technologies such as big data, cloud computing and AI,

5.5.2 *A Model of Intelligence-Led Smart City SSM*

Based on the cycle model of intelligence-led urban SSM operation, this section probes into the essence of urban SSM, clarifies its inherent logic, explores methods and approaches for urban SSM under the concept of smart cities, and constructs a model of intelligence-led urban SSM model (as shown in Fig. 5.12). The connotations of this model are analyzed as follows.

5.5.2.1 Essence of Urban SSM

From the perspective of SSM, the goal of urban SSM is to address the various SS challenges that cities face during their ongoing development through monitoring,

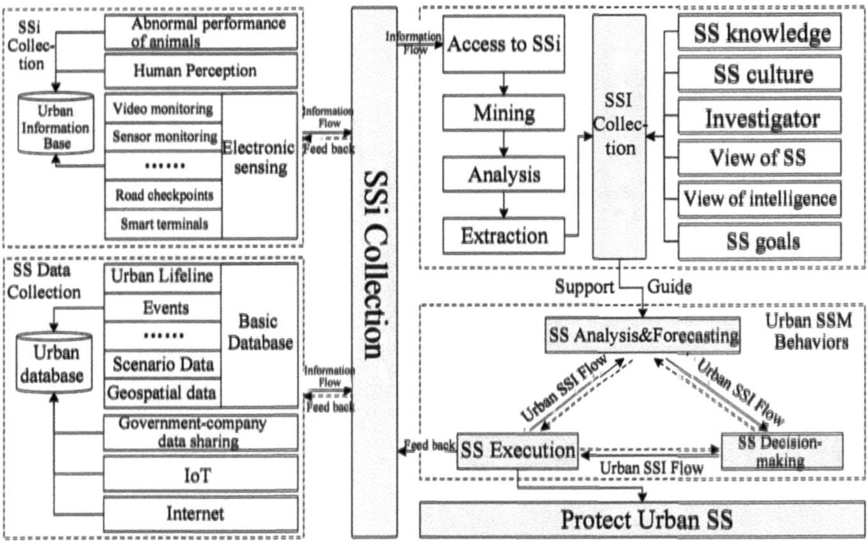

Fig. 5.12 Model of intelligence-led smart city SSM

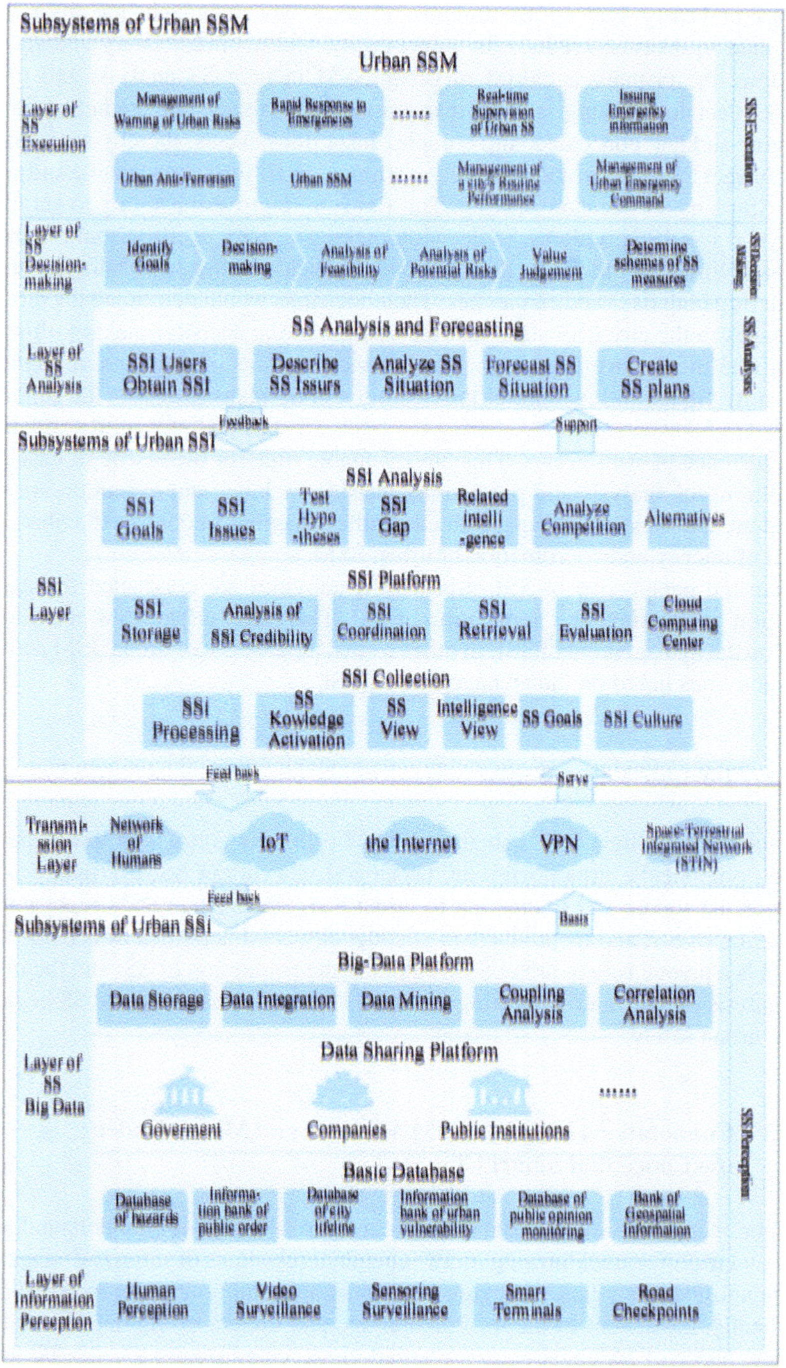

Fig. 5.13 System of intelligence-led smart city SSM

analysis, and early warning. By doing so, urban SSM can enhance the city's ability to respond to disasters, reduce its vulnerability, and maintain safety and security. Therefore, the primary activities of urban SSM include SS analysis and prediction, SS decision-making, and SS execution. SS analysis and prediction involve analyzing sudden, concealed, and complex hazardous events, thereby assessing the city's current SS state, providing timely warnings, and offering reasonable and appropriate SS plans for decision-making. The aim is to improve the ability to anticipate, detect, and avoid SS problems. SS decision-making requires decision-makers to make choices in line with the values that are generally held to offer satisfactory solutions, considering both risks and challenges. Their decisions would determine the direction of changes in the city's SS state, and the success of the decision-making ultimately depends on them, thus they are required of increased SS literacy and unprejudiced SS identity. SS execution requires the execution of SS plans to disrupt the conditions that lead to SS events, thereby altering the city's SS state and preventing or delaying the occurrence of such events. This buys valuable time for the rescue of casualties. In addition, SS analysis and prediction, decision-making, and execution are interdependent and must coordinate with each other, keeping in contact to enhance the overall efficiency and coordination of urban SSM.

From the perspective of SSI, urban SSM activities are conducted through the collection, processing, and application of SSI. Therefore, urban SSM activities are based on the management of SSI. In other words, the essence of urban SSM activities is the activities based on the management of SSI.

5.5.2.2 Intrinsic Logic of SSM

The model of intelligence-led smart city SSM primarily consists of three interconnected components: urban SSi data, urban SSI, and urban SSM. Urban SSi data serves as the source and foundation of urban SSI, while urban SSM represents the application of SSI and its products to conduct city SSM activities. It is evident that SSI plays a pivotal role in urban SSM, guiding the perception of SSi and the mining and analysis of data, and supporting urban SSM to achieve established SS goals and ensure urban safety.

5.5.2.3 Connotations of Urban SSM Methods and Means Under the Concept of Smart City

Information and data are not just resources; they are also means of understanding the world and tools for transforming it. The abundant resources of information and data have quantified many fields, playing an increasingly important role in urban SSM. Specifically, the advent of the information age and the era of big data has ushered in a new phase of SS construction of smart cities, emphasizing a more digital and intelligent approach. High-tech methods such as big data and cloud computing are used to collect and analyze vast amounts of urban SSi data, and activate them into

SSI. SSI products provide a solid foundation for precise analysis, early warning, and timely and effective control of SS events; they also offer new ideas and tools for mitigating urban SS risks and establishing a timely, effective, and intelligent system of urban SSM operations. As a result, urban SSM has become more logical, refined, and intelligent. It can be said that the method of intelligence-led urban SSM fully leverages the characteristics of the information age, providing a broad operational space for ensuring urban SS and making it possible to achieve smart city SSM.

5.5.3 A System of Intelligence-Led Smart City SSM

The city SS platform provides a crucial support for the entire urban SS system and critical scenarios for fulfilling roles of monitoring and early warning, interconnecting, and intelligence-aided decision-making. The analysis is as follows:

First, various technical means are employed to conduct comprehensive real-time monitoring of urban risk targets and sources of danger. The goal is to gather SSi and refine it into valuable SSI, thereby preventing the occurrence of various emergencies and providing a natural "protective umbrella" for urban SS.

Second, models and scientific theories of smart SSM are utilized to analyze, assess, and assist in decision-making on the basis of urban SSI. This provides effective SS measures and controlling strategies, enabling the quick and efficient management of SS events.

Third, based on intelligence-aided decision-making, SS measures are implemented to interrupt the occurrence of SS events and minimize casualties and property losses. Therefore, the quality of the urban SSM system affects the effective operation of urban SSM and has a direct impact on the prevention and control of SS events.

From the above understanding, the construction of the system of intelligence-led smart city SSM is depicted in Fig. 5.13:

The system of intelligence-led smart city SSM is built upon technical subsystems of the spatial deployment model of smart cities, aiming at making urban SSM more precise and intelligent, and improving SS decision-making and urban emergency command through the extensive use of SSI. The urban SSM system is divided into three subsystems (urban SSI subsystem, urban SSI subsystem, and urban SSM subsystem) and seven levels (SSI perception, SS big data resources, information transmission, SSI, SS analysis, SS decision-making, and SS execution). Among these, SSI perception, resources of SS big data, information transmission, and SSI form the support system and foundation for urban SS; SS analysis, decision-making, and execution constitute the management system for urban SS. Here the subsystems are illustrated.

5.5.3.1 Urban SSI Subsystem: Foundation of the Urban SSM System

SS perception involves collecting real-time data from various aspects of a city, providing a wealth of information for early warning and decision-making in urban SSM. The sources of SSi can be classified into three categories:

First, the SSi derived from human perception, such as detecting smoke, colored gases, or unusual lights.

Second, the SSi coming from platforms of the Internet of Things (IoT), including systems of video surveillance, audio frequency collection, GPS positioning, sensor detection, and smartphone terminals.

Third, the SSi obtained from SS databases, such as databases of urban hazard sources, banks of urban information of vulnerability, databases of city lifelines, case libraries, libraries of event information, and model libraries.

The SSi from the first and second sources is dynamic, providing a broader range of information. However, this dynamic SSi can also pose challenges, such as randomness, irrelevant data, missing of crucial information, and potential SS deceptions. The third source offers static SSi, which aids in the internal storage and sharing of SSi. It can be integrated with other foundational data for related analysis and effective synthesis, thereby enhancing the intelligence of the smart city SS system. Predicting urban SS situations using available SS data and historical cases is a valuable strategic resource. However, the risk of errors may increase because these predictions are not entirely grounded in factual data.

5.5.3.2 Urban SSI Subsystem: Leading and Supporting SSM

As mentioned earlier, urban SSM requires valuable SSI derived from a series of processed SSi, rather than a vast quantity of undifferentiated SSi. In other words, the goal is to convert SSi into actionable intelligence, which forms the foundation for SS analysis, decision-making, and execution. Thus, SSI operations are the cornerstone of urban SSM.

The collection of SSI is often constrained by factors such as the collector's knowledge and views of SS, ideas of intelligence, and SSI culture. Therefore, it is crucial to establish an assessment system for these influencing factors. Additionally, a comprehensive processing platform including storage, credibility analysis, retrieval, and evaluation is needed for processing the collected SSI. The advent of cloud computing has largely resolved the storage and processing challenges associated with SSI.

Most errors in SSI are not caused by a lack of raw data, including SSI, SS data, or intelligence, but rather an overwhelming abundance of such data, which makes it difficult to analyze and draw proper conclusions. Objectively, SSI can be categorized into two types: correct and incorrect—"signal" and "noise" in intelligence terms, with "noise" comprising the majority. Therefore, the core task of SSI analysis is to extract signals, eliminate noises, and conduct a thorough assessment of SSI.

5.5.3.3 Subsystem of Urban SSM

Based on accurate SSI, the fundamental task of further analysis is to provide essential information about hazardous objects, report the trends of development of people, objects, and the environment of the city, explain the reasons for changes in SS matters, analyze the consequences, predict the SS state of urban development, assess the potential outcomes of dangers, issue timely warnings, and propose SS countermeasures for decision-making. In short, SS analysis should not only explain the current SS state but also answer the question "what will happen next?" to provides a crucial buffer time for decision-makers to take necessary measures to ensure the safety and security of the city.

The SS decision-making process is a concentrated reflection of the upstream management capabilities within the urban SSM system. It includes two aspects. The first is risk decision-making, which involves assessing the necessity of take SS measures according to the development trends of urban SS and determining on the proper measure for eliminating threats. The second is opportunity decision-making, which focuses on the development trends of SS and the beneficial or balanced aspects of urban interests after taking measures. This type of decision-making is usually based on the current SS state and the decision-maker's value orientation.

SS execution is the critical link that directly determines the success or failure of management. The preparation of emergency plans before the occurrence of an event, emergency responses during the event, and reconstruction after the event are execution of SS behaviors. As effective execution requires a well-prepared operating environment, it is essential to establish execution platforms such as urban underground SS space (Chen Zhuo and She Lian 2009), time dimension urban SS space, "organic structures" capable of adapting to and recovering from disasters (Project Team of Strategic Research on the Construction and Promotion of China's Smart Cities 2015), and contingency response systems capable of rapid peak-identification.

In summary, the system of intelligence-led smart city SSM can be applied to various fields, including urban counter-terrorism, public security, traffic safety, food safety, emergency management, and daily management services. It is a comprehensive management system and mechanism that integrates accident prevention, disaster response, contingency management, and disaster recovery, ensuring the safety and secure of smart cities.

5.5.4 *Conclusions*

In the context of smart cities, it is essential to conduct SSM because urban SS is a top priority in the development of smart cities. With SSI serving as the foundation and catalyst for achieving SSM in smart cities, conducting research on intelligence-led SSM in smart cities is both theoretically and practically significant. This section firstly analyzed the cycle of intelligence-led SSM operation in smart cities. There are four closed loops in the cycle: SSI perception, SS analysis and prediction, SS decision-making, and SS execution, all supported by the cycle of SSI. The model of intelligence-led smart city SSM is then constructed with its connotation explained. Finally, the system of intelligence-led smart city SSM is proposed. It is found that while intelligence-led urban SSM is the fundamental method and tool for achieving smart city SSM, the system of intelligence-led smart city SSM becomes the foundation and guarantee for achieving smart city SSM. This system is constructed from the component technical systems of the spatial deployment model of smart cities, and can be divided into three subsystems (subsystem of urban SSI, subsystem of urban SSI, subsystem of urban SSM) and seven levels (SSI perception, big data of SS, information transmission, SSI, SS analysis, SS decision-making, and SS execution). The content of this section provides a theoretical basis and methodological guidance for achieving SSM in smart cities.

5.6 Intelligence-Led Emergency Prevention and Control

Emergencies pose a serious threat to human life and property, socio-economic development, and even national security (Wang Lingling and Zhou Limin 2016). Therefore, emergency prevention and control is crucial and has always been highly valued by the state, the government, and all sectors of a society. In 2007, The Emergency Response Law of the People's Republic of China was issued and implemented. With the increasing frequency of global emergencies, the recent years have seen a rise in emergency incidents in China (Liu Qian et al. 2018). The next 20 years represent a strategic opportunity period for China's socio-economic development, during which the society and economy will undergo transformation and rapid growth. Drawing from the experience of developed countries, this period is also likely to be characterized by a high incidence of emergencies. Therefore, it can be predicted that China will face more severe challenges posed by emergencies. Research on emergency prevention and control is urgently needed and holds significant theoretical and practical importance.

The science of emergency prevention and control is an interdisciplinary field that requires collaboration among researchers from various disciplines. Since emergencies are the primary focus of SS science, SSI can serve as "the eyes and ears, or the pioneer and adviser" of emergency prevention and control, thus intelligence is essential for managing emergencies (Wu Xiaotao and Shen Chen 2016; Liu Jianzhun et al. 2019; Wang Dan and Feng Weiguo 2019; Yang Feng et al. 2019; Guo Chunxia et al. 2019; Tang Mingwei et al. 2019; Wang Jingru and Song Shaocheng 2016). In recent years, the study of emergency intelligence has gained significant attention (Liu Jianzhun et al. 2019; Wang Dan and Feng Weiguo 2019; Yang Feng et al. 2019; Guo Chunxia et al. 2019; Tang Mingwei et al. 2019; Wang Jingru and Song Shaocheng 2016). The current research in this area primarily focuses on emergency response to incidents (Wu Xiaotao and Shen Chen 2016; Liu Jianzhun et al. 2019; Wang Dan and Feng Weiguo 2019; Yang Feng et al. 2019; Guo Chunxia et al. 2019; Tang Mingwei et al. 2019; Wang Jingru and Song Shaocheng 2016), and there is a notable absence of research on the entire process of emergency prevention and control, which has considerably limited the depth and scope of the field. Additionally, while numerous studies of emergency intelligence have been conducted, no practical methods have yet been proposed for effectively utilizing intelligence in emergency prevention and control, which has hindered the application of research findings (Wang Bing et al. 2020). Therefore, future research in this area needs to address the full cycle and whole procedure of emergency prevention and control, and to develop feasible methods for applying intelligence in related operations.

In light of this, in this section the theories and methods of intelligence-led SSM are applied to emergency prevention and control. Research on intelligence-led emergency prevention and control is conducted through the full cycle and whole procedure to promote innovation in emergency management and lay a theoretical foundation, thereby providing guidance for future research and practice and enhancing the capability of emergency prevention and control.

5.6.1 Analysis of Relationship Between Intelligence and Emergency Prevention and Control

5.6.1.1 Connotation of Emergency Prevention and Control

Emergency prevention and control refers to a series of management activities aimed at preventing, controlling, and responding to emergencies, with the emergency itself as the objective of management. Seen from various perspectives, emergency prevention and control activities can be categorized in several ways. The academic and practical communities generally classify them on the basis of two dimensions (the time dimension and the logical dimension), as detailed below:

1. From the time dimension, according to the entire life cycle of emergencies (before, during, and after) (Qu Yaogen 2009), emergency prevention and control can be divided into three stages: pre-event prevention and control, mid-event prevention and control, and post-event prevention and control. Pre-event prevention and control emphasize precautions and early warnings; mid-event prevention and control focus on response and control; and post-event prevention and control highlights the recovery and reconstruction of organizations or individuals affected by emergencies, as explained in detail in Table 5.3 below.

2. From a logical perspective, according to the general logical process of crisis management and SSM (Wang Bing and Wu Chao 2019g; MBA Think Tank 2019; Lu Wengang and Wang Yaxuan 2019), emergency prevention and control can be divided into six aspects: environment determination, emergency monitoring, emergency analysis, emergency decision-making, emergency response and evaluation, and feedback and learning. The specific explanations are provided in Table 5.3.

Table 5.3 Connotation of dimensions of the spatial structure of emergency prevention and control

No	Dimension	Specific link	Connotation
1	Time dimension	Pre-event prevention and early warning	This refers to all the preparatory actions of preventive and early control taken by emergency responders before the occurrence of an emergency. There are many factors that can lead to emergencies, including accidental and sudden elements. However, the occurrence of the vast majority of emergencies is a gradual process. If emergency responders can identify the risk factors and signs of emergencies in advance, establish an effective prevention and early-warning mechanism for emergencies, and take effective preventive measures in time, the occurrence of emergencies can be avoided. Therefore, pre-event prevention and early warning of emergencies is the primary link and ultimate task of emergency prevention and control. This is also a specific manifestation of the principle stated in Article 5 of *Emergency Response Law of the People's Republic of China* that "The principle of giving priority to prevention and combining prevention with emergency response shall be adhered to in coping with emergencies." The focus is on the risk management of emergencies

(continued)

Table 5.3 (continued)

No	Dimension	Specific link	Connotation
		In-event emergency response and control	This refers to the emergency and control actions taken by emergency responders during the occurrence of an emergency (mainly starting the emergency response procedure and implementing the emergency plan for emergencies). In short, the in-event emergency response and control refers to the actions taken by emergency responders to deal with the threats of emergencies that are happening, with the goal of responding to the threats of emergencies in a timely, swift, and effective manner
		Post-event recovery and reconstruction	This refers to the actions taken by emergency responders after the occurrence of an emergency to reduce the damage and impact of the incident (preventing the spread of the incident and the expansion of losses), and to use various measures and resources for recovery and reconstruction. The goal is to eliminate the negative impact of emergencies, restore the normal operation of organizations affected by the incidents as much as possible and as soon as possible, and help individuals affected by the incidents resume normal life and production
2	Logic dimension	Determination of environment	This refers to understanding and clarifying the factors of environmental background of emergency prevention and control, such as the objects, needs, boundaries, principles, resources, requirements, influencing factors of constraining, and key points of work for emergency prevention and control

(continued)

Table 5.3 (continued)

No	Dimension	Specific link	Connotation
		Emergency monitoring	This refers to the use of various information monitoring facilities, equipment, and technologies to monitor data, information, and intelligence related to emergencies (such as risk factors, signs, occurrence and evolution processes, and state changes of emergencies). In short, the monitoring of emergencies aims to systematically and comprehensively scan and collect various data, information, and intelligence related to emergencies, in order to provide data, information, and intelligence services and support for emergency prevention and control
		Emergency analysis	This refers to the in-depth analysis of the risk factors of emergencies based on the data, information, and intelligence collected through monitoring, confirming the occurrence of emergencies, diagnosing the current state and developing trend of emergencies, assessing the influencing factors of current emergency prevention and control, as well as strengths, opportunities, and threats, in order to assist in the decision-making of emergency prevention and control

(continued)

Table 5.3 (continued)

No	Dimension	Specific link	Connotation
		Emergency decision-making	This refers to the process of formulating specific plans for emergency prevention and control by emergency responders under the constraints of limited time, resources, etc. Generally speaking, pre-event decision-making for emergencies is mainly based on routine and procedural decision-making, while in-event and post-event decision-making for emergency prevention and control is often typical non-routine decision-making, carried out under conditions of high uncertainty of information and urgency, which is a challenging and difficult decision-making process
		Emergency response	This refers to the activities and actions carried out by emergency responders in accordance with specific emergency plans to deal with emergencies, such as pre-event information monitoring and prevention and early warning of emergencies, emergency response and control operation during the incident, and post-event recovery and reconstruction. In short, the response to emergencies is the implementation and execution of emergency plans by emergency responders

(continued)

Table 5.3 (continued)

No	Dimension	Specific link	Connotation
		Evaluation, feedback, and learning	This refers to the process in which emergency responders review, evaluate, and examine the emergency response plans, strategies, and measures taken, as well as their implementation effects, summarize strengths and experiences, identify problems and deficiencies, and organize them into learning materials for emergency prevention and control. These materials are then fed back to relevant departments and personnel for learning and reflection, in order to further improve the level of emergency prevention and control
3	Intelligence dimension	Emergency intelligence planning	This refers to determining the intelligence needs of emergency responders (i.e., intelligence users) and emergency prevention and control, that is, to clarify the scope, needs, content, and tasks of intelligence services and activities for emergency prevention and control
		Collection of emergency intelligence	This refers to identifying and collecting various sources and channels of data and information related to emergencies according to the intelligence needs of emergency responders and emergency prevention and control, through various monitoring facilities, equipment, and technologies for emergencies (including intelligence collection facilities, equipment, and technologies)

(continued)

Table 5.3 (continued)

No	Dimension	Specific link	Connotation
		Emergency intelligence analysis	This refers to the process of converting the collected data and information related to emergencies into intelligence that serves the related emergency prevention and control activities (mainly emergency decision-making) through various data and information mining and analysis facilities, equipment, and technologies, by integrating, mining, evaluating, comprehensively analyzing, processing, and interpreting the data and information, in order to meet the known or expected intelligence needs for emergency prevention and control
		Production of emergency intelligence	This refers to the process of organizing and "packaging" intelligence to produce real "emergency intelligence products" (mainly referring to the identification, analysis, and evaluation results of emergency risks or signs, as well as the formulated suggestions for emergency prevention and control), and supplying them for use by emergency responders

(continued)

Table 5.3 (continued)

No	Dimension	Specific link	Connotation
		Transmission of emergency Intelligence	This refers to the process of delivering (supplying) emergency intelligence products to relevant emergency responders according to their intelligence needs, so that they can use the intelligence products to carry out related emergency prevention and control activities, such as implementing emergency prevention and control plans
		Feedback of emergency intelligence	This refers to the process in which emergency responders provide feedback on the application and implementation effects of emergency intelligence products, and propose suggestions and requirements for revising and improving intelligence operation. The purpose of the feedback of emergency intelligence is to modify and improve the original intelligence or produce new intelligence

5.6.1.2 Reflections on Intelligence of Emergency Prevention and Control

Based on the connotation of emergency prevention and control and the intelligence from the management perspective, the connotation of emergency prevention and control from the intelligence perspective is given. From the intelligence perspective, emergency prevention and control is an organic integration of management and intelligence operation. It takes emergencies as the management objects, focuses on the prevention, control, and response to emergencies based on relevant emergency information (including emergency data). The information is processed, analyzed, and transformed into useful intelligence for emergency prevention and control (i.e., emergency intelligence), and used to serve and support emergency management and intelligence operation. In short, the essence of emergency prevention and control is to take relevant practices by obtaining and harnessing emergency intelligence, which is an indispensable component of the process.

As a typical project of a complex giant system, the emergency prevention and control needs theoretical basis and methodological guidance from the system engineering and its theories. Hall's three-dimensional (including time dimension, knowledge dimension and logical dimension) structural model (Hall 1969) is an important system engineering methodology. It is considered as a unified ideological methodology to solve problems in the planning, organization, and management of large complex systems [for example, it has been applied to intelligence-led urban SSM research (Wang Bing and Wu Chao 2019f)]. Similarly, it can provide inspiration and methodological guidance for emergency prevention and control. Based on the two dimensions in emergency prevention and control (time dimension and logical dimension) and drawing on Hall's three-dimensional structural model (Hall 1969), its third important dimension (intelligence dimension) is proposed according to the meaning of emergency prevention and control from the intelligence perspective. The intelligence dimension of emergency prevention and control shows that emergency prevention and control needs support from the intelligence of emergency prevention and control, which is similar to the connotation of the knowledge dimension of Hall (which embodies that solving system problems requires knowledge from various disciplines and sectors to provide support). According to the general process (cycle) of intelligence operation (Wang Bing and Wu Chao 2019g; Li Guoqiu and Lv Bin 2012), emergency intelligence operation embraces six important links: emergency intelligence planning, emergency intelligence collection, emergency intelligence analysis, emergency intelligence production, emergency intelligence transmission, and emergency intelligence feedback. In addition, the connotations of the time dimension and logic dimension of emergency prevention and control (separately reflecting the stages of emergency prevention and control and the workflow or steps of them) agree with that of Hall's. Based on this, a three-dimensional spatial structure of emergency prevention and control is constructed (Fig. 5.14), and the specific meanings of each dimension are explained in Table 5.3.

From Fig. 5.14 and Table 5.3, it can be concluded that the three dimensions are interconnected, forming a complete system of emergency prevention and control. The relationship between the three dimensions in Fig. 5.14 is explained as follows:

1. The relationship between the intelligence dimension and the time and logic dimensions of emergency prevention and control: (1) Since the intelligence dimension acts as the foundation and safeguard for the time and logic dimensions, providing essential support for both, the operation of emergency prevention and control cannot function properly without the support of intelligence. Emergency intelligence and its operation run through the time and logic dimensions, and are crucial to the functioning of both. The absence of intelligence is often the primary cause of failure in emergency prevention and control; (2) Since intelligence is derived from data and information (in other words, data and information are the fundamental resources for generating intelligence), the various types of information (including data) generated within the time and logic dimensions serve as raw materials for the intelligence dimension, influencing emergency prevention and control in this regard; (3) As shown in Table 5.3, there is a close relationship

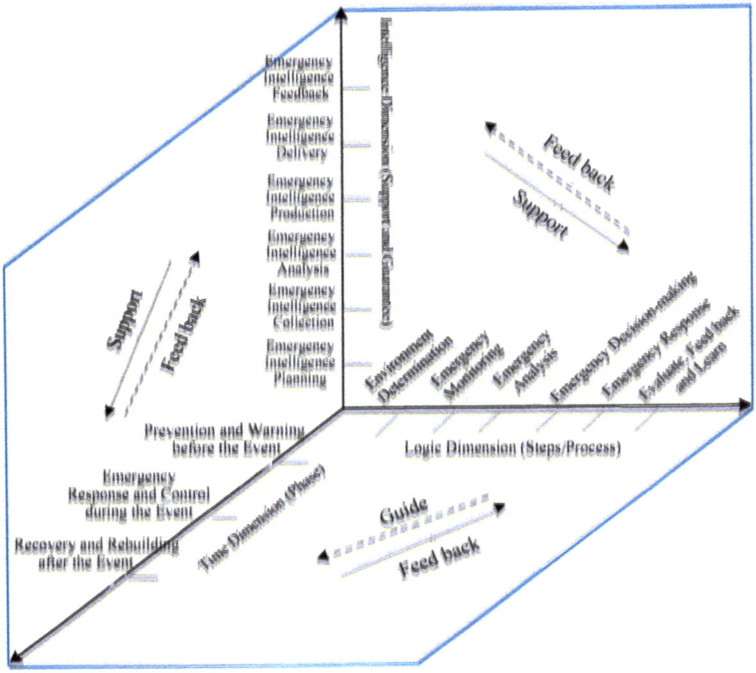

Fig. 5.14 Three-dimensional spatial structure of emergency prevention and control

and mutual influence between the intelligence and logic dimensions (Fig. 5.15). For detailed information, please refer to Table 5.3.

2. The relationship between the time dimension and the logic dimension of emergency prevention and control: As the three stages of the time dimension are connected to the six components of the logic dimension in a natural order, the procedure (process) of the logic dimension can provide methodological guidance for each stage of the time dimension. Conversely, the time dimension can offer feedback to the logic dimension, influencing its various tasks.

5.6.2 Proposal of Intelligence-Led Emergency Prevention and Control

Intelligence is the basis and support of management (especially decision-making). If the value of intelligence in management practices is to be shown at the largest extent, advocating and conducting the intelligence-led management becomes a must. Intelligence-led management refers to managers, based on their own management skills and experience, rigorously, accurately, and effectively harnessing the available intelligence to lead management activities according to expectations, resources, and

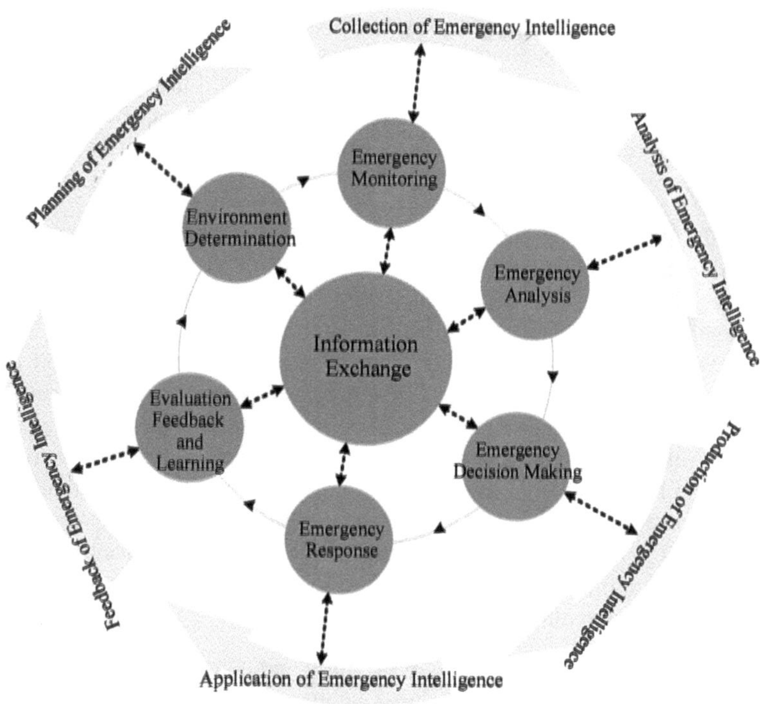

Fig. 5.15 Relationship between the intelligence dimension and the time dimension of emergency prevention and control

other restrictions of the activities, aiming to formulate and implement the best (specifically the most reliable and most effective) plans and measures to solve management problems.

Based on this understanding, intelligence-led emergency prevention and control can be understood as: emergency managers, leveraging their professional skills and experience, develop and implement optimal plans and measures with consideration of the circumstances of the emergency and the needs of prevention and control. The primary goal is prevention and control, with a strong emphasis on addressing the lack of intelligence in emergency prevention and control. This approach takes into account the actual demands and conditions, and is fundamentally based on the quality of emergency intelligence. In essence, intelligence-led emergency prevention and control is a comprehensive process that guides prevention and control efforts with intelligence, centered around the goal of effectively managing emergencies. It is based on three key elements: emergencies, emergency intelligence, and emergency managers with effective interaction among these elements serving as the foundation, utilizing emergency intelligence to steer and oversee the entire process of emergency prevention and control. Drawing from the conceptual model of intelligence-led management (Liska 2015; Wang Bing and Wu Chao 2019g), a conceptual model for intelligence-led emergency prevention and control is proposed, as illustrated in Fig. 5.16. In summary, this approach highlights the leading and dominating role of emergency intelligence in prevention and control efforts. Furthermore, the connotation of intelligence-led emergency prevention and control can be analyzed from the perspectives of the operation of emergency intelligence and the operation of emergency prevention and control.

1. From the perspective of emergency intelligence operation: This involves the comprehensive collection, analysis, and mining of various emergency data to produce actionable intelligence. The resulting intelligence products are then packaged and applied to prevention and control efforts.
2. From the perspective of the operation of emergency prevention and control: Emergency intelligence is placed at the center of emergency prevention and control efforts, serving as an integral component. The analysis and interpretation of emergency intelligence form the foundation of these efforts. Therefore, to effectively harness the guiding role of intelligence in the prevention and control of emergencies, it is essential to place a strong emphasis on the work of emergency intelligence. Continuous efforts must be made to enhance the capacity of emergency intelligence and its endeavor to serve, support, and ensure the success of emergency prevention and control efforts.

Traditional methods of emergency prevention and control have numerous shortcomings, which would result in failures in managing emergencies. In contrast, intelligence-led emergency prevention and control has made significant improvement, especially in seven aspects: cognitive understanding of prevention and control, foundational basis for prevention and control, sources of intelligence, methods of intelligence gathering, analysis and utilization of intelligence, modes of prevention and control, and distinctive features (see Table 5.4).

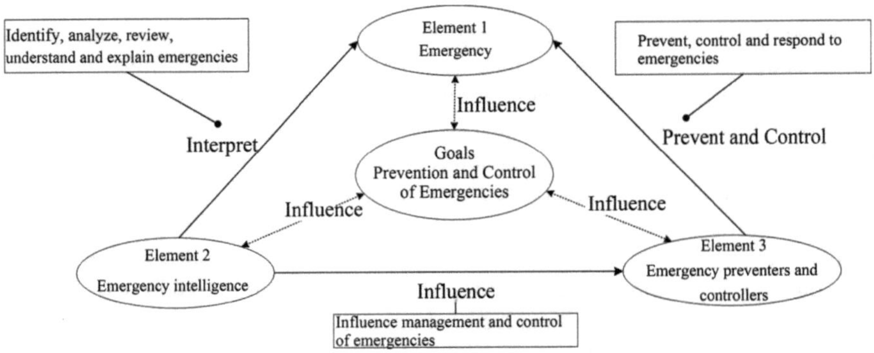

Fig. 5.16 A conceptual model of intelligence-led emergency prevention and control

Table 5.4 Comparison of intelligence-led emergency prevention and control and traditional emergency prevention and control

No	Dimension of Comparison	Intelligence-Led Emergency Prevention and Control	Traditional Emergency Prevention and Control
1	Understanding of prevention and control	The understanding the failure of emergency prevention and control has shifted from the perspective of information deficiency to that of intelligence deficiency. Therefore, it aims to address the issue of intelligence gaps in emergency prevention and control	The reason for the failure in emergency prevention and control is the loss of information; hence, it aims to address the issue of information deficiency in the prevention and control of emergencies
2	Basis for prevention and control	Currently available high-quality intelligence about emergency	Laws, regulations, and standards related to emergencies, as well as experience, etc
3	Intelligence sources	Considering comprehensive sources (all sources) of intelligence related to emergency prevention and control	Focusing on intelligence related to emergency prevention and control from certain sources
4	Intelligence acquisition	Systematically and comprehensively acquiring intelligence about emergencies	The acquisition of intelligence about emergencies is not comprehensive and lacks a systematic approach
5	Analysis and utilization of intelligence	Emphasizing	Not emphasizing
6	Model of prevention and control model	Comprehensively considering emergencies, intelligence about emergencies, and emergency responders	Centered around the emergency responders
7	Features of prevention and control	Scientific, proactive, intelligent, and precise	Characterized by being reactive and lacking in intelligence and specificity

5.6.3 Establishment of the Intelligence-Led Emergency Prevention and Control System

A system of intelligence-led emergency prevention and control is built according to the relationship between intelligence and the intelligence-led emergency prevention and control, and the connotation of the intelligence-led emergency prevention and control. It is shown in Fig. 5.17.

As depicted in Fig. 5.17, the system comprises three primary levels that influence and interact with each other: the foundation, the methodology, and the service. The level of foundation provides support for the methodology level, while the level of

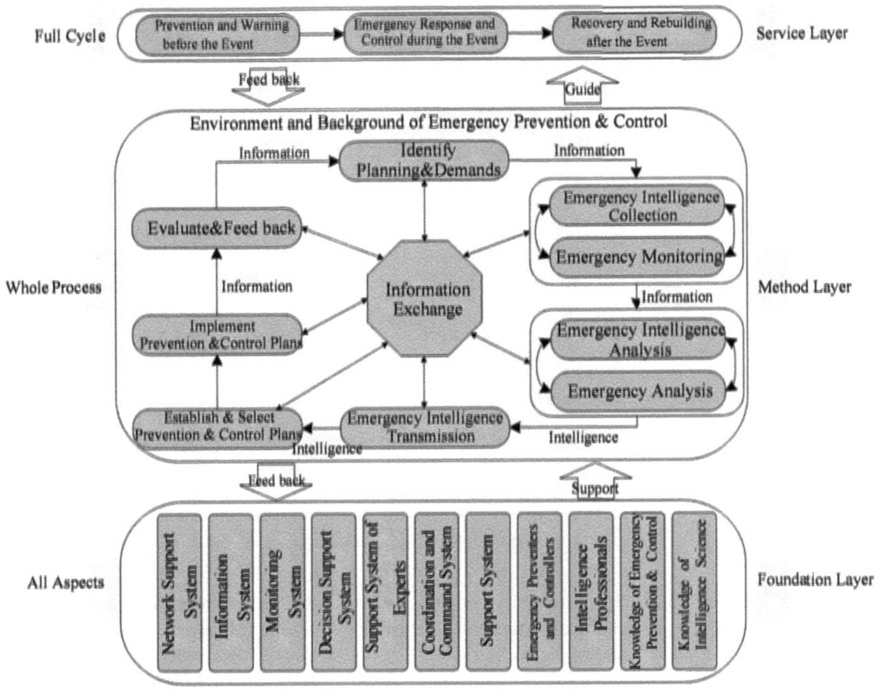

Fig. 5.17 System of intelligence-led emergency prevention and control system

methodology offers guidance for the level of service. The feedback mechanism allows each level to influence its adjacent levels from the top to the bottom. Based on Fig. 5.17, the connotations of each level of the system are briefly analyzed as follows:

5.6.3.1 Connotation of the Level of Foundation

This level presents that the implementation of intelligence-led emergency prevention and control depends on a variety of resources, including hardware and software, human resources, and domain-specific knowledge. Typically, the software and hardware resources embrace network support systems (such as the Internet, the mobile Internet, and the IoT, etc.), information support systems (such as platforms for collecting, mining, and analyzing emergency data), monitoring systems, support systems of decision-making, expert support systems, systems of coordination and command (aimed at unified command and coordination of relevant departments and personnel to jointly conduct emergency prevention and control efforts), and support systems (primarily providing logistical support, such as financial and resource assistance). Human resources mainly consist of emergency prevention and control professionals and intelligence specialists, who must collaborate to leverage their complementary strengths. Domain-specific knowledge primarily includes knowledge of emergency prevention and control (involving crisis management, risk management, SSI, emergency management, sustainable development, etc.) and intelligence knowledge (mainly knowledge of SSI science). The implementation of intelligence-led emergency prevention and control requires numerous foundational conditions, necessitating comprehensive development across all areas.

5.6.3.2 Connotation of the Level of Method

This level outlines how to integrate intelligence operations with the prevention and control process (once integrated, it mainly consists of seven links. The connotations of each link are detailed in Table 5.4). This integration ensures that intelligence operations are effectively incorporated into the entire process of emergency prevention and control, enhancing prevention and control efforts. It is important to note that the core of the intelligence-led emergency prevention and control process is information communication and exchange. Specifically, through communication and exchange of information between emergency preventers, emergency intelligence professionals, and each link of the implementation process, a consensus can be reached on the risks associated with emergencies and their related factors, as well as on the plans of emergency prevention and control.

5.6.3.3 Connotation of the Service Level

This level describes how intelligence-led emergency prevention and control can serve the life cycle of emergency prevention and control operation, or the temporal dimension of such operation. The entire process includes pre-event prevention and early warning, mid-event emergency response and control, and post-event recovery and reconstruction.

5.6.4 Conclusions

Emergency prevention and control must rely on intelligence and its operation. Therefore, discussing intelligence-led emergency prevention and control is of both theoretical and practical significance. At the same time, intelligence-led emergency prevention and control represents the future of prevention and control operation in the era of informatization and intelligence. It is an advanced concept and method of emergency prevention and control as well as a guarantee for its effectiveness. This section begins from a theoretical perspective, discussing the foundation for constructing intelligence-led emergency prevention and control (that is, the relationship between intelligence and emergency prevention and control) on the basis of the fundamental meanings of intelligence and emergency prevention and control. On this basis, it proposes intelligence-led emergency prevention and control and establishes a comprehensive, whole-process, full-cycle system for intelligence-led emergency prevention and control. It is found that intelligence-led emergency prevention and control emphasizes the application of emergency intelligence in leading the overall situation and process of prevention and control. The intelligence-led emergency prevention and control system consists of three levels: the level of foundation, the level of method, and the level of service.

The research in this section may promote the innovation of the operational mode of emergency prevention and control, and can lay a theoretical foundation and provide theoretical guidance for future research and practice. Based on the research in this section, a series of new topics are extended in discussion, such as the principles of its implementation, influencing factors, effectiveness evaluation, research on intelligence-led emergency prevention and control in the big data environment, and application research in various emergency events (such as natural disasters, accidents, public health events, and social SS events of all kinds).

5.7 Intelligence-Led SSM of National Mineral Resources

Since the 19th National Congress of the Communist Party of China incorporated "a holistic approach to national security" into the basic line and policy of socialism with Chinese characteristics in the new era, research on national SS has garnered significant attention from the academic community (Gao Jinhua 2019a). Mineral resources are an essential material foundation for the national economy and the survival and development of society, providing crucial support for the survival and development of a country (Long Ruyin and Yang Jiahui 2018). The safety and security of mineral resources is closely linked to the overall national security, for it is not only the core of resource SS, it is also associated with other SS fields, such as economic SS, ecological SS, military SS, political SS, and social SS, among others (Yu Hongyuan and Yu Bowen 2017). For instance, mineral resources provide key raw materials for national defense and the military industry, supply fundamental materials for China's industrialization and urbanization, and are essential for the implementation of national strategies like *Made in China 2025* and *The Strategy for Energy Production and Consumption Revolution (2016–2030)*. The legal status of resource SSM was officially established with the promulgation of *The National Security Law of the People's Republic of China* on July 1, 2015, making national mineral resource SSM one of the important topics in national SS research, particularly in the area of resource SS. In recent years, China's national mineral resource SS faces significant challenges from international competition and a complex landscape, as the strategic position of mineral resources supporting emerging industries is becoming increasingly prominent as countries compete for economic high ground (Long Ruyin and Yang Jiahui 2018; Yu Hongyuan 2018). The high dependency on foreign mineral resources, which is a fundamental aspect of China's national conditions, is unlikely to change significantly in the foreseeable future. The situation of national resource SS is severe, and the modernization of China's mineral resource SSM system faces huge challenges (Yu Hongyuan 2018). Therefore, researching the sustainable development of China's mineral resources currently holds strong urgency and significant theoretical and practical importance.

In recent years, research and practice in information science have increasingly focused on major SS issues, with particular attention given to the SSM challenges in key industries (such as core technologies, major innovations, significant national projects, critical national infrastructure and equipment, and industries like nuclear, aerospace, resources, science and technology, information, and major emergency events) (Wang Bing and Wu Chao 2019g; Liska 2015; Zhang Jianian and Ma Feicheng 2018). Some SS scholars have also begun to explore major SS issues from a perspective that integrates information science and SS science (Yao Leyuan and Fan Wei 2014; Dong Yin and Liu Qianli 2016; Wang Bing and Wu Chao 2019c). A consensus has emerged: SSI plays a crucial role in SSM, for it serves as the core support and fundamental resource for SSM and is essential for achieving smart SSM (Wang Bing and Wu Chao 2019g; Liska 2015; Zhang Jianian and Ma Feicheng 2018; Yao Leyuan and Fan Wei 2014; Dong Yin and Liu Qianli 2016). On this basis, the concepts, strategies, and methods of "intelligence-led SSM" have been developed, particularly for key industries and fields (Liska 2015; Zhang Jianian and Ma Feicheng 2018; Wang Bing and Wu Chao 2019g; Yao Leyuan and Fan Wei 2014; Dong Yin and Liu Qianli 2016). As mineral resource SS is regarded as the core of resource SS and national SS, and a significant issue in national development, the concepts, strategies, and methods of "intelligence-led SSM" can bring inspiration into national resource SSM and provide a direction for its reform and innovation, specifically through the implementation of strategies of intelligence-led national mineral resource SSM. It is worth noting that the application of big data has become a global trend to promote economic growth and enhance management capabilities. In the context of mineral resource management, discussions have been held on the application of big data thinking and methods, such as mineral resource evaluation (Xiao Keyan et al. 2015) and mineral resource planning (Xu Gepei 2017). Similarly, big data is a powerful tool for mineral resource SSM. Therefore, the theories, thinking, methods, and technologies of big data can offer new opportunities for implementing strategies of intelligence-led national mineral resource SSM, fulfilling the vision of intelligence-led SSM of national mineral resources.

In summary, it is imperative to explore the paradigm of intelligence-led mineral resource SSM for developing national mineral resource SSM in China. The reasons are: firstly, the impact of intelligence and big data on national mineral resource SSM is both substantial and realistic; secondly, the fulfillment of strategies and models of intelligence-led national mineral resource SSM rely on the big data; thirdly, incorporating an "intelligence-led" approach into current management practices is conducive to clarifying the essential role that big data plays in national mineral resource SSM. More specifically, the primary value of big data in mineral resource SS lies in its ability to generate intelligence for mineral resource SS, rather than in leading the management process itself. It is crucial to avoid blindly embracing big data, as this could potentially confuse efforts in national mineral resource SSM. However, it is regrettable that there is currently limited attention given to intelligence-led national mineral resource SSM within the context of big data. In light of this, this section integrates insights from big data and intelligence to approach fundamental and forward-looking issues of intelligence-led mineral resource SSM in the environment of big data from the theoretical perspective, aiming to promote the reform and innovation of national mineral resource SSM in the new era, particularly in the era of big data and intelligence.

5.7.1 *Basic Concepts*

5.7.1.1 SS of Mineral Resources

Currently, the view of resource system is one of the most central and mainstream views in resource studies (Wu Qing and Sha Jinghua 2015). The science of system SS represents an advanced stage in the development of modern SS science. It advocates for conducting SS research from a systemic perspective with its distinct advantages and becomes the dominant research paradigm in SS science (Wang Bing and Wu Chao 2018b). Based on the existing definitions of resource SS (including mineral resource SS) (Long Ruyin and Yang Jiahui 2018; Yu Hongyuan and Yu Bowen 2017; Yu Hongyuan 2018; Winzer C 2012; Gu Shuzhong, et al. 2002) and the definition of SS from a systemic perspective (which refers to the state of a system being free from unacceptable adverse effects of internal and external factors), the definition of mineral resource SS from the perspective of the mineral resource system is as follows: Resource SS refers to the ability of a country or region's mineral resource system to supply the required mineral resources and related products in an acceptable state. The logical and mathematical expressions are:

$$\begin{cases} S = (X_0, +\infty) \\ X = f(x_1, x_2) = f(x_{11} + x_{12}, x_{21} + x_{22}) \end{cases} \tag{5.11}$$

In the formula, S is the degree of SS of the mineral resource system; X is the ability of the mineral resource system to supply the required mineral resources and mineral resource products to the country or region where it is located; X_0 is the minimum acceptable ability (threshold) of the system to supply the required mineral resources and mineral resource products to the country or region where it is located; x_1 is the internal SS risks of the system; x_2 is the external SS risks of the system; x_{11} is the deliberate internal SS risks of the system; x_{12} is the accidental internal SS risks of the system; x_{21} is the deliberate external SS risks of the system; x_{22} is the accidental external SS risks of the system.

Here, "capability is in an acceptable state" refers to the ability of a country or region's mineral resource system to supply the required mineral resources and their products of high quality in a timely and continuous, stable and reliable, and economical and reasonable manner. The "degree of acceptability of the capability" can be measured according to the "SS risks of acceptable mineral resource." From the above definition, it is evident that the connotation of mineral resource SS can be understood from four aspects:

First, mineral resource SS is manifested as a crucial capability of the mineral resource system. It represents the ability to provide mineral resource guarantee to the country or region where it is located (i.e., the ability to support socio-economic development in that country or region). The quantity, quality, and structure of the mineral resources within a country or region's mineral resource system directly determine the SS capability of that system (i.e., the SS state of the system).

Second, mineral resource SS is a relative concept, similar to SS in other fields, for there is no absolute safety and security. The acceptable level of SS risk in the mineral resource system varies among different countries and regions.

Third, mineral resource SS risks encompass both "safety" and "security." According to Piètre-Cambacédès and Chaudet (2010), "safety" risks generally refer to accidental or internal risks within the mineral resource system, while "security" risks typically refer to deliberate or external risks within the system.

Fourth, the subjects of mineral resource SS include both countries and regions. SS generally involves the subjects of SS, and mineral resource SS is no exception. Based on different subjects, mineral resource SS can be divided into national mineral resource SS and regional mineral resource SS; there are fundamental differences in terms of the interest subjects and the orientation of interests between the two, and national interest is not simply the sum of regional interests. In terms of mineral resource SS, regional interests should serve national interests. In this discussions, more emphasis is placed on national mineral resource SS.

5.7.1.2 Big Data on Mineral Resources SS

According to the definition of SS data (which records and describes the SS state of a system and its patterns of change) (Wang Bing and Wu Chao 2017c) and the definition of mineral resource SS provided earlier, mineral resource SS data can be defined as follows: the record and description of the SS state of the mineral resource system and its patterns of change (such as quantity, quality, type, price, and the supply and demand dynamics of mineral resources and their patterns of change). It can be seen as an abstract representation of the SS state of the mineral resource system and its patterns of change.

Based on the definition of SS data and the most significant characteristic of big data (i.e., its massive scale), the big data of mineral resource SS can be defined as a vast dataset used to record and describe the SS state of the mineral resource system and its patterns of change. The logical and mathematical expression is as follows:

$$D \Leftrightarrow H(S) \tag{5.12}$$

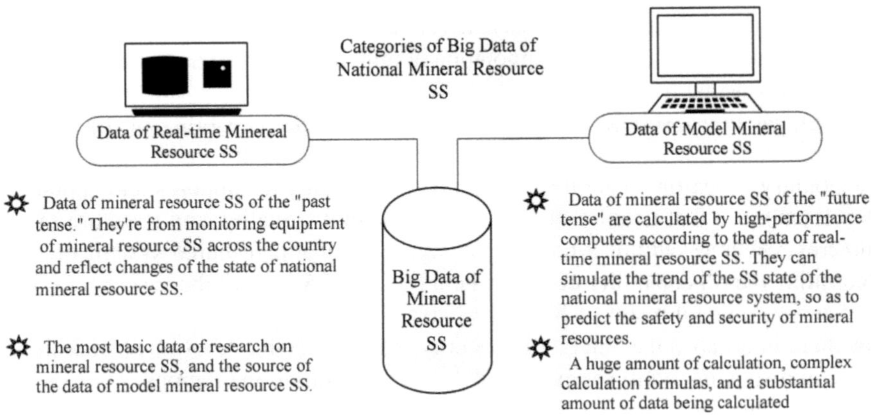

Fig. 5.18 Categories of big data of national mineral resource SS

In the formula, *D* represents the big data of mineral resource SS; *H(S)* represents the SS state of the mineral resource system and their changes, which is a comprehensive reflection of the SS degree (*S*) of the system.

Further explanations of the above definition are provided in four aspects: (1) This definition indicates that mineral resource big data belong to a specific category of big data, thus they possess the general characteristics of big data, particularly the 4 V characteristics, namely the massive scale (volume), the variety of mineral resource SS data (variety), the extremely fast flow (velocity), and the relatively low value density (value) (Wang Bing and Wu Chao 2017c); (2) The "mineral resource system" in this definition embraces all factors or elements related to mineral resource SS within the system; (3) The "SS state" in this definition refers to the relatively stable SS conditions and trends exhibited by the system in specific times and spaces; (4) The "ways of change of SS state" in this definition refers to the dynamic patterns of changes in the SS state of the mineral resource system as time and space changes. The big data of mineral resource SS, though complex, can be divided into two categories: "data of live mineral resource SS" and "data of model mineral resource SS" (for specific explanations and characteristics, see Fig. 5.18). Among them, data of live mineral resource SS serve as the source for mineral resource SSM (especially for predicting mineral resource SS), while data of model mineral resource SS are the essence of the big data of mineral resource SS, with their big data characteristics being more pronounced.

5.7.1.3 SSI of Mineral Resources

From the perspective of system SSM, SSI refers to the intelligence that affects system SSM, while SSI refers to the self-display of the system's SS state and its changes. In terms of affiliation, mineral resource SSI is a subset of SSI. Therefore, based on the definition of SSI, mineral resource SSI can be defined as: the SSI that affects the SSM of mineral resources in a country or region. In simpler terms, mineral resource SSI refers to the useful SSI for mineral resource SSM. This definition addresses two fundamental questions about mineral resource SSI: first, what is it (the essence of mineral resource SSI is a type of SSI. In other words, all mineral resource SSI originates from SSI, but not all SSI is mineral resource SSI); second, what is the use of mineral resource SSI (mineral resource SSI serves mineral resource SSM, with the goal of providing effective intelligence services for SSM).

However, there are obvious shortcomings in the definition of mineral resource SSI. For example, the differences between mineral resource SSI and related SSI are not clearly delineated, nor is the method for collecting mineral resource SSI and its specific role in mineral resource SSM. To address this problem, a more detailed definition is provided here: feasible mineral resource SSM recommendations obtained through a complete SSI processing process that affect mineral resource SSM. According to this definition, the connotation of mineral resource SSI is further explained as follows:

First, mineral resource SSI is a product derived from the processing of mineral resource SSi. Generally, the processing of mineral resource SSi involves planning, collection, analysis, interpretation, dissemination, and information storage. Mineral resource SSI primarily includes intelligence products obtained through analysis in the following areas: mineral resource SS evaluation, threats (risks) to mineral resource SS, management capabilities, the pros and cons of mineral resource SSM, and current landscapes and trends. In other words, as a product, mineral resource SSI operates as "feasible recommendations for mineral resource SSM."

Second, mineral resource SSI is a process. It is a systematic process aimed at collecting, analyzing, and managing mineral resource SSi that influences mineral resource SSM planning, prediction, decision-making, and execution. In other words, the process of mineral resource SSI involves collecting, analyzing, and harnessing mineral resource SSi to meet the short-term and long-term needs of mineral resource SSM in a country or region.

Third, mineral resource SSI provides "recommendations for mineral resource SSM" to support and optimize mineral resource SSM in a country or region. Mineral resource SSM recommendations refer to the evidence for mineral resource SSM. The actual purpose of mineral resource SSM is to collect and process mineral resource SSi (data) to provide feasible mineral resource SSI (i.e., evidence for mineral resource SSM). In other words, mineral resource SSI focuses on past and present mineral resource SSi which is utilized to produce SSI that motivates a country or region's mineral resource SSM.

Fourth, mineral resource SSI focuses on identifying and resolving issues of mineral resource SS. The relevance and context of these issues are determined through the analysis of mineral resource SSi. Additionally, according to Wang Bing and Wu Chao's understanding of SSI (Wang Bing and Wu Chao 2019c), mineral resource SSI aims to address issues of mineral resource SS, that is, to improve mineral resource SSM. Therefore, in the practice of mineral resource SSM, high-quality SSI about mineral resources should be provided to corresponding SS managers at all levels to help identify and resolve SS issues.

In summary, mineral resource SSI is both a process and a product for it informs mineral resource SS managers of all the information they need before they give directions or address SS issues.

5.7.2 *A Paradigm of Intelligence-Led SSM of Mineral Resources*

5.7.2.1 Meaning of Intelligence-Led SSM of Mineral Resources

In the SSM of mineral resources, the mineral resource SSI plays a significant role for being a fundamental resource and tool. To effectively leverage its role of mineral resource strategy, it is imperative to implement the strategy of "intelligence-led SSM of mineral resources." Based on the concept of intelligence-led SSM (Wang Bing and Wu Chao 2019g), the meaning of intelligence-led mineral resource SSM is derived, with a detailed explanation shown in Fig. 5.19.

According to Fig. 5.19, referring to the basic elements of intelligence-led SSM (Wang Bing and Wu Chao 2019g), the three fundamental elements of intelligence-led SSM of mineral resources are summarized: SSI of mineral resources, SS issues of mineral resources, and SS managers of mineral resources. These elements interact through information flow (for specific relationships, see Fig. 5.20), collectively promoting the continuous improvement of the efficiency and quality (performance) of mineral resource SSM.

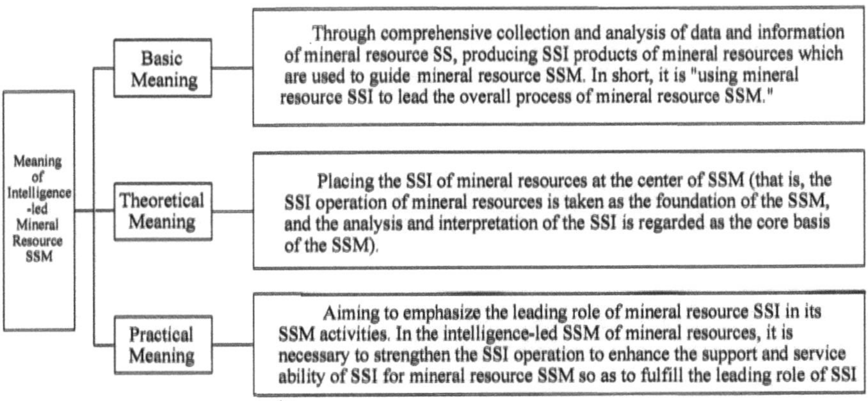

Fig. 5.19 Meaning of intelligence-led SSM of mineral resources

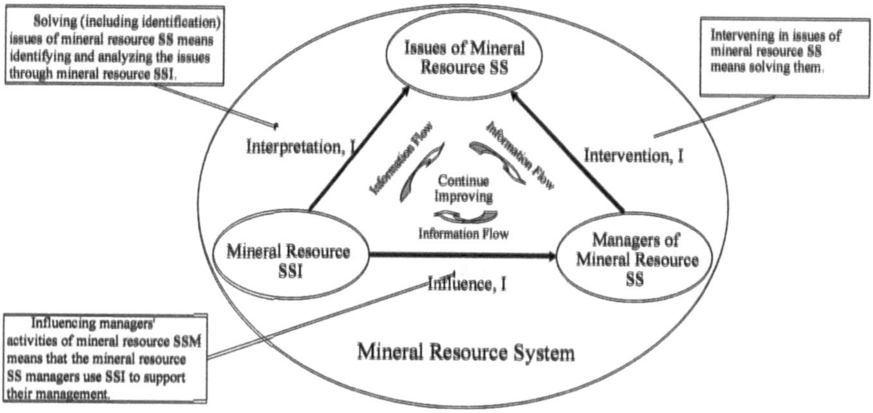

Fig. 5.20 Basic elements of intelligence-led SSM of mineral resources

5.7.2.2 A Model of Intelligence-Led SSM of Mineral Resources in the Big Data Environment

The big data of mineral resource SS can provide substantial raw materials for producing the corresponding SSI. Based on the "relationship among data, information and intelligence of mineral resources," "the meaning of intelligence-led mineral resource SSM," and "the meaning of mineral resource security big data," a model of intelligence-led SSM of mineral resources in the big data environment can be built, as presented in Fig. 5.21. This figure shows that the model in the big data environment primarily includes "data-driven" and "intelligence-led" components, which are closely linked. The former serves as the foundation for producing SSI products of mineral resources, while the latter acts as the means or tool, primarily using intelligence to serve and support management. The connotation of this model is analyzed in Fig. 5.21.

Firstly, the connotation of the "data-driven" segment. The basic steps of this segment mainly include: "collection of mineral resource SS data (using various data collection technologies, channels, and methods, with the assistance of the Internet, the mobile Internet, cloud computing, and the IoT platforms to collect various mineral resource SS data, especially data of live mineral resource SS)—accumulation of mineral resource SS data (creating a large database)—mining of big data of mineral resource SS—analysis of big data of mineral resource SS—extraction of valuable mineral resource SSI—production of mineral resource SSI products." It can be concluded that the essence of the "data-driven" segment is the task of mineral resource SS big data and the operation of mineral resource SSI.

Secondly, the connotation of the "intelligence-led" segment. The central task and goal of this segment is to guide the entire process of mineral resource SSM (all the inks and activities of mineral resource SSM) with mineral resource SSI, in order to achieve the desired effect. The links and activities of mineral resource SSM mainly include: prediction of mineral resource SS (identifying SS issues), decision-making

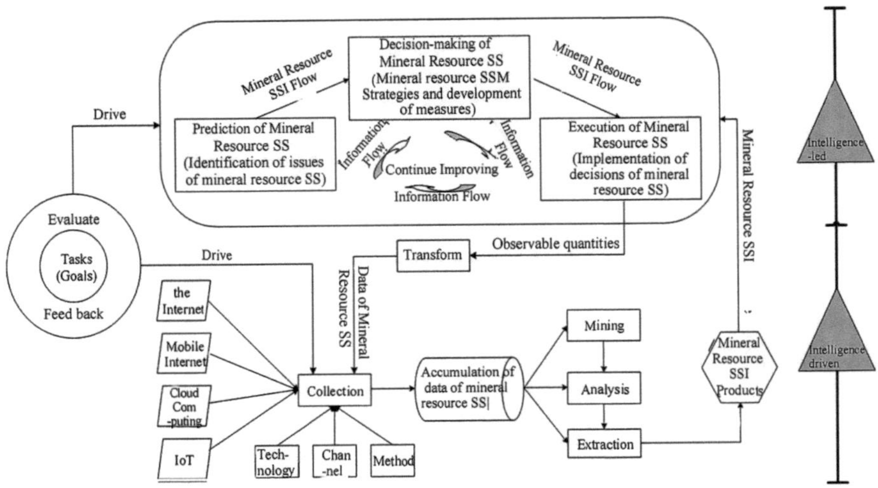

Fig. 5.21 Model of intelligence-led SSM of mineral resources in the big data environment

of mineral resource SS (formulating SSM strategies and specific measures), and execution of mineral resource SS (implementing mineral resource decisions, i.e., using the formulated strategies and specific measures of mineral resource SSM to address the related SS issues). In summary, the essence of the "intelligence-led" segment is the effective utilization of intelligence in mineral resource SSM.

Thirdly, other implications. While serving mineral resource SSM, the mineral resource SSI generates observable quantities (changes in the SS state of the mineral resource system, which represent the performance of mineral resource SSM) at the end of the mineral resource SSM (i.e., the execution of mineral resource SS). These changes can be transformed into mineral resource SS data through a series of processes. Clearly, new mineral resource SSI is produced with the obtained data as raw materials. Based on this repetitive cycle, a mechanism of continuous improvement and refinement for mineral resource SSM can be established, thereby achieving ongoing enhancement in the performance of mineral resource SSM. Additionally, the collection of mineral resource SS data and the operation of mineral resource SSM should be driven by specific tasks (goals) of mineral resource SSM to enhance the relevance, accuracy, effectiveness, efficiency, and economy of the endeavors.

It can be seen that the intelligence-led mineral resource SSM in the big data environment is essentially an organic combination of mineral resource SSI operation in the big data environment with traditional methods, as well as the fusion of "data-driven" and "intelligence-led" mechanisms, aiming to establish the mechanism and methods for mineral resource SSM that "generates intelligence with data and leads management with intelligence."

5.7.3 *Platform Support for Intelligence-Led SSM of National Mineral Resources in the Big Data Environment*

The big data of mineral resources serve as raw materials for producing mineral resource intelligence, while the big data cloud of mineral resource SS is the foundation for collecting and processing the data. In this vein, to achieve the intelligence-led SSM national mineral resources, it is essential to strengthen the construction of the big data cloud of national mineral resource SS. This also involves creating a comprehensive and authoritative online data warehouse of mineral resource SS and a cloud computing platform for the mining and application of mineral resource SS data. These platforms will support and promote the intelligence-led SSM of national mineral resources. Based on the mineral resource SSM at the national level and following the concept of building "proprietary clouds + public clouds" of mineral resource SS, a comprehensive and integrated data cloud of national mineral resource SS is formed, connecting all levels and departments, as shown in Fig. 5.22.

As shown in Fig. 5.22, the platform comprises a national center for primary/backup big data of mineral resource SS, several provincial big data nodes of mineral resource SS, and a public cloud of data repository of mineral resource SS (the hardware and software systems of the big data cloud of mineral resource SS are deployed in the proprietary "national-provincial" cloud, while a big data platform for mineral resource SS based on the public cloud is established at the national level. All levels and components are interconnected, forming a national layout of big data services of mineral resource SS. In summary, the big data cloud of "national-provincial" mineral resource SS, through a unified metadata management system of mineral resource SS, has achieved interconnectivity, mutual backup, and mutual service.

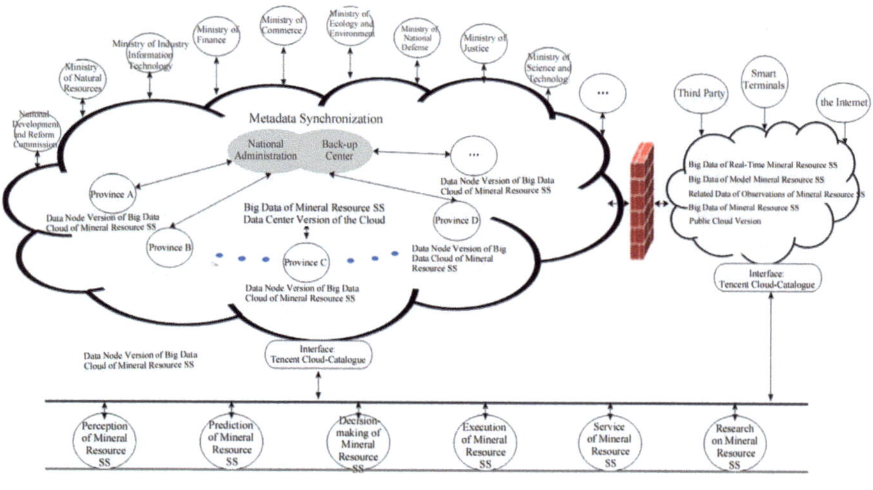

Fig. 5.22 Framework diagram of the big data cloud of national mineral resource SS

This provides support of data intelligence for the perception, prediction, decision-making, execution, service, and scientific research concerning mineral resource SS, thereby assisting intelligence-led SSM of national mineral resources and smart SSM of national mineral resources. Additionally, as depicted in Fig. 5.22, the construction of the big data cloud of national mineral resource SS should be led by the Chinese government, in collaboration with relevant departments of national and provincial mineral resource SSM (such as the Development and Reform Commission, Ministry of Natural Resources, Ministry of Industry and Information Technology, Ministry of Finance, Ministry of Commerce, Ministry of Ecology and Environment, Ministry of National Defense, Ministry of Justice, and Ministry of Science and Technology), for unified design and development. This ensures the authority, standard uniformity, and connectivity of mineral resource SS data.

The big data cloud of national mineral resource SS is centered on the demands of the mineral resource SSM, which should be both the starting point and the ultimate goal of constructing this platform. In the era of big data, various types of SS data of mineral resource are collected and transmitted via private networks, government extranets, the Internet, and IoT. By harnessing cloud computing platforms and incorporating cutting-edge information technologies like big data, machine learning, and artificial intelligence, the capacities of extracting and analyzing data pertaining to mineral resource SS are to be enhanced. These endeavors pave the way for the creation of superior intelligence products of mineral resource SS, promote the efficiency, precision, user-friendliness, and sophistication of the SSM of national mineral resources, and effectively address the requirements of the SSM of national mineral resources. Additionally, with the big data cloud of mineral resource SS at the center of the SSM business of national mineral resources, a "cloud + endpoints" flat SSM process of national mineral resources can also be established and operated in business (see Fig. 5.23).

5.7.4 Conclusions

In the information age, particularly in the era of intelligence and big data, it is an irreversible trend to utilize intelligence and big data to transform and develop new approaches for the SSM of national mineral resources, thus it is both inevitable and urgent to discuss intelligence-led SSM of national mineral resources in the context of big data. This section starts from a theoretical level, defining mineral resource SS, big data of mineral resource SS data, and mineral resource SSI. It also establishes a paradigm for intelligence-led SSM of mineral resources in the big data environment, and constructs a theoretical framework of the big data cloud of national mineral resource SS. It is evident that intelligence-led SSM of national mineral resources in the big data environment primarily comprises "data-driven" and "intelligence-led" components (the former being the foundation, the latter the means and tools), with the big data cloud of mineral resource SS as its basis. This section has explored some fundamental and forward-looking issues (including basic concepts and theoretical

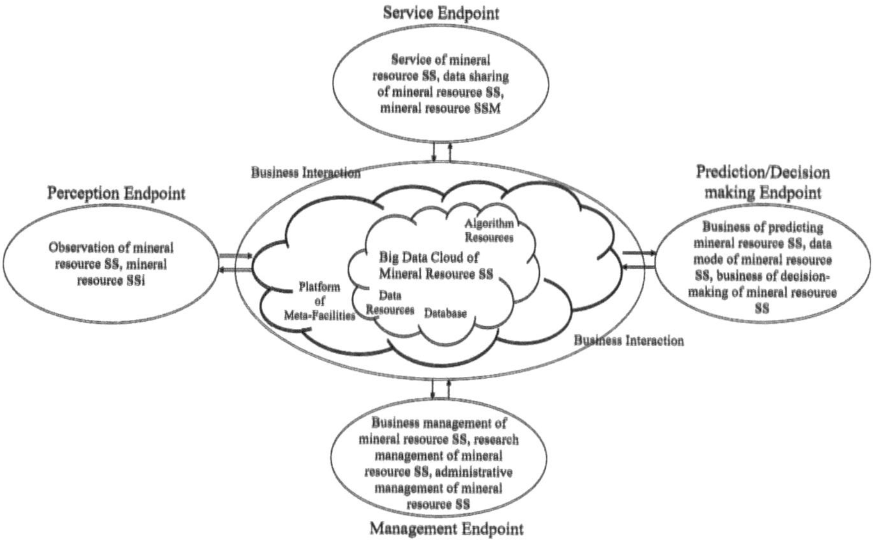

Fig. 5.23 Model of "cloud + endpoints" business interaction with the big data cloud of mineral resource SS at the core

frameworks) regarding intelligence-led SSM of mineral resources in the context of big data, paving the way for research and practice in the field of intelligence-led SSM of mineral resources.

5.8 Chapter Summary

Since safety and security are the cornerstones of national socio-economic growth, SSM is of paramount importance; SSI should also be actively promoted and implemented as it is an indispensable asset for SSM. The intelligence-led SSM has the potential to become the most influential revolution of SSM of the twenty-first century, representing a paradigm shift in the information age, particularly in the operational patterns of SSM in the future era of intelligence. This chapter primarily introduces intelligence-led SSM and its exploratory applications in some typical fields. The following are some conclusions or possibly helpful insights: (1) From the perspectives of SSM theory and intelligence theory, the research focuses on the mechanism and value of SSI in the entire SSM process, providing a universally applicable theoretical basis and reference for management research and practice; (2) There is a solid foundation for establishing intelligence-led SSM methods. Intelligence-led SSM refers to utilizing SSI to guide the SSM process. The conceptual model and implementation model of intelligence-led SSM offer comprehensive theoretical and methodological guidance for its research and practice, and intelligence-led SSM effectively promotes the integrated development of SSI and SSM in both theory and practice; (3) The method of intelligence-led SSM demonstrates remarkable applicability and effects in fields like urban SSM (including smart city SSM), emergency prevention and control, and mineral resource SSM. It brings new ideas and changes to the aforementioned SSM field, and is worth promoting and applying. Admittedly, studying and promoting a new SSM concept and related methods is no easy task, and much work remains to be done, as research on the methods of intelligence-led SSM and its typical application is still in the preliminary stage. In the future, more in-depth research concerning intelligence-led SSM and its application is required, upon the basis of the theoretical framework established and application explored in this chapter.

Postscript

Perhaps many people are similar to me, reading a book from its "Foreword" to "Postscript." Generally speaking, the text of a scientific research is often a relatively rational product from the author's "brain" while the postscript is often a comparatively casual expression of true feelings presented from the author's "heart." It is necessary to read the postscript in order to fully understand a scientific work, and comprehend the thoughts of the author (especially an author who is considered introverted and not quite good at talking). It is for this reason that at the very beginning of writing, I planned and thought about how I would wantonly use my words to savor the past, reflect on the present, and give an outlook for the future in the postscript, but as this book is ready to be printed, the feeling of expressing myself becomes intense. I want to record it in order to show readers the journey of my mind and also to provide them with some necessary "intelligence." Now is the winter vacation, in daytime I perform duties of fatherhood—taking care of my children and spending time with them; in the night time, I write these words down in the quietness of the night after they go to bed. It is a moment of enjoyment for me, because I am more likely to fall into the depths of thinking in the quietness, and be carried away by romantic thoughts about the beauties in life. It is a moment to express how I truly feel.

The process of writing is long and arduous. After hundreds of days and nights of hard work, and after many discussions and corrections, my work is finally about to be printed. Given that I have checked off one of the things on my bucket list, I breathe a long sigh of relief. Here I have packed "six shares" of feelings for those who helped me: I feel immensely grateful for my teacher (the second author of this book, Professor Wu Chao); I feel deeply thankful to my family; I feel truly grateful and thankful to all my colleagues, relatives, and friends who have given me their care, support, and encouragement; I feel pride of my own persistence, hard work and efforts; I have a deep love and passion for SSI research; I am impressed by editors of this book for their diligence and conscientiousness. The assistance I receive from these wonderful people energize me, which adds fuel to my impulse, concentration and passion for SSI research.

B. Wang and C. Wu, *An Introduction to Safety and Security Intelligence Science*,
https://doi.org/10.1007/978-981-96-7261-5

It must be destiny or a good luck that I became a student of Professor Wu Chao, that we share our understandings and pursuits, and that we can study and write together! What is even more fortunate is that this book will be the 9th book we co-authored. "Nine" is a lucky number in the traditional Chinese culture, meaning "eternity," "countless," and "keep pushing the boundaries"; it always goes on. I hope that we can produce more desirable academic results with this book as the base. I believe that our destiny and good luck would go on perpetually, and those expectations can be realized. At first, I could not think of right words to express how I felt about our student–teacher relationship, then I thought it couldn't be better to quote Professor Wu Chao's words from his blog on the Science Net and his emails to me: "Although we are 34 years apart in age—he is a '90s and I am a 50s', I am both his teacher and friend," "Haha, we share the same interest in scientific research. You are very persistent!We are kind of best research partners," "There is no academic 'generation gap' between us, for we are a very compatible pair of '*May to December*' in doing academic research."

In recent years, due to my personal academic interest and theoretical innovation's need of SS science in the new era, I continued to carry out basic theoretical research on SSI, so my entrance into the field of library, information, and archives management, especially the field of information science, might be accidental. I might have got my "first pot of gold" in this field by publishing a series of research papers in journals of library archives and information science. At present, I am leading several vertical research programs funded by The National Social Science Fund of China, and this book is the final result of the later-stage program of the fund (under the discipline of library, information, and archives management). Now, I would like to share with you what drove me to enter the field of intelligence studies. It is mostly due to my understanding, ideal, and pursuit of interdisciplinary research. People like to say that they are engaged in interdisciplinary research, but what exactly does an interdisciplinary researcher do? I think there are two indicators: one is that the content of the research should be interdisciplinary; the other is that the researcher should be able to publish research papers in academic journals of his/her own discipline and other disciplinary fields. Because the SSI research I am engaged in belongs to the cross field of security science and information science, I hope that my SSI research results will be published in academic journals both in the fields of SS science and information science. Because of this understanding, ideal, and pursuit of interdisciplinary research, I am not satisfied with my publication in academic journals in the field of SS science, but determined to publish my research results in journals in the field of information science, and to be recognized or criticized.

At the same time, I am willing to share my personal research experiences and gains from this journey. If you have read the Foreword of this book, you know that the SSI research series and this book are the results of my endeavors. I have gained a lot from writing this book, and a precious understanding comes with it: when someone regards his/her profession and work as happiness, piety comes from the bottom of his/her heart. Here are my research experiences and gains:

First, simplicity is required for academic research. The famous Taoist saying "Great truths are often simple" tells us that there is great wisdom and beauty in extreme simplicity. I am the kind of person who pursuits a simple state in life, with very few interests and fantasizing ideas, and I don't have much interest about assessment indicators or "titles" one could gain for his/her outstanding performance in research. I yearn for freedom, and my greatest enjoyment is to read and write alone when listening to the happy tunes of square dance. Perhaps it is this understanding of simplicity that prompted me to complete this book quickly.

Second, tranquility is required for academic research. I remember reading a widely acclaimed article "Why Do We Lack A Maverick Attitude towards Life." I resonate with it because it is not easy to leave some "quiet time" for ourselves! We are accustomed to the hustle and bustle in life, but one day we will find that what helps us to grow is mostly our own "quietness."

Third, dedication is required for academic research. As the saying goes, "Rome wasn't built in a day." From the beginning of my doctoral study to the completion of this book, I have been thinking about it, studying it, and enriching it almost every day (including most of the holidays) from day to night.

Fourth, lasting interest is required for academic research. I firmly believe that interest is the driving force and essence of scientific research activities, thus my research is based more on the exploration with interest. Since this kind of exploration is not restrained by one's fame or wealth, I find it quite joyful and valuable. For this reason, this book was written not with much passive "persistence" but with the time I am willing to devote to writing.

Fifth, belief and perseverance are required for academic research. The background of my academic discipline is security science and engineering, which is set up in the Polytechnic College where I am considered as a "special" or "peculiar" researcher. Because the basic theoretical research in security science (including SSI) I am engaged in is more of social science in nature, which is very different from the mainstream scientific research direction and assessment system, my research results hardly count. Sometimes, I even feel a bit aggrieved. In fact, it is a problem faced not only by me, but by other researchers of interdisciplinary research in universities of arts and sciences in China. Fortunately, I have insisted on my choice and firmly believe that "The basic theoretical research in security science is not only achievable, but also highly promising." Due to my perseverance in the pursuit in the field of SSI, I eventually harvested a lot of research fruits. It was only through this valuable experience that I understood the saying "One gains the sight because of his belief and perseverance."

Sixth, big-picture thinking is required for academic research. Researchers are required of big-picture thinking to yield high-level research results. This book also originates from big-picture thinking: a research vision that truly belongs to major SS, an academic faith of "daring to create new disciplines, take the lead in the international arena, and be the first comer in theoretical innovation," an academic courage of "sitting on the cold bench willingly, sticking to the interest, and spending years of time focusing on one matter," an academic pursuit of "making down-to-earth efforts while looking up to the stars," and a methodology of "making a blueprinting → producing series (results) → forming a system."

Seventh, authentic feelings are required for academic research. Whether it is for people, events, or objects, human beings are after all emotional beings. In my journey of scientific research, the phrase "safety and security" has always stayed with me as an intimate friend. I believe that any love is by no means a whim. Destiny may start a relationship, from making acquaintance to liking or loving someone, but strong feelings are only formed in a lengthy process—years and months of accumulation and cultivation, please think of the development of maternal love in the months of pregnancy. I confess that my love for the study of SSI has taken time to develop, and I believe this is an understanding shared by many readers. I still remember that the title of the acknowledgment of my undergraduate thesis was "Security, it is happy to love you." And as time slipped by, my love for security has become even purer. I am willing to take the trouble and time to explore SSI because I love it truly.

People pursue value in everything they do. I hope that the thinking, theories, and methods of SSI science conveyed by this book can contribute to the development of SS and intelligence studies, the construction of a higher level of safe and secure China, the cause of national SS, and the pursuit of the SS and well-being of all mankind. I would like to dedicate this book to all my colleagues who are passionate about the cause of SS and intelligence studies (especially SSI). I hope we can live up to our ambitions by aiming at the stars and making great plans. Our aspiration and passion could be our steeds. Yes, we see because we believe. I will remain passionate about SSI and dedicate my life to the field of SSI. It is my hope that in the course of research and exploration, I will be able to publish a textbook entitled *A Detailed Treatise on SSI* or *SSI* as soon as possible, so as to fulfil the expectations of my readers.

Considering that a redundant text only fatigues its readers, I conclude the recording and sharing of my personal experiences and thoughts in SSI research, thus my readers can listen to the voice deep within my heart.

This concludes the postscript.

Wang Bing

Midnight on February 15, 2022

References

Treatises

Ba, Z. Urban Development and National Security: Proceedings of the 6th China National Security Forum. Current Affairs Press, Beijing, 2008.
Bao, C. Preface to the Information Analysis Series. In Sun, Z., et al. Introduction to Information Analysis. Tsinghua University Press, Beijing, 2007.
Gao, J. The Great Blunder: Mysteries of Major Intelligence Wars in the 20th Century. Jiangsu People's Publishing House, Jiangsu, 1998.
Gao, J., Wu, X. A History of Intelligence Thought. Jin Cheng Publishing House, Beijing, 2014.
Guo, S., Hou, Y., & Wang, X. Information Economics. Tsinghua University Press, Beijing, 2014.
Han, C., Zhu, H., Duan, Z., et al. Multi-source Information Fusion. Tsinghua University Press, Beijing, 2006.
Li, L., Ren, H. Infectious Diseases. People's Medical Publishing House, Beijing, 2013.
Liang, H. Research on Public Security Information Science Discipline System. China Legal Publishing House, Beijing, 2017.
Ma, F., Zhao, Z. An Introduction to Intelligence Engineering. Scientific and Technical Documentation Press, Beijing, 2019.
Qian, X. On Cognitive Science. Shanghai People's Publishing House, Shanghai, 1986.
Qin, D. Theory of Information Literacy. Nanjing University Press, Nanjing, 2012.
Qiu, J., Wang, Y. Bibliometric Content Analysis Method. National Library of China Publishing House, Beijing, 2008.
Tian, S., Jing, G. Safety Management (2nd Edition). China Machine Press, Beijing, 2016.
Wang, C. Strategic Management. Tsinghua University Press, Beijing, 2013.
Viktor Mayer-Schönberger. Big Data: A Revolution That Will Transform How We Live, Work, and Think. Translated into Chinese by Zhou, T. Zhejiang People's Publishing House, Hangzhou, 2012.
Wu, C. Methodology of Safety Science. Science Press, Beijing, 2016.
Wu, C., Wang, B. New Disciplines of Safety Science. Science Press, Beijing, 2018.
Xu, Z., Jiang, X. Safety Systems Engineering (3rd Edition). China Machine Press, Beijing, 2016.
Yan, Y. Introduction to Information Science (Revised Edition). Wuhan University Press, Wuhan, 1994.
Ye, Y., Wu, Y. An Elementary Coursebook on Information Science (Second Edition). Science Press, Beijing, 2012.
Zhang, X. A Study of U.S. Military Intelligence Theory. Military Science Publishing House, Beijing, 2007.
Zhao, R. Competitive Intelligence Science. Science Press, Beijing, 2017.

B. Wang and C. Wu, *An Introduction to Safety and Security Intelligence Science*,
https://doi.org/10.1007/978-981-96-7261-5

Project Team for Strategic Research on Construction and Promotion of China's Smart Cities. Strategic Research on Safety Development of China's Smart Cities. Zhejiang University Press, Zhejiang, 2015.

Chinese Theses

Published Theses

An, L., Wu, Y., Li, G. The Framework of Counter-Terrorism Intelligence and Information Work Capability. Research on Library Science. 2018 (17).
Ba, Z., Li, G., An, L., et al. Information Integration of Big Data for National Security: Application Architecture and Realization Methods. China Soft Science. 2018 (7).
Bai, M., Zhu, Q., Guo, H., et al. Planning and Design of Intelligent Pension Information System Based on Cloud Computing. Journal of Shandong University of Finance and Economics. 2017 (3).
Bao, C. The Quiet Dawn—A Reconsideration of Intelligence and Information Science in China. Library and Information Service. 2009 (8).
Bao, C., Jin, X., Zhang, J., et al. The Construction of Discipline System of Intelligence Studies in China. Journal of Intelligence. 2018 (10).
Cai, S. The Fusion of Intelligence in America Homeland Security. Journal of Intelligence. 2019 (1).
Cao, R., Shi, J., Guo, H. Research on Intelligent Decision Based on Competitive Intelligence: Connotation, Mechanism and Process. Information Studies: Theory and Application. 2017 (12).
Chen, C. The Essence of Intelligence. Competitive Intelligence. 2017 (2).
Chen, C., Zeng, Q. The Path of Intelligence Capability Building of Think-tank in View of Intelligence Research. Library and Information Service. 2018 (21).
Chen, M., Chen, F. An Analysis on the Path of Information Awareness for Science and Technology Security. Information Science. 2019 (5).
Chen, M., Wang, Y. Analysis on Conditions of Information Awareness. Information Studies: Theory and Application. 2018 (8).
Chen, M. The Fourth Paradigm: Data-Intensive Scientific Discovery. Computer Education. 2013 (9).
Chen, M., Wang, Q. The Intelligence Security in the Holistic Review of National Security. Journal of Wuhan University of Technology (Social Sciences Edition). 2018 (5).
Chen, X., He, Y. Construction of Evaluation Index System of the University Library Service Capacity under the Background of "Double First-Class" Construction. Journal of Academic Library and Information Science. 2018 (5).
Chen, Z., She, L. Research on City Vulnerability to Safe Development—Based on an Angle of Underground Space Utilization. Journal of Huazhong University of Science and Technology (Social Science Edition). 2009 (1).
Cui, S., Yu, X. Security Governance: From the Perspective of Non-traditional Security Management. World Economy and Politics. 2010 (1).
Dai, L. Application of Social Networks in Open Source Security Intelligence. Information Security and Communications Privacy. 2016 (1).
Deng, K. An Analysis of Russian Security Intelligence and Early Warning Mechanism of Threats. Journal of Jiangnan Social University. 2018 (3).
Ding, L., Xu, K., Li, X. Analysis on the Influencing Factors for Quality of Intelligence Collection in Think Tanks Based on Grounded Theory. Library and Information Service. 2019 (21).
Ding, X., Su, X. Financial Security Intelligence Analysis Based on Blockchain Driven Trustable Big Data and AI. Journal of the China Society for Scientific and Technical Information. 2019 (12).
Dong, K., Qiu, J. Research on the Influence of Big Data Environment on the Development of Informatics. Journal of the China Society for Scientific and Technical Information. 2017 (9).

Dong, Y., Liu, Q. Research on the Intelligence-led Supply Chain Risk Management. Information Studies: Theory and Application. 2016 (4).
Du, L., Tang, Y. Research on Evaluation Index System of Scientific and Technical Information Service Ability in University Library from the Perspective of Supply Chain. Journal of Library and Information Science in Agriculture. 2019 (12).
Fan, B. On the Transmission Mechanism of US National Security Intelligence. Journal of International Security Studies. 2019 (2).
Fan, B., Li, J. On Reengineering Method for Collaborative Emergency Information System. Journal of Intelligence. 2011 (7).
Feng, Q., Zhu, X. Research on Application of Artificial Intelligence in Information Work. Information Studies: Theory and Application. 2019 (11).
Gan, Y., Wang, L., Huang, J., et al. Intelligence Analysis and Technologies Development Exploring in Big-data and Artificial Intelligence Era. Telecommunication Engineering. 2018 (5).
Gao, J. National Security Intelligence Science as a Discipline. Information Studies: Theory and Application. 2019a (1).
Gao, J. On the National Security Intelligence and Research Objects of National Security Intelligence Studies. Journal of Intelligence. 2019b (1).
Gao, J. On the Approaches to Reforming the System of National Security Intelligence. Journal of Public Security Science. 2019c (2).
Gao, J. Evolution of Intelligence Analytic Methodology: From A Perspective of National Security Intelligence Analysis. Scientific Information Research. 2020 (1).
Ge, J. The Management Pattern, the Problems and Countermeasures of Wisdom Agriculture. Ecological Economy. 2017 (11).
Gu, S., Yao, Y., Shen, L., et al. Conceptual Framework and Research Focus of Resource Security. Journal of Natural Resources. 2002 (3).
Gu, J., Tang, X., Zhu, Z. A Survey of Wuli-Shili-Renli System Approach. Journal of Transportation Systems Engineering and Information Technology. 2007 (6).
Guan, L., Hu, G., Wang, Z. Research on Network Security Situational Awareness Technology Based on Big Data. Netinfo Security. 2016 (9).
Guang, X., Wang, B., Wu, C., et al. A Study of Intelligence-Led Model and System for the Safety and Security Management of Smart Cities. Journal of Intelligence. 2020 (2).
Guo, C., Xu, C., Chu, J. Preliminary Exploration of Information Analysis Talents Training for Emergency Management in Big Data Era. Library and Information Service. 2019 (5).
Guo, C., Zhang, J. Research on the Construction of Rapid Response Intelligence System for Emergency Decision-making. Information Studies: Theory and Application. 2016 (5).
Guo, H., Su, X., Deng, S. Research on City Emergency Management Information System under the Background of "Smart City". Library and Information Service. 2016 (15).
Guo, J., Huang, S. A Study of Architecture of Big Data Based on Military Intelligence Analysis and Service System. Journal of China Academy of Electronics and Information Technology. 2017 (4).
Guo, R., Tang, H. Research on Development and Trend of the Research of Urban Security in China. Urban Development Studies. 2013 (11).
He, H., Liu, S. The Transfer Paths of France's Security Intelligence. Journal of Jiangnan Social University. 2017 (3).
He, Z. An Overview of Artificial Intelligence. Technology and Economic Guide. 2018 (33).
Hong, L., Zhou, L., Chen, L. Research on the Construction of Data-driven Smart Information Service System in Libraries. Library & Information. 2018 (2).
Hu, S. Methodology of Safety Science Research. China Safety Science Journal. 2003 (9).
Hu, Y., Bai, R., Pan, B. Analysis and Evaluation of Intelligence-led Policing Model in Foreign Countries. Journal of Modern Information. 2014a (6).
Hu, Y., Liu, Q., Pan, B. Research on Prediction of Intelligence for Science and Technology Security. Journal of Intelligence. 2014b (9).

Hua, B. Multi-source Information Requires Multi-party Fusion. Journal of the China Society for Scientific and Technical Information. 2013 (3).
Hua, B., Li, G. Wisdom Intelligence in Data Era from the Perspective of Multi-dimensional. Information Studies: Theory and Application. 2016 (5).
Hua, B., Li, G. Architecture Design and Key Techniques of Intelligence Analysis System Based on Artificial Intelligence. Library & Information. 2017 (6).
Hua, B., Zheng, Y. The Theory of Information Transformation (II): Transformation from Information into Intelligence. Information Studies: Theory and Application. 2012 (4).
Huang, L., Wu, C., Jia, N. A Study of Methodology of Safety Theoretical Model Construction. China Safety Science Journal. 2016 (12).
Jiang, H. On the Evolution of Relationship between National Security and Intelligence Work. Journal of Intelligence. 2015 (12).
Jiang, X., Wang, B. A Preliminary Study of the Big Data Analytics and Its Adaptability in Intelligence Studies. Library & Information. 2014 (5).
Jiang, D. Research on the Development and Management of Intelligence Personnel. Science and Technology Innovation Herald. 2013 (16).
Jiao, J., Zhu, L., Zhang, M. Analysis Model and Application of Military Intelligence Value in Decision-Making. Journal of Modern Information. 2008 (3).
Jin, S. Intelligence Characteristics and Intelligence Value. Journal of Modern Information. 1996 (3).
Jin, H. On the Construction of National Intelligence Law Mechanism under the Concept of Overall National Security—Taking the "Three Levels" and "Three States" as the Framework. Journal of Intelligence. 2018 (11).
Lai, J., Yan, X., Deng, L., et al. A Comparative Study of the Concept of "Intelligence" in China, Japan and Korea. Journal of Intelligence. 2018 (4).
Lan, X. A Study of the Strategy to Bridge China's Urban-Rural Digital Divide Based on the Quantitative Method of SWOT. Information Science. 2016 (2).
Li, B., Wang, F., Wu, L., et al. An Urgent Need for Bio-safety Education in Higher Vocational School and Junior College. China Medical Herald. 2009 (4).
Li, C., Wu, C. Tacit Knowledge and the Training of Intelligence Analysts. Information Studies: Theory and Application. 2019 (8).
Li, D., Gong, J., Shao, Z. From Digital Earth to Smart Earth. Geomatics and Information Science of Wuhan University. 2010 (2).
Li, G., Li, Y. Research on the Emergency Decision-making from the Perspective of Intelligence. Information Studies: Theory and Application. 2015 (8).
Li, G., Li, Y. Construction of Emergency Decision-making Intelligence System Against the Background of Smart City. Journal of Library Science in China. 2016 (3).
Li, G., Lyu, B. On Intelligence Cycle. Library Journal. 2012 (1).
Li, H., Liu, Z. Anti-Terrorism Intelligence Information Work under the Holistic View of National Security: Issues and Practices. Administration and Law. 2018 (11).
Li, H., Dong, F., Guo, R. Key Points in Studies of Medical Informatics from the Visual Angle of Information Chain and Their Future Development Direction. Chinese Journal of Medical Library and Information Science. 2015 (1).
Li, H., Zhang, H., Hou, Y., et al. Research on Service Capabilities of Scientific and Technical Information in Intelligence 3. 0 Era. Information Studies: Theory and Application. 2017 (3).
Li, K., Xue, L. Non-traditional Security Theories: Conceptions, Schools and Characteristics. The Journal of International Studies. 2012 (2).
Li, M., Wu, C. A Study of Methodology of Safety Humanology. China Safety Science Journal. 2015 (3).
Li, M., Sun, J., Zhang, Z. Frame of Industrial Information Security Intelligence Analysis Based on Big Data. Cyberspace Security. 2019 (4).
Li, X., Gong, H. A Survey on Big Data Systems. Scientia Sinica Informationis. 2015 (1).

Li, Y., Jin, P. Review of Theoretical Research on the Undergraduate Talent Training Mode of Higher Education in China. Jiangsu Higher Education. 2003 (5).

Li, Y. Consideration on the Construction of Intelligence Capability Oriented to Urban Emergency Management. Journal of Modern Information. 2019a (5).

Li, Y. System Construction and Implementation Approach for Anti-Terrorist Intelligence Information Service in a New Era. Documentation, Information & Knowledge. 2019b (4).

Li, Y., Li, G. Practice and Application of Intelligence Engineering Oriented to the Emergency Decision-making under the Background of Smart City. Library and Information Service. 2016 (11).

Li, Y. Win a New Victory in Preventing and Mitigating Potential Risks and Dangers. Retrieved from http://csr.mos.gov.cn/content/2019-03/13/content_74905.htm [2019–03–13].

Liang, G. Intelligence Service—The Library Service in the Knowledge Economy Era. Research on Library Science. 2011 (6).

Liang, Z. An Inquiry into Some Issues Concerning Information Science. Information Studies: Theory and Application. 2003 (3).

Liao, Q. The Concept of "Intelligence" in Japan and Its Impact on Intelligence Work. Information Studies: Theory and Application. 1992 (5).

Liu, J., Tang, X., Shi, M., et al. Research on Intelligence Intervention and Fusion Model in Emergency Management. Library and Information Service. 2019 (18).

Liu, J., Huang, D. Application of Big-Data in Information Security Field. Secrecy Science and Technology. 2013 (9).

Liu, Q., Gan, Y. A Discussion on the Enterprise-Oriented Collaborative Training Mode for Competitive Intelligence Talents. Research on Library Science. 2015 (17).

Liu, L., Wang, C., Liu, Y. A Study of Triangle Transformation Model of "Data- Information-Intelligence". Journal of Modern Information. 2015 (2).

Liu, Q., Yang, Y., Yin, J. A Study of Communication Features and Rules of Public Opinions in Public Emergency: Taking the "Problematic Vaccine Event of Changsheng Biotechnology Cooperation" as an Example. Journal of News Research. 2018 (21).

Liu, R., Wu, C., Liu, Y., et al. History and Development of Scientific and Technical Intelligence in China. Journal of the China Society for Scientific and Technical Information. 2019 (1).

Liu, X. Safety Ethics and the Construction of Models of "Moral" Safety Management. Reform of Economic System. 2007 (6).

Liu, Y. Science and Technology and National Security. Journal of North China Electric Power University (Social Sciences). 2000 (4).

Liu, Z. Review of the "Macroscopic Intelligence" Idea. Information Studies: Theory and Application. 1999 (2).

Liu, Z. On the Connotation of the Idea of "Macroscopic Intelligence". Journal of Intelligence. 2000 (3).

Long, R., Yang, J. Research Status and Prospect of National Mineral Resource Security. Resources Science. 2018 (3).

Lu, W., Wang, Y. On Emergency Governance in Hospital-Patient Mass Conflicts Based on the System of the Whole Life Cycle and the PPRR Theory: Ten Typical Cases. Journal of Guangzhou University (Social Science Edition). 2019 (2).

Luo, L., Li, G. Smart Intelligence Service and Knowledge Fusion. Information and Documentation Services. 2019 (2).

Luo, T., Wu, C. Basic Issues of Safety Information Science. Science & Technology Review. 2018 (6).

Luo, Y. Analysis and Insights on President Xi's Strategic Thoughts for Public Safety and Emergency Management. Retrieved from https://mp.weixin.qq.com/s/9E7HtQpmN43im-kfnYvYXw [2018–05–17].

Luo, Y., Huang, X., Xu, M. Exploration on Development and Trend of Scientific Management on Work Safety. Journal of Safety Science and Technology. 2016 (10).

Lyu, X. A Research on the Model of Intelligence-led Policing. Journal of People's Public Security University of China (Social Sciences Edition). 2008 (4).
Ma, D. On the Construction of "Big Intelligence" System for Public Security Bureau with Emphasis on the Culture of Intelligence Sharing. Journal of People's Public Security University of China: Social Sciences Edition. 2011 (3).
Ma, D., Su, Y. Research on Several Basic Issues of Public Security Intelligence Studies in China from the Perspective of "Intelligence Studies". Information Studies: Theory and Application. 2013 (5).
Ma, F. A Review of Yovits' Views on Intelligence Theory. Technology and Market. 1984 (4).
Ma, F. Historical Review of the Development of Information Science with Proposing Frontier Topics. Documentation, Information & Knowledge. 2013 (2).
Ma, G. Research on the Construction of the Training Mode of Innovative Talents. Higher Agricultural Education. 2001 (4).
Ma, H., Zhang, B. Analysis on the Diffusion of Security Information Laws and Regulations and Policies in China. Journal of Intelligence. 2019 (7).
Ma, W. Science and Technology Security: Definition, Intension and Extension. Studies in International Technology & Economy. 1999 (2).
Mo, J. Some Reflections on Speeding up the Construction of National Bio-safety Legal System. Journal of Xinjiang Normal University (Edition of Philosophy and Social Sciences). 2020 (4).
Nie, H. Discussion on Data Mining and Artificial Intelligence Technologies. Technology and Market. 2019 (5).
Ouyang, Q., Wu, C. A Study of Influencing Factors of Safety Big Data Sharing and its Model Construction. Journal of Safety Science and Technology. 2017 (2).
Ouyang, Q., Wu, C., Huang, L. Research on Basic Principles of Applications of Big Data in Field of Safety Science. China Safety Science Journal. 2016 (11).
Pan, W., Shen, G. The Evaluation Model of Emergency Intelligence System Service Capability. Information Science. 2019 (11).
Pang, N. Cognitive Informatics: New Opportunities of Information Analysis in Big Data Era. Information Studies: Theory and Application. 2018 (12).
Peng, Z. Intelligent Safety Management: A New Mode for University Safety Management in the Big Data Era. Legal System and Society. 2019 (11).
Peng, Z. Research on the Intelligence Process: Review and Reflection. Journal of The China Society for Scientific and Technical Information. 2016 (10).
Peng, Z. Research on the Integration of Public Security Intelligence Analysis and Big Data Analysis. Information Science. 2017a (10).
Peng, Z. A Probe into Public Security Information Science: Discipline Attributes and Changes under Big Data Environment. Information and Documentation Services. 2017b (5).
Qian, X., Yu, J., Dai R. A new discipline of science—The study of open, complex, giant system and its methodology. Chinese Journal of Nature. 1990 (1).
Qin, D., Zhang, Y. Information Literacy: The Core Element of Information Security Theory. Information Studies: Theory and Application. 2015 (4).
Qin, D., Zhang, Y. Construction and Practice of Three-level Information Literacy Theory. Information Studies: Theory and Application. 2017 (6).
Qin, T. On the Basic Attributes of Intelligence. Journal of the China Society for Scientific and Technical Information. 1991 (1).
Qu, Z., Lan, Y., Xia, Y. The Construction Research of Emergency Intelligence Analysis Model Based on the Big Data. Journal of Modern Information. 2017 (1).
Shang, P. Construction Outline of National Security Intelligence Studies: Research Object, Discipline Characteristics, System and Research Methods. Journal of Intelligence. 2018 (8).
Shen, J., Zhang, K. An empirical analysis of factors leading to typical urban problems in China. Progress in Geography. 2020 (1).
Shi, Y., Chen, X., Xie, W. Characteristic, Task and Tools of Intelligence 3.0 Period. Journal of Intelligence. 2017 (10).

Su, X. The Rise of Intelligence Studies in the Age of Big Data. Journal of the China Society for Scientific and Technical Information. 2018 (5).
Su, Y., Liu, R. The Rapidity of Earthquake Disaster Intelligence. Journal of Intelligence. 2001 (2).
Sun, C., Li, Q. Public Library Service Ability Evaluation Index System Construction and Empirical Research. Information and Documentation Services. 2015 (6).
Sun, J., Li, Y. On Several Issues about the "Smart" Development of Intelligence Studies and Intelligence Work. Journal of Information Resources Management. 2019 (1).
Sun, L., Wang, Y. A Multi-Source Competitive Intelligence Fusion Method Based on Knowledge Element. Journal of Intelligence. 2017 (11).
Sun, R., Ma, H. Research on the Construction of National Intelligence Work Security System with Chinese Characteristics under the View of Overall. Information and Documentation Services. 2019 (1).
Sun, Y. Accelerating the Modernization of National Biosafety Governance Capacity. Security Work. 2020 (4).
Tan, Z., Wu, S., Han, X. References and Enlightenments from the Public Health Emergency Response Systems of the United States of America and Japan—Pneumonia Epidemic Response to Novel Coronavirus Pneumonia in China. Health Economics Research. 2020 (3).
Tan, B. Model and Validation: Construction and Application of University Students' Information Literacy Evaluation System. Information Research. 2017 (12).
Tan, Q., Miao, D., Liu, J. A Study of Event Tree Analysis Method Based on Reliability Theory. Journal of Safety Science and Technology. 2015 (6).
Tan, X., Wu, C. Research on Rapid Intelligence Production Line System Based on Workflow. Information Studies: Theory and Application. 2019 (11).
Tang, C. A Study of the Construction of Public Security Intelligence Legal System in the Guidance of Holistic View of National Security. Journal of Intelligence. 2017 (4).
Tang, M., Su, X., Zhang, Y. The Big Data Oriented Emergency Information Collection with Internet of Things. Information Science. 2018 (3).
Tang, M., Su, X., Wang, H. A Case Study of Emergency Rapid Response Intelligence System—Base on the Public Safety Emergency. Information Science. 2019 (1).
Tang, X., Zheng, D., Tan, M. Construction of Enterprise Competitive Intelligence System Model Integrating Information Methodology and Artificial Intelligence Technology. Information Science. 2019 (7).
Wang, G. Uncertainty and Intelligence Services. Journal of Library and Information Science. 2007 (1).
Wang, B. Computational Safety and Security Science. Modern Occupational Safety. 2020a (1).
Wang, B. Bio-Security & Safety Intelligence: A Significant New Topic in Security & Safety Intelligence Science. Journal of Intelligence. 2020b (10).
Wang, B., Liu, H., Wu, C. A Study of Intelligence-led Emergency Prevention and Control. Journal of Information Resources Management. 2020 (1).
Wang, B., Wang, Y. Analysis of Influencing Factors of Safety and Security Intelligence Failure in Safety and Security Management. Journal of Intelligence. 2021a (2).
Wang, B., Wang, Y. A Study of the Cause Model of Safety and Security Intelligence Failure. Information Studies: Theory and Application, 2021b (1).
Wang, B., Wu, C., Huang, L. Intelligence-Led National Mineral Resource Safety and Security Management in the Big Data Environment: Paradigm and Platform. Journal of Intelligence. 2019 (10).
Wang, B., Wu, C. An Outline of the Research of System Safety Science under the Guidance of Safety Information. Journal of Intelligence. 2017a (10).
Wang, B., Wu, C. Structure and Deduction of the System Safety Model of Safety-related Information—Safety-related Behavior (SI-SB). Journal of Intelligence. 2017b (11).
Wang, B., Wu, C. A Study of the Innovation Research of Safety Science Based on the Safety Big Data. Science and Technology Management Research. 2017c (1).

Wang, B., Wu, C. A Study of the Background and Foundation for the Establishment of Safety and Security-Related Intelligence Science. Journal of Intelligence. 2018a (10).
Wang, B., Wu, C. Thoughts about the Three Key Questions of Safety Management Informatization on the Science Level: Basic Connotation, Theoretical Motives and Focus Changes. 2018b (7).
Wang, B., Wu, C. The Outline of Safety Informatics. Journal of Intelligence. 2018c (2).
Wang, B., Wu, C. A New Approach of System Safety Management Based on the Evidence and Risk: ERBS. Journal of Intelligence. 2018d (9).
Wang, B., Wu, C. Safety-Related Information Literacy: An Important Concept in the Interdisciplinary Field of Library & Information Science and Safety Science. Information Studies: Theory and Application. 2018e (7).
Wang, B., Wu, C. Modern Urban Safety and Security from the Perspective of System Safety Science—Safety and Security Issues of a Typical Complex Giant System. Shenzhen Security. 2018f (6).
Wang, B., Wu, C. The Origin, Evolution Trends and Meaning of Safety and Security-related Intelligence. Library and Information Service. 2019a (11).
Wang, B., Wu, C. A Study of Some Basic Issues of Safety and Security-related Intelligence Science under the Guidance of the Big Safety and Security Concept. Journal of Intelligence. 2019b (3).
Wang, B., Wu, C. Some Problems and Thinking on the Discipline Construction of Safety and Security Intelligence Science. Library and Information Science. 2019c (24).
Wang, B., Wu, C. Some Thoughts on National Safety and Security Science—A Speculation from the Safety and Security Science Researcher. Journal of Intelligence. 2019d (5).
Wang, B., Wu, C. Function Mechanism and Value of Safety and Security-related Intelligence in Safety and Security Management. Information Studies: Theory and Application. 2019e (2).
Wang, B., Wu, C. An Approach to the Acquisition and Analysis of Safety and Security-Related Intelligence: R-M Method. Journal of Intelligence. 2019f (1).
Wang, B., Wu, C. Intelligence-Led Safety and Security Management (ILSM): Basis, Meaning and Model. Information Studies: Theory and Application. 2019g (6).
Wang, B., Wu, C. A Study of the Theoretical Framework of Safety &Security-Related Intelligence System (SIS). Journal of Modern Information. 2019h (1).
Wang, B., Wu, C. Change and Development of Safety and Security Intelligence Science in the Big Data Environment. Library and Information Service. 2020a (10).
Wang, B., Wu, C. A Study of Intelligence-Led City Safety and Security Management. Information Studies: Theory and Application. 2020b (4).
Wang, B., Wu, C. Big Data-driven and Intelligence-led Intelligent Safety and Security Management: Meaning and Research Framework. Library and Information Service. 2020c (11).
Wang, D., Feng, W. Based on Emergencies: Intelligence Needs Analysis and Risk Prevention and Control in Large-scale Sports Events. Journal of Intelligence. 2019 (8).
Wang, F. Intelligence 5.0: Parallel Intelligence System in the Parallel Era. Journal of the China Society for Scientific and Technical Information. 2015 (6).
Wang, J., Song, S. Construction of the Multimodal Crisis Intelligence Fusion System of the Emergency Management. Information Science. 2016 (12).
Wang, J., Xu, M., Zhu, Z. Research on Application of Methods in Information Analysis. Information Studies: Theory and Application. 2020 (1).
Wang, K., Feng, X., Liu, X. Competitive Intelligence and Enterprise Crisis Early Warning. Information Science. 2012 (3).
Wang, L., Zhou, L. Public Safety, Risk Prediction and Management Strategies: A Case Study of Stampede at the Bund in Shanghai on 31, December. Journal of Guangzhou University (Social Science Edition). 2016 (15).
Wang, M., Zuo, J. Chen, Y. The Analysis and Construction of International Professional Training Mode in Information Management in Big Data Environments. Library. 2015 (4).
Wang W., Ma, J. Discussion on Information Talents Training in China Based on the Demand of the Job Market. Information Studies: Theory and Application. 2017 (6).

Wang, X. Cyber Biosecurity: A Special Frontier in the Rivalry of Great Powers Rivalry. Scientific Chinese. 2019 (7).
Wang, X., Cheng, Z. The Demands Analysis of Counter-intelligence for Enterprise Crisis Early Warning. Information Science. 2017 (1).
Wang, X., Wang, X. Research on Intelligent Service System of Industrial Competitive Intelligence. Competitive Intelligence. 2019 (3).
Wang, Y. Particularity—Insights in the Distinctive Development of Library and Information Science. Information Studies: Theory and Application. 2019 (10).
Wang, Y., Zhao, K., He, F. Analysis of the Relationship of Information, Intelligence and Wisdom. Information Studies: Theory and Application. 2016 (2).
Wang, Y. Research on the Problems in the Development of Information Science in China. Library and Information Studies. 2020 (1).
Wang, Y., Zou, B., Song, X. Intelligence Analysis Knowledge Base Driven by Big Data and Its Architecture Design. Information Studies: Theory and Application. 2019 (1).
Wang, Z. Chen, Q. Characteristics and Role of Competitive Intelligence in Companies. Journal of Intelligence. 2004 (1).
Wang, Z., Ge, L. Comparison between SWOT Model and BCG Matrix in Competitive Intelligence. Library & Information. 2013 (3).
Wu, X., Yang, Z. The Concept of Smart City and Future City Development. Urban Development Studies. 2010 (11).
Wu, C., Wang, B. Research Review on the Trends and Theoretical Progress of Safety Science in Recent Years. Journal of Safety and Environment. 2018b (2).
Wu, C., Huang, L., Jia, N. A Study of the Construction of Generalized Safety Model. Science and Technology Management Research. 2018a (1).
Wu, C., Wu, L. Thoughts on Safety and Security Management Modes from the Perspective of Safety and Security-related Intelligence. Journal of Guangzhou University (Social Science Edition). 2020 (2).
Wu, C., Yang, M., Wang, B. Safety Definition, Intention, Extension and Inferences from Scientific View. Journal of Zhengzhou University (Engineering Science). 2018b (3).
Wu, C., Li, H., Fu, H. Intelligence Service Towards the Era of 3.0. Information Studies: Theory and Application. 2015 (9).
Wu, C., Tang, X. Changes and Challenges Faced by National Security Intelligence in the Big Data Era. Journal of Intelligence. 2020 (6).
Wu, Q. On the Quality and Cultivation of Scientific and Technological Intelligence Personnel. Information Research. 2012 (7).
Wu, Q., Sha, J. A Study of Regional Resource Security Index System of Resource System View Perspective. Economic Research Guide. 2015 (18).
Wu, Q., Wu, C., Liu, R. Construction of Intelligence Service System under Intelligence 3. 0. Information Studies: Theory and Application. 2018 (11).
Wu, S. Risks and Policy Choices in the Big Data Era. Outlook. 2013 (32).
Wu, S., Chen, Y., Wang, K. On the "Target-Centered" Intelligence Analysis Process of U.S. Journal of Intelligence. 2013 (4).
Wu, X., Shen, C. New Features and Enlightenments of the US Emergency Prevention System in the New Phase. Journal of Wuhan University of Technology (Information & Management Engineering). 2016 (2).
Xi, C., Deng, S., Zheng, H. Artificial Intelligence and Intelligence Collection Analysis. Security Work. 2018 (4).
Xiao, K. Sun, L., Li, N. Mineral Resources Assessment under the Thought of Big Data. Geological Bulletin of China. 2015 (7).
Xiao, L., Meng, T., Wang, W. Entity Recognition of Intelligence Method Based on Deep Learning: Taking Area of Security Intelligence for Example. Data Analysis and Knowledge Discovery. 2019 (10).

Xiao, P., Su, Y., Zhang, R. Research and Construction of a Centralized Control and Management Platform for Security Intelligence of Power Grid Information. Network Security Technology & Application. 2017 (1).

Xie, X. Challenges and Solutions: An Analysis of China's Intelligence Studies Based on Overall National Security. Journal of Public Security Science. 2019 (4).

Xie, X., Gao, J. History of National Security Intelligence Studies in China (1949–1999): Intelligence Theory Dominated by Historical Paradigm. Information Studies: Theory and Application. 2020 (4).

Xiong, Y., Hu, J. Postgraduate Innovation &Startup Ability and Interdisciplinary Cultivating Mode—Analysis Based on Survey to Postgraduates at School. Journal of Graduate Education. 2017 (2).

Xu, G. Research on Mineral Resource Planning from the Perspective of Big Data Visual Analytics. Technology Innovation and Application. 2017 (26).

Xu, Y. Research on Talent Training Mode of Library and Information Science Under the Background of Big Data—A Case of GSLIS and SILS. Journal of Modern Information. 2016 (10).

Xu, C. Rapid Information Flow is Everything When Responding to Major Emergencies. http://www.aisixiang.com/data/120092.html. [2020–02–10].

Xu, H., Xu, R. Research on the Application of Cloud Computing in Military Intelligence Integration. Journal of Xi'an University of Arts Science (Natural Science Edition). 2014 (1).

Xue, M. Space-Based Modeling of the Public Security Intelligence Analysis. Geomatics & Spatial Information Technology. 2011 (3).

Yan, H. A Review of Research Hotspots in Chinese Library, Information and Archives Management Fields in 2019. Information and Documentation Services. 2020 (2).

Yan, Y. Relationship Between Safety and Security and Emergency: Analysis Based on the Branches of Production, Research and Education. Journal of Intelligence. 2019 (9).

Yang, F., Zhang, Y., Yao, L. The Method for Intelligence Awareness in an Emergency Based on Scenario Similarity. Journal of the China Society for Scientific and Technical Information. 2019 (5).

Yang, G., Li, P. Deepening Intelligence Work under the Context of the Overall Outlook on National Security. Journal of Intelligence. 2018 (5).

Yang, J. Rethinking the Philosophical Foundation of Information Science. Journal of the China Society for Scientific and Technical Information. 2020 (3).

Yang, S. Industry 4.0 and Industrial Internet: Comparison, Enlightenment, and Response Strategies. Contemporary Finance and Economics. 2015 (8).

Yang, Y. A New Concept of Intelligence: Information Science Providing One-to-one Service. Information Research. 2013 (8).

Yang, Z., Xie, H. Defining Intelligence through the Method of Scientific Induction. Journal of China Society for Scientific and Technical Information. 1988 (1).

Yao, L., Fan, W. A Study of Intrinsic Mechanism of Intelligence in Emergency Management. Library and Information Service. 2014 (23).

Ye, Y., Ma, F. Relation between the Arising of Data Science and Information Science. Journal of the China Society for Scientific and Technical Information. 2015 (6).

Yu, H. Mineral Security and Resource Strategy in the Belt and Road Initiative. Pacific Journal. 2018 (5).

Yu, H., Yu, B. Resource Independence and Global Governance: An Analysis of the EU's Mineral Resources Security Strategy. Chinese Journal of European Studies. 2017 (2).

Yu, J. Development and Application of System Science and System Engineering. Scientific Decision Making. 2017 (12).

Yu, J., Liu, Y. Complexity research and system science. Studies in Science of Science. 2002 (5).

Yuan, L., Yao, L. Exploration of "Data-Resources-Application" Intelligence Fusion Model in Emergency Management. Library and Information Service. 2014 (23).

Yuan, Y. Probe into Qian Xuesen's Academic Ideology of Information Research. Information Studies: Theory and Application. 2013 (9).

Zeng, Q., Chen, C. Construction of Counterterrorism Intelligence Analysis System Based on Meta-synthesis Method. Journal of Intelligence. 2018 (4).
Zeng, W., Li, H., Li, R. Exploration on Analysis and Application of Intelligent Information from the Perspective of Data Engineering. Information Studies: Theory and Application. 2018a (7).
Zeng, W., Li, H., Li, X. Research on Scientific and Technological Information Serving for National Science and Technology Security. Information Studies: Theory and Application. 2018b (6).
Zeng, X. An Investigation of HI Education in the Big Data Era. Library Tribune. 2020 (2).
Zeng, Z., Huang, C. Research on the Intelligence System of Public Health Emergencies with an Epidemic Control Orientation. Journal of Science. 2017 (10).
Zhang, J. On the Fusion between Resilience and Intelligence from the Perspective of Safeguarding National Security. Journal of Intelligence. 2017 (1).
Zhang, J., Ma, F. A Study of Construction and Operation of Chinese National Security Intelligence System. Information Studies: Theory and Application. 2015a (8).
Zhang, J., Ma, F. Research on the Structure and Operation of American National Security Intelligence System. Information Studies: Theory and Application. 2015b (7).
Zhang, J., Ma, F. On the System and Construction of the National Security Intelligence of Science and Technology. Journal of the China Society for Scientific and Technical Information. 2016 (5).
Zhang, J., Ma, F. Intelligence Service in the New Era: the New Connotation, New Challenges, New Opportunities and New Functions Based on the Overall National Security Concept. Information Studies: Theory and Application. 2018 (7).
Zhang, J., Wang, G. Natural Property and Social Property of Safety. China Safety Science Journal. 2001 (5).
Zhang, J. A Brief Discussion on Cultivation of Nuclear Intelligence Personnel and User Services in the Big Data Environment. China Nuclear Science and Technology Report (Volume 5)—Proceedings of the 2017 Annual Meeting of the China Nuclear Society, Volume 9 (Nuclear Intelligence Sub-volume, Nuclear Technology Economy and Modernization Management Sub-volume, Nuclear Electronics and Nuclear Detection Technology Sub-volume)
Zhang, L. An Analysis Model for the Value of Decision Information. Journal of Shenzhen University (Science & Engineering). 2002 (2).
Zhang, Q., Tang, C. Research on the Development of Intelligence Subject under the Guidance of Overall Outlook on National Security. Journal of Intelligence. 2015 (12).
Zhang, X. Intelligence, Intelligence Science, and National Security: An Interview with Mr. Bao Changhuo. Journal of Intelligence. 2017 (5).
Zhang, X. Mai, F., Li, C. Research on the Connotation of and Common Technology System for Enterprise Competitive Intelligence Dynamic Information Processing Based on Big Data. Information Studies: Theory and Application. 2014 (3).
Zhang, Y., Jin, P. Several Key Issues of Informatics Theory Research at Abroad: on the Perspective of Social Public Security. Information Research. 2014 (8).
Zhang, Y., Wu, C., Wang, B. Safety and Security Intelligence Literacy: The Required Literacy of Security Professionals under the Background of Overall National Security Idea. Journal of Intelligence. 2019 (3).
Zhao, B. On Intelligence-Dominated Competition, Journal of Intelligence. 2014 (1).
Zhao, F. Emergency Intelligence Research in China's Emergencies—A Literature Review. Journal of Modern Information. 2020 (2).
Zheng, T., Huang, P. Shen, B. Status and Prospect of Global Biosecurity. Military Medical Sciences. 2012 (10).
China Society for Scientific and Technical Information, Chinese Information Society of Social Sciences. Nanjing Consensus on Intelligence Science and the Development of Intelligence Operation. Library and Information Service. 2018 (1).
Zhou, B. Research on the Concepts of Intelligence. Journal of Modern Information. 1997 (3).
Zhou, G. City and Its Region: A Typical Open Complex Huge System. Urban Planning Forum. 2002 (2).

Zhou, J., Liu, R., Zhao, F. Remodeling Big Intelligence View in the New Era. Information Studies: Theory and Application. 2019 (8).
Zhou, Z. Construction of National Security Intelligence Legal System: From the Perspective of Constitutional “National Security” Provisions. Journal of Intelligence. 2020 (3).
Zhu, F. Analysis of “Non-traditional Security”. Social Sciences in China. 2004 (4).
Zhu, Z. The WSP Approach in the International Systems Community. Systems Engineering, Systems Science and Complexity Research--Proceeding of 11th Annual Conference of Systems Engineering Society of China. 2000.
Zi, S. Choice and Security of a Nation. SDX Joint Publishing Company, Shanghai, 2005.
Zou, P., Dong, Y., Li, X. Comparative Research on Crisis Management in Library and Information Science at Home and Abroad in the View of Scientific Knowledge Map. Information Science. 2018 (12).

Master’s and Doctoral Dissertations

Cao, W. Research on Biosafety and Intelligence Analysis Based on Multilayer Heterogeneous Complex Networks. Tianjin University for the Degree of Master. 2017.
Dong, L. Design and Implementation of Intelligent Tourist Attraction Information Management Service System. Zhejiang University of Technology for the Degree of Master. 2017.
Gao, M. Discussion for Intelligence Work for Security Issues in Medical Field. Nanjing University for the Degree of Master. 2019.
Huang, L. Research on Enterprise Competitive Intelligence Service Model Based on Cloud Computing. Nankai University for the Degree of Doctor. 2019.
Huang, X. Studies on the Methodology of Complexity Sciences. Tsinghua University for the Degree of Doctor. 2005.
Jiang, X. Research on Cyber Security Intelligence Work and Countermeasures under the Overall National Security Concept. Nanjing University for the Degree of Master. 2019.
Li, P. Research on the Causes of Intelligence Failure and the Evasion Strategies under the Perspective of Intelligence Studies. People’s Public Security University of China for the Degree of Master. 2019.
Li, X. The Research on Constructing Innovative CI Personnel Cultivation Mode. Fuzhou University for the Degree of Master. 2010.
Qu, Y. Research on Whole Process Emergency Response Management of Chemical Industry Park Public Emergency—Taking Shanghai Chemical Industry Park as Case. Shanghai Jiao Tong University for the Degree of Master. 2009.
Wei, Y. Research on Information Science Curriculum in Security Field. Nanjing University for the Degree of Master. 2017.
Xiang, Q. Campus Network Security Intelligence Management System. Beijing Forestry University for the Degree of Master. 2019.
Zhang, X. Research on the System Construction and Capability Evaluation of University Library Think-tank Service, Jilin University for the Degree of Doctor. 2019.

Newspapers

China Business News
Legal Daily (The newspaper was renamed *Legal Daily* in 2020)
Guangming Daily
China Military
Science and Technology Daily
People’s Daily
Study Times

Foreign Literature

Asthana, N. C., Nirmal, A. 2008. Intelligence and Security Management. Aavishkar Publishers, Distributors, vii.

Atkinson, M., Belayeva, J., Zavarella, V., et al. 2010. News Mining for Border Security Intelligence. IEEE International Conference on Intelligence & Security Informatics. IEEE.

Barnes, P., Charles, M. B., Branagan, M., et al. 2007. Intelligence and Anticipation: Issues in Security, Risk and Crisis Management. International Journal of Risk Assessment & Management, 7(8): 1209–1223.

Benzidane, K., Alloussi, H. E., Warrak, O. E., et al. 2016. Toward A Cloud-Based Security Intelligence with Big Data Processing. NOMS 2016–2016 IEEE/IFIP Network Operations and Management Symposium. IEEE.

Betts, R. K. 1978. Analysis, War, and Decision: Why Intelligence Failures are Inevitable. World Politics, 31(1): 61–89.

Brookes, B. C. 1980. The Foundations of Information Science: Part II. Quantitative Aspects: Classes of Things and the Challenge of Human Individuality. Journal of information science, (2): 209–221.

Bruenisholz, E., Wilson-wilde, L., Ribaux, O., et al. 2019. Deliberate Fires: From Data to Intelligence [J]. Forensic Science International, 301: 240–253.

Brummer, H. L., Badenhorst, J. A., Neuland, E. W. 2006. Competitive Analysis and Strategic Decision-Making in Global Mining Firms. J. Global Bus. Technol. 2(2): 26–38.

Burnhein, R. 1992. Information Literacy: A Core Competence.Australian Academic & Research Libraries, 23(4): 188–196.

Cates, S. 2015. The Evolution of Security Intelligence. Network Security, (3): 8–10.

Clark, R. M. 2010. Intelligence Analysis: A Target-Centric Approach. Washington, DC: CQ press.

Devlin, K. 2000. The Role of Conceptual Structure in Human Evolution. //GANTER, B, MINEQU, G. W., (Eds). Conceptual Structures: Logical, Linguistic, and Computational Issues, 8th International Conference on Conceptual Structures. Berlin: Springier Verlag.

Donoho, D. 2017. 50 Years of Data Science. Journal of computational & graphical statistics, 26(4): 745–766.

Dulles, A W. 1965. The Craft of Intelligence. New York: Harper & Row, 52.

Fingar, T. 2011. Reducing Uncertainty: Intelligence Analysis and National Security. California: Stanford University Press.

Fleisher, C., Bensoussan, B. 2003. Strategic and Competitive Analysis: Methods and Techniques for Analyzing Business Competition. Upper Saddle River, NJ: Prentice Hall.

Florea, G, Popa M. 2012. Safety and Security Integration in LPG Tank Farm Process Control. IFAC Proceedings, 45(623–25): 1828–1831.

Fraser-Arnott, M. 2017. Competencies for Information Specialists in Emerging Roles. Library Management, 38(1): 65–76.

Frevel, B. 2013. Managing Urban Safety and Security in Germany: Institutional Responsibility and Individual Competence. European Journal of Criminology, 10(3): 354–367.

Garai, H., Cochrane, P. 1997. Managing Information. New York: Ashgate Publishing.

George, F., William A. 1997. Artificial Intelligence: Structures and Strategies for Complex Problem Solving. Six Edition. USA: Addison-Wesley Longman Publishing Co., Inc.

Hale, A. 2003. Safety Management in Production. Human Factors and Ergonomics in Manufacturing & Service Industries, 13(3): 185–201

Hale, A. 2005. Safety Management, What do We Know, What Do We Believe We Know, and What Do We Overlook. Tijdschrift voor toegepaste Arbowetenschap, 18 (3): 58–66

Hall, A. D. 1969. Three-Dimensional Morphology of Systems Engineering. IEEE Transactions on System Science and Cybernetics, SSC-5(2):156–160.

Handel, M. I. 1989. War, Strategy and Intelligence. Psychology Press.

Hey, T., Tansley, S., Tolle, S. 2020.The Fourth Paradigm: Data—Intensive Scientific Discovery. [2020–02–28]. http://www.amazon.de/The-Fourth-Paradigm-Data-Intensive-ebook/dp/B00318D9Y2#reader_B00318D9Y2.

Huang, L., Wu, C., Wang, B., et al. 2018a. A New Paradigm for Accident Investigation and Analysis in the Era of Big Data. Process Safety Progress, 37(1): 42–48.

Huang, L., Wu, C., Wang, B., et al., 2018b. Big-Data-Driven Safety Decision-Making: A Conceptual Framework and its Influencing Factors. Safety Science, 109: 46–56.

IARPA. 2020. Research Programs. [2020–04–10]. https://www.iarpa.gov/index.php/research-programs.

Johnson, L. K. 2012. National Security Intelligence. Oxford: Polity Press.

Kahaner, L. 1996. Competitive iintelligence. Simon and Schuster, New York.

Kent, S. 1951. Strategic Intelligence for American World Policy. New Jersey: Princeton University Press.

Kim, Y., Kim, I., Park, N. 2014. Analysis of Cyber Attacks and Security Intelligence. Lecture Notes in Electrical Engineering, 274: 489–494.

Liew, A. 2013. DIKIW: Data, Information, Knowledge, Intelligence, Wisdom and their Interrelationships. Business Management Dynamics, 2(4): 49–62.

Liska, A. 2015. Building an Intelligence-Led Security Program. Amsterdam: Elsevier.

Lowenthal, M. M. 2009. Intelligence from Secrets to Policy (4th ed.). CQ Press, Washington, D.C.

Luhmann, N. 1984. The Self-Description of Society: Crisis Fashion and Sociological Theory. Comparative Sociology, 25(1): 59–72.

Lynch, C. 2008. Big Data: How do your Data Grow? Nature, 455(7209):28–29.

Marshall, P., Hirmas, A., Singer, M. 2018. Heinrich's Pyramid and Occupational Safety: A Statistical Validation Methodology. Safety Science, 101 (1): 180–189.

Meyerson Laura, A., Reaser Jamie, K. Biosecurity: Moving toward a Comprehensive Approach. Bioscience, 52(7):593–600.

Ouyang, Q., Chao, W., Lang, H. 2018. Methodologies, Principles and Prospects of Applying Big Data in Safety Science Research. Safety Science, 101(1): 60–71.

Patriarca, R., Gravio, G. D., Cioponea, R., Licu, A., 2019. Safety Intelligence: Incremental Proactive Risk Management for Holistic Aviation Safety Performance. Safety Science, 118: 551–567.

Peterson, M. 2005. Intelligence-Led Policing: The New Intelligence Architecture. Bureau of Justice Statistics, 20: 310–325.

Piètre-Cambacédès, L, Chaudet, C. 2010. The SEMA Referential Framework: Avoiding Ambiguities in the Terms "Security" and "Safety". International Journal of Critical Infrastructure Protection, 3(2): 55–66.

Pramanik, M. I., Lau, R. Y. K., Yue, W. T., et al. 2017. Big Data Analytics for Security and Criminal Investigations. Wiley Interdisciplinary Reviews: Data Mining and Knowledge Discovery, 7(4): e1208, https://doi.org/10.1002/widm.1208.

Rankin, M. 1986. National Security: Information, Accountability, and the Canadian Security Intelligence Service. University of Toronto Law Journal, 36(3): 249–285.

Rausand, M. 2014. Risk Assessment Theory, Methods, and Application. New York: John Wiley & Sons Inc.

Richardson, J. H. 1999. Biosafety in Microbiological and Biomedical laboratories. Government Printing Office Report on tb Laboratory Services, 4(4): 206–207.

Riel, A., Kreiner, C., Messnarz, R., et al. 2018. An Architectural Approach to the Integration of Safety and Security Requirements in Smart Products and Systems design. CIRP Annals, 67(1): 173–176.

Rosa, E. A. 1998.Metatheoretical Foundation for Post-normal Risk. Journal of Risk Research, 1(1): 15–44.

Rozenfeld, O., Sacks, R., Rosenfeld, Y., et al. 2010. Construction Job Safety Analysis. Safety Science, 48(4): 491–498.

Schlegelmilch, J., Albanese, J. 2014. Applying Business Intelligence Innovations to Emergency Management. Journal of Business Continuity & Emergency Planning, 8(1): 31–40.

Shannon, C. E. 1948. A Mathematical Theory of Communication. Bell Labs Technical Journal, 5(4): 3–55.
Shulsky, A. N., Gary, J. S. 2002. Silent Warfare: Understanding the World of intelligence. Washington, DC: Brassey's.
Simon, H. A. 1959. Theories of Decision-Making in Economics and Behavioral Science. The American Economic Review, 49(3): 253–283.
Song, I., Zhu, Y. 2016. Big Data and Data Science: What Should we Teach?. Expert systems, 33(4): 364–373.
Trim, P. R. J. 2003. Disaster Management and the Role of the Intelligence and Security Services. Disaster Prevention & Management, 12(1): 6–15.
Turban, E., Sharda, R., Aronson, J. E., King, D. N., 2008. Business Intelligence: A Managerial Approach. Upper Saddle River, NJ: Pearson Prentice Hall.
Turner, M. A. 2005. Why Secret Intelligence Fails. Dulles: Potomac Books,Inc..
Varey, P. 1999.Process Plants: A Handbook for Inherently Safer Design (Review).
Viktor, M. S., Kenneth, C. 2013. Big Data: A Revolution that will Transform How We Live, Work, and Think. London: Hodder & Stoughton.
Wang, B., Wu, C., Huang, L., et al. 2019. Using Data-Driven Safety Decision-Making to Realize Smart Safety Management in the Era of Big Data: A Theoretical Perspective on Basic Questions and their Answers. Journal of Cleaner Production,210: 595–1604.
Wang, B., Wu, C., Shi, B., et al. 2017. Evidence-Based Safety (EBS) Management: A New Approach to Teaching the Practice of Safety Management (SM). Journal of Safety Research, 63 (2): 21–28.
Wang, B., Wu, C. 2019. Demystifying Safety-Related Intelligence in Safety Management: Some Key Questions Answered from a Theoretical Perspective. Safety Science, 120(12): 932–940.
Wang, B., Wu, C. 2020. Safety Informatics as a New, Promising and Sustainable Area of Safety Science in the Information Age. Journal of Cleaner Production, 252: 119852.
Wang, B. 2020. Safety Intelligence as an Essential Perspective for Safety Management in the Era of Safety 4.0: From a Theoretical to a Practical Framework. Process Safety and Environmental Protection, 120(12): 932–940.
Wang, R., Ji W., Liu M., et al. 2018. Review on Mining Data from Multiple Data Sources. Pattern Recognition Letters, 109: 120–128.
Weller, G. R. 2010. The Canadian Security Intelligence Service Under Stress. Canadian Public Administration, 31(2): 279–302.
Winge, S., Albrechtsen, E., Arnesen, J., 2019. A Comparative Analysis of Safety Management and Safety Performance in Twelve Construction Projects. Journal of Safety Research, 71: 139–152.
Winzer, C. 2012.Conceptualizing Energy Security. Energy Policy, 46(3): 36–48.
Roberta, W. 1962. Pearl Harbor: Warning and Decision. Stanford University Press.
Xu, R., Wu Y., Luan, J., et al. 2020. Consumer-Perceived Risks of Genetically Modified Food in China. Appetite, 147.
Yovits, M. C., Kleyle, R. M. 1993. The Average Decision Maker and Its Properties Utilizing the Generalized Information System Model. Journal of the American Society for Information Science, 44(6): 352–363.

MIX
Papier aus verantwortungsvollen Quellen
Paper from responsible sources
FSC® C105338

If you have any concerns about our products,
you can contact us on
ProductSafety@springernature.com

In case Publisher is established outside the EU,
the EU authorized representative is:
Springer Nature Customer Service Center GmbH
Europaplatz 3, 69115 Heidelberg, Germany

Printed by Libri Plureos GmbH
in Hamburg, Germany